Circular Economy and Sustainability

This book series aims at exploring the rising field of Circular Economy (CE) which is rapidly gaining interest and merit from scholars, decision makers and practitioners as the global economic model to decouple economic growth and development from the consumption of finite natural resources. This field suggests that global sustainability can be achieved by adopting a set of CE principles and strategies such as design out waste, systems thinking, adoption of nature-based approaches, shift to renewable energy and materials, reclaim, retain, and restore the health of ecosystems, return recovered biological resources to the biosphere, remanufacture products or components, among others.

However, the increasing complexity of sustainability challenges has made traditional engineering, business models, economics and existing social approaches unable to successfully adopt such principles and strategies. In fact, the CE field is often viewed as a simple evolution of the concept of sustainability or as a revisiting of an old discussion on recycling and reuse of waste materials. However, a modern perception of CE at different levels (micro, meso, and macro) indicates that CE is rather a systemic tool to achieve sustainability and a new eco-effective approach of returning and maintaining waste in the production processes by closing the loop of materials. In this frame, CE and sustainability can be seen as a multidimensional concept based on a variety of scientific disciplines (e.g., engineering, economics, environmental sciences, social sciences). Nevertheless, the interconnections and synergies among the scientific disciplines have been rarely investigated in depth.

One significant goal of the book series is to study and highlight the growing theoretical links of CE and sustainability at different scales and levels, to investigate the synergies between the two concepts and to analyze and present its realization through strategies, policies, business models, entrepreneurship, financial instruments and technologies. Thus, the book series provides a new platform for CE and sustainability research and case studies and relevant scientific discussion towards new system-wide solutions.

Specific topics that fall within the scope of the series include, but are not limited to, studies that investigate the systemic, integrated approach of CE and sustainability across different levels and its expression and realization in different disciplines and fields such as business models, economics, consumer services and behaviour, the Internet of Things, product design, sustainable consumption & production, bio-economy, environmental accounting, industrial ecology, industrial symbiosis, resource recovery, ecosystem services, circular water economy, circular cities, nature-based solutions, waste management, renewable energy, circular materials, life cycle assessment, strong sustainability, and environmental education, among others.

K. Mathiyazhagan • Vernika Agarwal
Snigdha Malhotra • Veronica Scuotto
Editors

Humanizing Circular Supply Chain Management

Springer

Editors
K. Mathiyazhagan
Thiagarajar School of Management
Madurai, Tamil Nadu, India

Vernika Agarwal
Apeejay School of Management
Dwarka, Delhi, India

Snigdha Malhotra
Fortune Institute of International Business (FIIB)
Delhi, Delhi, India

Veronica Scuotto
University of Napoli Federico II
Department of Economics, Management, Institutions
Napoli, Italy

ISSN 2731-5509 ISSN 2731-5517 (electronic)
Circular Economy and Sustainability
ISBN 978-3-031-86561-9 ISBN 978-3-031-86562-6 (eBook)
https://doi.org/10.1007/978-3-031-86562-6

This Springer imprint is published by the registered company Springer Nature Switzerland AG
The registered company address is: Gewerbestrasse 11, 6330 Cham, Switzerland

Preface

The global manufacturing landscape is at a crossroads, facing the dual challenges of ensuring environmental sustainability and prioritizing human welfare. The pressing need to align economic objectives with ecological and social responsibilities forms the core of this book, *Humanizing Circular Supply Chain Management*.

In the introductory chapter, we set the stage by exploring the transformative potential of human-centered circular supply chain management and circular economy principles. These frameworks address critical concerns such as resource efficiency, waste minimization, and workplace safety while fostering ethical practices and stakeholder collaboration. The subsequent chapters delve into specific dimensions of this paradigm. Chapter 2 presents a comprehensive framework illustrating how supply chain practices can effectively integrate circular economy principles to enhance reusability, recycling, and waste management, while Chap. 3 emphasizes the pivotal role of training and development in equipping the workforce with the skills necessary to drive sustainable change. Exploring sector-specific applications, Chap. 4 focuses on the healthcare industry, revealing how human-centric skills and technological innovations shape the performance of circular supply chains. Chapter 5 shifts attention to consumer behavior, unraveling the psychological and social drivers behind garment recycling in the fast fashion industry. The challenges of embedding humanized approaches within circular supply chains are scrutinized in Chap. 6, which identifies ethical dilemmas, stakeholder engagement issues, and technological barriers. Chapters 7 and 8 employ innovative methodologies like ISM, MICMAC, and working capital management frameworks to analyze and optimize sustainable practices in diverse industries, including pharmaceuticals. Skills for CSCM remain a recurring theme, as Chap. 9 highlights the competencies essential for operationalizing circular practices across sectors. Chapter 10 explores the intersection of entrepreneurship and supply chain management, showcasing the influence of human creativity and leadership on sustainability. Finally, Chap. 11 addresses the integration of digital technologies in humanized supply chains, demonstrating how tools like the f-DEMATEL approach can enhance operational intelligence while upholding human involvement.

The later chapters further extend the discourse by tackling the nuances of supply chain resilience and innovation. Chapter 12 introduces a life cycle assessment and multi-criteria decision-making approach to identify sustainable packaging solutions in the beverage industry, aligning with Industry 5.0 objectives. In Chap. 13, entrepreneurial strategies from India exemplify how businesses navigate challenges and spearhead sustainability efforts within circular supply chains. The exploration of essential skills for employees continues in Chap. 14, identifying new roles like green procurement specialists and strategies for bridging the circular economy skills gap. Chapter 15 offers a comparative analysis of global sustainability frameworks, providing organizations with a structured guide for effective decision-making. Dissecting job roles further, Chap. 16 proposes updated performance evaluation metrics tailored to a circular economy. Chapter 17 delves into the concept of friendshoring, emphasizing the importance of human capital in fostering resilient and sustainable global supply chains. Chapter 18 examines the hindrances in the procurement of used products within the electrical and electronic equipment sector, offering actionable strategies to overcome consumer perception challenges and technological constraints. Chapter 19 analyzes the challenges to humanizing circular supply chains, prioritizing resistance to change, organizational culture, and cross-sector collaboration as focal areas for improvement. Expanding on the human touch in supply chain management, Chap. 20 provides a comprehensive review of key terminologies and emerging research directions in human circular supply chain management. Finally, Chaps. 21 and 22 apply lean tools and frameworks to enhance operational efficiency in aeronautic firms within the MENA region, highlighting the importance of integrating circular economy principles into supply chain operations.

This book aspires to bridge theoretical insights and practical applications, equipping businesses, policymakers, and researchers with the knowledge to transition toward sustainable supply chains that are both circular and humanized. By integrating technical, behavioral, and systemic perspectives, it lays the foundation for a resilient future where economic growth harmonizes with environmental stewardship and social equity. We hope this work inspires meaningful dialogue and actionable strategies to meet the demands of the circular economy while upholding the values of human dignity and collective progress.

Madurai, Tamil Nadu, India K. Mathiyazhagan
Dwarka, Delhi, India Vernika Agarwal
Delhi, Delhi, India Snigdha Malhotra
Napoli, Italy Veronica Scuotto

Contents

Chapter 1
Introduction to Humanizing Circular Supply Chain Management

Youssef Raouf, Zoubida Benmamoun, and Hanaa Hachimi

Abstract The global manufacturing industry is confronting intensifying challenges relating to environmental sustainability and safety. Human-centered circular supply chain management (CSCM) and circular economy fundamentals are gaining acceptance as a transformative approach. These concepts are encouraged through enhanced safety and lowered hazard in waste concentration, resource efficiency and minimizing waste through continuous use and circulation of materials in closed loops. Combining the circular economy with CSCM promotes a comprehensive strategy addressing environmental, social, and economic sustainability. Redesigning products to be recyclable, suitable for manufacturing, and capable of adding value through waste helps cut emissions and reduce the need for virgin resources. The integration of circular economy with CSCM leads organizations to a holistic approach toward environmental, social, and economic sustainability. By redesigning for recyclability, re-manufacturing, and waste-value creation, emissions are reduced, and virgin resources are spared. By adhering to these guidelines, businesses can minimize pollution, decrease resource consumption, and prevent workplace safety incidents when working conditions are improved. As a concept, humanizing CSCM focuses on ethical practices, participation within the community, and labor welfare throughout the supply chain. It encourages interactions between producers, suppliers, policy makers and consumers to overcoming technology, regulatory and behavioral barriers. This chapter introduces the notion of Humanizing Circular Supply Chain Management and illustrate how through this approach the manufacturing industry can transform to overcome the current sustainability and safety challenges.

Y. Raouf (✉) · H. Hachimi
Laboratory of Advanced Systems Engineering, National School of Applied Sciences ENSA, Ibn Tofail University, Kenitra, Morocco
e-mail: youssef.raouf@uit.ac.ma; hanaa.hachimi@uit.ac.ma

Z. Benmamoun
College of Engineering and Computing, Liwa University, Abu Dhabi, United Arab Emirates
e-mail: zoubida.benmamoun@lc.ac.ae

K. Mathiyazhagan et al. (eds.), *Humanizing Circular Supply Chain Management*, Circular Economy and Sustainability,
https://doi.org/10.1007/978-3-031-86562-6_1

Keywords CSCM · Circular economy · Humanization · Sustainable supply chain · Economic performance

1.1 Introduction

Companies today face several challenges, each of which can affect their daily business and even business model. To maintain a sustainable competitive advantage, the manufacturing sector must prioritize both environmental and commercial sustainability. The population of the globe is growing at an astounding rate, making it more important than ever to make the most use of the resources at our disposal in order to maintain the viability of this world (Liu and Kringos 2024). This emphasis on environmental issues has grown in parallel with an approach to sustainable resource management, which is seen as a way to address these issues. Concerns about climate change, reducing resources, and rising environmental pollution have also contributed to this focus. Furthermore, companies face major challenges in reaching the "triple bottom line," which is a combination of sustainability in three dimensions. There are three components to the triple bottom line: social, environmental, and economic (Eisenreich et al. 2022). The approach has been embraced by several businesses in order to assess their operations and generate more value. In this regard, the business and every phase of the circular supply chain (CSC) it employs should integrate social, economic, and environmental sustainability. Businesses should thus be prepared for the potential that management would pursue goals other than increasing profits and simultaneously decreasing expenditures (Ayati et al. 2022). From this angle, the most comprehensive and relevant approach to supply chain transformation in the era of Industry 4.0 is the circular economy, which is a strategy that can be put into practice to decrease the number of resources used up and waste that is produced (Kadio 2023b). Creating a highly technologically advanced production environment may now include ecologically conscious resource management. Circular Supply Chain incorporates sustainable resource management to efficiently safeguard social, economic, and environmental benefits. This aligns with the goals of the 12th Sustainable Development Goal (SDGs) set out by the United Nations for the 2030 outlook (Fatimah et al. 2024).

In the transition to a circular economy, the implication for human factor is one research area that has been often overlooked. The present chapter is split into two portions to fill in this gap. The first part elaborates Circular Supply Chain Management (CSCM), concepts on Circular Economy (CE), fundamental principles of circular supply chain management, and differences between linear and circular economy models. Part two delves into the why behind humanizing circular supply chain management.

1.2 Understanding Circular Supply Chain Management

1.2.1 Circular Economy

Researchers have applied the idea of the circular economy in a variety of settings. The most popular theory is based on the 3R framework, which uses the phrases Reduce, Reuse and Recycle to describe the circular economy (Calzolari et al. 2021). This framework has been in use for more than 10 years, though the exact origins of this framework remain unclear. Implementation addresses the (How-to) part and is heart of to the circular economy. From a systems perspective, another classification of the circular economy describes it across multiple levels (micro, meso, and macro) (Theeraworawit et al. 2022). However, a more crucial concept for understanding this framework is the waste hierarchy principle, which is rooted in the Cradle to Cradle (C2C) philosophy. This principle emphasizes treating waste as a resource, respecting diversity, and using renewable solar energy (Prajapati et al. 2022).

There are also many other circular economy concepts that have been advocated for by different schools of thought, including the Performance Economy, Regenerative Design, Industrial Ecology, Blue Economy, Permaculture, Natural Capitalism, and Industrial Symbiosis. These theoretical frameworks can be viewed as complementary and together establish the relationship of sustainability with the circular economy (Huo et al. 2017). Collectively, those perspectives suggest that sustainability, and sustainable development, are fundamentally linked to the core concepts of the circular economy. They may emphasize social justice and well-being, preservation, defense, and/or restoration of environmental quality, and the protection of future generations.

In addition to the three Rs, a number of recommended practices may help create a more circular economy. These are referred to the 9 Rs (Recover, Recycle, Repurpose, Remanufacture, Refurbish, Repair, Reuse, Reduce, Redesign/Rethink, and Refuse). The importance of these practices, in contrast to the 3 Rs, has not received formal government recognition. Supply chain participants, however, can see these methods as means of expanding the range of circularity opportunities (Angelis et al. 2017).

The importance of the circular economy in enacting systemic change is becoming more widely acknowledged. Although there is ongoing debate on the precise definition, the idea is generally understood to refer to an economic structure that completely separates economic operations from their effects on the environment. It achieves this goal by:

- Reducing the pace at which materials are used by the economy: This is done by making durable items last longer.
- Regenerating natural systems entails avoiding the use of non-renewable energy and materials, protecting or enhancing renewable resources (returning valuable nutrients to the soil to support natural ecosystems, for example, or using

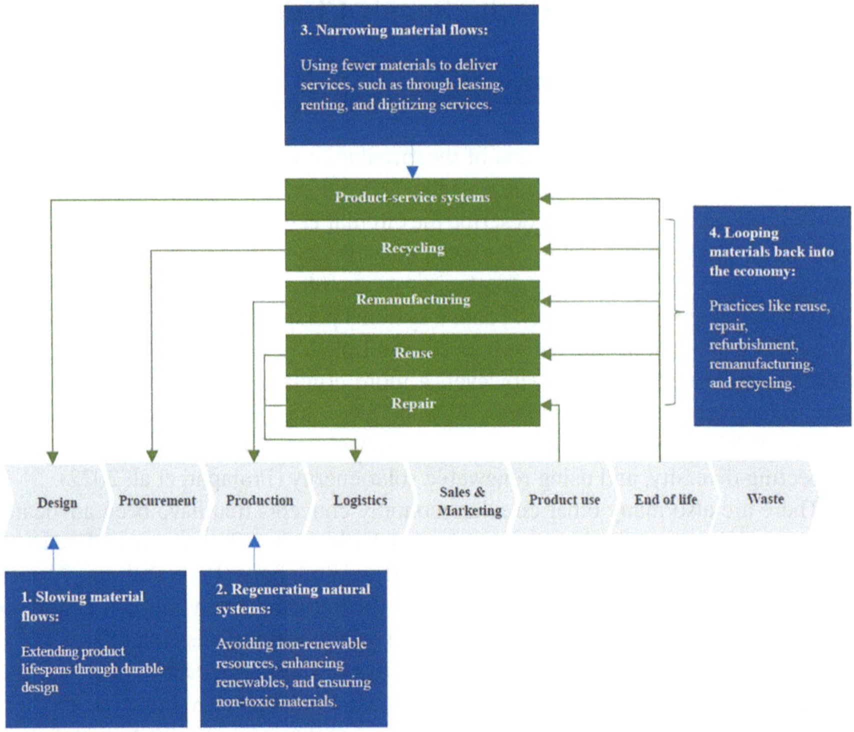

Fig. 1.1 The circular economy's overview

renewable energy in place of fossil fuels), and making sure that all materials moving through the economy are safe and non-toxic (Benmamoun et al. 2023a).
- Narrowing material flows: This refers to delivering the same service with less materials by altering consumption habits, for as by renting, leasing, or digitizing services.
- Bringing resources back into the economy at the end of their useful lives through recycling, remanufacturing, reuse, and refurbishment (Skalli et al. 2023) (Fig. 1.1).

1.2.2 Fundamental Principles of Circular Supply Chain Management

The fundamental principles of Circular Supply Chain Management (CSCM) are the essential policies and strategies that businesses use to make the shift to a circular economy model (Farooque et al. 2019a). The management of materials and products throughout their lifespan is guided by these principles in order to minimize waste, maximize resource efficiency, and lessen the impact on the environment. Important components that these concepts usually contain are:

- **Design for Circularity**
- From the beginning, designing for circularity entails taking into account a product's full lifespan. This entails using recyclable materials, building goods with readily replaceable or updated modular components, and making sure they are durable to prolong their useful life. By adding these components, businesses may produce goods that use less resources and produce less waste while still being sustainable (Kadio 2023a). The product's end-of-life possibilities are much improved by this proactive approach throughout the design phase, making recycling, repurposing, and remanufacturing simpler.
- **Efficiency of Resources**
- The goal of resource efficiency is to generate goods and services with the least amount of waste possible by making the most use of water, energy, and materials. This entails employing energy-efficient technology, streamlining manufacturing procedures to minimize material waste, and putting water-saving strategies into action (Montag 2023). Business perceived resource efficiency by which businesses can decrease their cost and environmental footprint. Rethinking supply chain and logistics practices are likely to decrease unnecessary transportation, hence lessening emissions that have a direct relationship in the direction of sustainability objectives (Jebbor et al. 2024). It therefore remains applicable for preserving the interaction between business and environment in favor of more sustainable overall outcomes, not only at operational level but also over time on a larger societal benefits perspective (Remme and Jackson 2023).
- **Producer Responsibility Extension**
- Manufacturers are also responsible for the total product lifecycle under the Extended Producer Responsibility (EPR) regulatory framework including recycling, take-back and disposal. By taking the onus of accountability away from consumers and governments, this incentivizes companies to manufacture more easily recyclable or reusable products (Hartley et al. 2024). EPR schemes often involve setting up infrastructure for the collection of end-of-life products, ensuring they are recovered in an environmentally sound manner and raising awareness among consumers about the value of returning stuff to be re-used or recycled.
- **Closed-Loop Logistics Systems**
- A closed-loop supply chain involves creating practices that allow resources and products to be regenerated within the manufacturing process. This means that things are reprocessed, reimplemented, rejuvenated or remedied to make new ones instead of being thrown when they come at the end of their useful life. Closed-loop system waste in general is reduced, for both environmental pollution and material used as raw materials (Lopes De Sousa Jabbour et al. 2019). Closed-loop supply chains can help businesses develop more resilient and sustainable operational approaches, inspire innovation as well as saving wasted cost in terms of waste management or raw material acquisition.

- **Cooperation and Transparency**
- For circular SCM to succeed, it requires tight collaboration with transparency among manufacturers, suppliers, retailers and customers. All should be on the same page in terms of sustainable practices and goals, for open communication information exchange (Singh 2022). Collaboration might lead to innovation and improve efficiency in driving supply chain, whilst the openness helps instill accountability as well as confidence. If this is handled together different hurdles resulting from technology, laws and consumer behavior could be overcome leading to a supply chain that sustains the environment and at the same time works efficiently.
- **Value Preserving Mechanisms**
- Preservation value focuses on retaining assets or goods at the height of their best worth for as long as possible. It involves reusing, repurposing, remanufacturing and repairing products before they get stuck in a recycling process. By extending the services life of their goods, businesses can prevent waste and thus reduce new needs for resources (Roshan et al. 2024). The same operations extend the life of these things and increase cost savings by diminishing costs as well as developing remanufacturing in addition to refurbished markets.
- **Reverse Logistics**
- Reverse logistics is the handling and transporting of returned goods, such as unsold products or for reuse pieces from production sites to end consumers which also involves taking back materials used in manufacturing processes (Benmamoun et al. 2017). This involves formalizing systems for collecting used items, categorizing to determine whether it should be recycled, refurbished or reused and optimizing logistics in order to reduce costs and environmental impact (Elma Sibonghanoy Groenewald 2024). Reverse logistics are extremely important to close the loop in supply chains, and essential for guaranteeing that valuable resources can be recovered and used again within manufacturing.
- **Sustainable Sourcing**
- Sustainable sourcing refers to buying products from vendors who work within moral labor and environmental conditions (Moktadir et al. 2018). Material derived from substances that are less harmful to the environment and renewable resources will get more preference. In order to ensure that the suppliers comply with sustainability requirements and advocate for a more socially responsible and sustainable supply chain, businesses need to conduct supplier audits regularly (Benmamoun et al. 2023b). Using sustainable resources, businesses can reduce their environmental footprint by doing more to protect animal welfare and support humane labor standards while preserving natural resources (Azmi et al. 2024).
- **Lifecycle Assessment**
- Life Cycle Assessment (LCA) is a systematic analysis of the environmental impact of products throughout their life cycle, starting with the extraction of raw materials to the disposal of the product (Jain et al. 2018). By identifying potential

sustainability hotspots, LCA can contribute to scale up of more sustainable practices and shrink the environmental footprint that a business creates. The result of LCA can be used to influence product and process redesign resulting in limited resource input, lower emissions and better environmental performance (Abdullah and Sanusi 2023).

- **Regenerative Approaches**
- Regenerative methods are about more than just sustaining existing environments; they involve enhancing and restoring the environment by revitalizing natural systems. This includes things like fossil fuel-free (or minimal) practices, tree planting and habitat restoration for essential nutrient cycling. By utilizing regenerative strategies, businesses can contribute to increased biodiversity and the restoration of natural ecosystems while also having a positive impact on the state of this planet. These methods also ensure that natural resources are maintained and replenished for future use, which means long-term economic feasibility (Farooque et al. 2019b). Together, these concepts form the basis for circular supply chain management principles that lead resilience and sustainability in business operations.

1.2.3 Differences Between Linear and Circular Economic Models

In many important respects, a circular economy differs from the current economic paradigm and provides insights into future societal shifts. Since the seventeenth century, industrial processes have served as the foundation for the linear economic model that largely shapes the modern economy. Usually promoting a linear production cycle 'take-make-waste', this paradigm frequently ignores the effects that product usage and resource consumption have on the environment (Lahane et al. 2020). It places a strong emphasis on market orientation and value creation through the development of new, wasteful, and sophisticated goods. Open loop production systems lose waste streams during operation or dispose of them in landfills at the end of their useful lives. As a result, materials and products are designed to be used just once, which feeds into a cycle made worse by deliberate obsolescence. The intentional reduction of product life, shown by cellphones and computers, hastens obsolescence and increases waste generation (Lahane and Kant 2021) (Table 1.1).

The circular economy, in contrast to the conventional linear economy, offers a more effective strategy for environmental preservation. The circular economic model makes it easier to turn waste or parts from end-of-life items into inputs rather than considering them as worthless. This conversion aids in closing loops within the manufacturing chain in addition to adding economic benefit (Kharola et al. 2022).

Table 1.1 Characteristics of a circular economy versus a linear economy

Linear economic model	Circular economic model
Characteristics: • Take-Make-Waste-Dispose (Marino and Pariso 2016) • A purely market-oriented strategy that promotes limitless consumer growth • Worthless trash • Minimal or no attention to the environment's finite resources	**Characteristics:** • 3 Rs: Reduce, Reuse, and Recycle ≥9 Rs (Colombo et al. 2022) • A new strategy for the market that separates growth from material consumption • Valuable waste • Prioritizing the prevention of waste pollution and the preservation of biodiversity
In an open-loop cycle: • Value is only created by manufacturing a large number of new items • Waste streams (input, output, and trash) are usually directed toward landfill disposal or leakage (Dieckmann et al. 2020)	**In a closed-loop cycle:** • Turn waste into inputs to create value • Divert waste streams into the manufacturing process to eventually produce zero waste (Didenko et al. 2018)
Product-level focus area: • Materials and products should only be used once throughout their lifespan (Sapan Kumar and Jhuma 2024)	**Product-level focus area:** • Multiple lifetime use of materials and products(Sørensen 2018)
Practical difficulties include: • Planned obsolescence: Things become outdated more quickly; this leads to a wasteful culture and a rise in e-waste from computers and cellphones with short lifespans • Using or importing virgin materials, such as minerals and other fundamental natural resources that are derived from their country of origin (Sariatli 2017)	**Practical difficulties include:** • Life-cycle approach: In contrast to planned obsolescence, items are kept for as long as feasible • The use of secondary materials: to encourage a greater amount of secondary material substitution for virgin material (Suchek et al. 2021)
Risks: • Increasing waste leaks in international supply chains, which have a significant impact on CO_2 emissions and lead to a lot of trash that might be avoided (Afteni et al. 2022)	**Risks:** • Adverse consequences of Rs practices: in a destination country with inadequate recycling capacity or an aging vehicle fleet, recyclable clothing trash or used automobiles may pose greater environmental problems (Zhou and Smulders 2021)

1.3 The Need for Humanizing Circular Supply Chain Management

1.3.1 Circular Economy Issues and Difficulties with Traditional Supply Chains

Numerous obstacles confronting traditional supply chains need a shift toward more human-centered and sustainable methods with a foundation in the social and environmental spheres (Abidi et al. 2014).

- **Resource Depletion**
- Conventional supply chains are a resource and environmental inefficient system that causes severe damage to the environment and natural resources (Chokshi et al. 2023). It involves reducing resource extraction, whilst ensuring that methods such as recycling, remanufacturing and repair are used to sustain the value of resources within economic systems.
- **Waste Generation**
- Linear supply chains are also notorious for the waste they generate in terms of surplus goods, and packing materials (Raouf et al. 2023). Here, on the other hand circular supply chains hold exceedingly more importance when it comes to reduction and management of waste. They aim to close material loops and therefore reduce waste through strategies such as product reuse, recycling or upcycling (Jebbor et al. 2023).
- **Lack of Transparency**
- Lack of transparency in existing supply chains makes it challenging to purchase and understand the sustainability aspects like, where products come from, how they are produced or its overall impact on the environment. Circular supply chains can be formed in which transparency and traceability of products are maintained (Khan and Ali 2022) for the stakeholders to take ethical decisions about life cycle knowledge of a product from its sustainability perspectives.
- **Social injustice**
- Traditional supply chains often compound social inequalities with exploitative working conditions and inadequate protection of worker rights, among other practices (Ghisellini et al. 2023). To improve work condition and the benefits of equal economic opportunity, humanizing circular supply chain management with respect to equitable labor practice for empowerment workers socially responsible within its chains (Matarneh et al. 2024).
- **Short Product Lifecycles**
- The linear economy model tends to promote short product lifecycles and consequently fast products obsolescence, hence high levels of waste production. Circular supply chain models carefully consider product durability, repairability and longevity by designing products to extend lifespan but reduce environmental impact at the same time (Skrypko et al. 2021).
- **Exorbitant Transportation Costs**
- Linear supply chains depend on long-distance transportation, an approach that significantly contributes to the overall operating costs as well as carbon emissions. Circular supply chains are based on localizing sourcing, production and distribution networks to reduce transport distances, environmental impacts thereby improving the resilience of the supply chain (Benchekroun et al. 2019).
- **Resistance to Change**
- Removing these barriers will be key to transitioning from linear, take-make-waste models to more circular operations within supply chains. Cultivation amongst stakeholders of a culture of innovation, cooperation in the implementa-

tion phase and continuous improvement can position circular strategies to be well suited for adapting to new paradigms towards lasting outcomes (Saroha et al. 2020).

- The humanizing circular supply chain management reconciles all these environmental, social and economic interests to target addressing the issues with traditional supply chains and driving forward sustainable business & societal practices (Mangla et al. 2018).

1.3.2 The Importance of Human Elements in CSCM

Understanding the importance of people in the supply chain is key to introducing human elements into construction supply chain management. Humanness must feature in Circular Supply Chain Management for a range of reasons.

- **Ethical Responsibility**
- A humanized supply chain recognizes fair labor standards, the dignity of work and is designed to ensure that those who may be at their most vulnerable are afforded the same opportunities available to others (Le 2022). This encourages moral accountability and brings companies into line with public expectations about Corporate Social Responsibility CSR (Upadhayay and Alqassimi 2019).
- **Employee Engagement and Satisfaction**
- CSCM improves employee engagement and satisfaction by emphasizing human factors including safe working conditions, chances for training and development, and inclusive workplace policies. Consequently, this raises productivity and lowers attrition rates (Graczyk-Kucharska and Hojka 2021).
- **Community Relations**
- Human-centered CSCM fosters good linkages with surrounding communities by promoting job opportunities, fostering community growth, and upholding cultural norms and values (Başar 2024).
- **Risk Mitigation**
- By addressing human factors in CSCM, risks related to labor conflicts, noncompliance with regulations, and reputational harm are reduced. Companies that put an emphasis on ethical supply chain procedures are better able to handle crises and keep the trust of their stakeholders (Šapalaitė and Kiaušienė 2023).
- **Long-Term Sustainability**
- As societal and environmental concerns change over time, resilience, creativity, and flexibility are fostered by sustainable corporate strategies, such as humanizing CSCM (Bailey 2023).

1.3.3 *Benefits of Humanized CSCM for Businesses and Society*

There are several advantages to humanizing CSCM for both enterprises and society.

- **Improved Brand Reputation**
- Companies that place a high priority on human factors in CSCM gain credibility as morally and socially conscious companies. This raises customer loyalty and trust, which boosts market competitiveness (Quevedo and Pradeep 2020).
- **Cost Savings**
- Humanized CSCM can save money on hiring, training, and operational interruptions by enhancing working conditions and lowering attrition rates. Increased productivity from employees also results in financial advantages (Benmamoun et al. 2023a).
- **Compliance and Risk Management**
- By adopting ethical labor practices and observing regulatory requirements, organizations can avoid legal and compliance problems (Hazen et al. 2021). When this human factor in CSCM is taken care of then chances are less of supply chain interruptions and are also fines.
- **Efficiency and Innovation**
- A humanized CSCM drive productivity for the management of resources, streamline supply chain procedures and opens ways to innovative product design. Increased efficiency and operational performance, along with reduction of waste are beneficial to businesses (Jaeger-Erben et al. 2021).
- **Positive Social Impact**
- This complete humanized CSCM endeavored to go beyond commercial benefits and also include the enhancement of societal welfare through promoting social justice, sustaining sustainable livelihoods as well as community development projects (Ashton et al. 2022).
- **Environmental Stewardship**
- Clearly, CSCM also bolsters more specific sustainability goals insofar as it is a reminder of the human connection. Ethical production and consumption behaviors are encouraged by businesses to support resource efficiency and environmental conservation.
- Incorporating human dimensions in CSCM enhances operational efficiency, inspires innovation and generates social benefits on top of assisting firms meet their legal and moral responsibilities (Villalba-Eguiluz et al. 2023). Companies that embed humanized product lifecycle management (PLM) as an enduring business practice, can secure a long-standing expansion and more importantly serve demonstrably to ensure the world economy has foundations which are ethical.

1.3.4 Key Components of Humanizing CSCM

Table 1.2 shows several essential factors involved in the CSCM humanization.

These key pillars layer together to create a responsible and sustainable ecosystem underpinning supply chain which fosters the success of corporations over the long term while simultaneously boosting employee welfare and driving social justice.

Table 1.2 Key components of humanizing CSCM

Components of humanizing CSCM	Insights
Ethical labor practices	Prioritizing fair wages and good workplace conditions and workers' rights throughout the supply chain. It involves ensuring that there is no discrimination, fostering equal opportunity, and following labor laws (Eyo-Udo et al. 2024b)
Worker well-being	Promoting physical and emotional wellness of employees by providing them access to health care, reasonable time off, and ergonomic workstations. These are some main components like programs for employee wellbeing and support systems (Ben Abdelaziz et al. 2024)
Training & development	Opportunities for career advancement, skill development, and ongoing training for workers at all levels of the supply chain This boosts productivity, encourages retention and enhances job satisfaction (Eyo-Udo et al. 2024a)
Community engagement	Engaging with neighboring communities to understand their needs; celebrating cultural diversity, and undertaking community development–oriented projects Helping local economies and integrating community input into supply chain processes (Castillo et al. 2018)
Transparency and accountability	Ensuring all lines of the supply chain are open, which means how the supplier works, how the goods are sourced, and how it impacts the environment. They are systems of accountability: tracking and recording social and ethical performance (Brun et al. 2020)
Supplier relationships	Cooperative, win-win partnerships with suppliers—Developed through mutual respect, trust, and shared values This involves fair contract terms, timely payment, and support to develop supplier capacity (Shrivastava 2024)
Human rights due diligence	Identifying and remedying human rights violations such as child labor, forced labor, and unsafe working conditions by assessing the risks to human rights throughout the supply chain. Implementing corrective action plans where necessary (Gonzalez 2014)
Continuous improvement	Vehicle for continuous improvements through stakeholder engagement, performance measurements, and feedback processes. It ensures that humanizing CSCM processes will evolve correctly in response to changing commercial needs and cultural norms (Berardi 2021)
Environmental considerations	Environmental sustainability can be internalized in humanizing CSCM by reducing the consumption of resources, promoting environmentally friendly behaviors and reducing carbon footprint. This involves sustainable sourcing methods alongside circular economy principles (Kazancoglu et al. 2021)
Corporate social responsibility	Connecting more general CSR programs that help the environment and society to humanizing CSCM. That means engaging with stakeholders, supporting local endeavors, and promoting ethical business practices (Haseli et al. 2024)

1.4 Conclusion

The research illustrates that integrating circular economy enablers with human-centred concepts is not only economically rewarding in improving cost-efficient outcomes and impacts but is central to transitioning business processes enabling them to address future challenges. Collectively, these data demonstrate that in the end, the operational resilience and trust businesses can gain from worker well-being and community-engaged labor practices ultimately results from ethical labor practices. It also minimizes typical risks posed by international supply chain disruptions, something that simply matter in today's stringently linked world economic situation.

The research indicates that integrating humanized circular supply chain management (CSCM) will not only yield significant cost savings due to enhanced resource efficiency and low waste, but it also makes the sustainability of profit possible over time through innovation incentives, as well as extension of product life. Also, because they enable greater human stewardship capability, the addition of human envoys provides a more powerful mechanism for future sustainability and social responsibility challenges.

Actually, those results are a reminder of the strategic reshape that should be integrated with a business model linking economic performance, social and environmental commitments in supply chain management. The companies using this framework are the ones positioning themselves as the key actors to achieving sustainable development and the transition toward a more resilient and equitable global economy. The ability of this study to shed light on the root causes of these disconnects is really quite important for meaningfully guiding improvements in practice in industries that will lead to a more sustainable future and that will also address some of the major global challenges we face.

References

Abdullah, N. H. N., & Sanusi, S. (2023). Circular supply chain management in manufacturing companies. In Y. Ramakrishna & S. N. Wahab (Éds.), Advances in finance, accounting, and economics (p. 395-411). IGI Global. https://doi.org/10.4018/978-1-6684-7664-2.ch019

Abidi H, De Leeuw S, Klumpp M (2014) Humanitarian supply chain performance management: a systematic literature review. Supply Chain Manag 19(5/6):592–608. https://doi.org/10.1108/SCM-09-2013-0349

Afteni C, Păunoiu V, Afteni M (2022) Study on the transition from the linear economy to the circular economy. Ann "Dunarea de Jos" Univ Galati, Fascicle V Technol Mach Building 39(1):49–55. https://doi.org/10.35219/tmb.2021.1.08

Angelis R, Howard M, Miemczyk J (2017) Supply chain management and the circular economy: towards the circular supply chain. Prod Plan Control 29:425–437. https://doi.org/10.1080/09537287.2018.1449244

Ashton WS, Fratini CF, Isenhour C, Krueger R (2022) Justice, equity, and the circular economy: introduction to the special double issue. Local Environ 27(10–11):1173–1181. https://doi.org/10.1080/13549839.2022.2118247

Ayati SM, Shekarian E, Majava J, Wæhrens BV (2022) Toward a circular supply chain: understanding barriers from the perspective of recovery approaches. J Clean Prod 359:131775. https://doi.org/10.1016/j.jclepro.2022.131775

Azmi FR, Roni M, Sa'at M (2024) Circular supply chain management in developing countries: challenges, opportunities and pathways to sustainability. Inf Manag Bus Rev 16(1(I)):105–115. https://doi.org/10.22610/imbr.v16i1(I).3666

Bailey SG (2023) Humanising the product-service system within a circular economy for product design and engineering students. In: Proceedings of the international conference on engineering and product design education, EPDE 2023. 25th International conference on engineering and product design education. https://doi.org/10.35199/EPDE.2023.90

Başar D (2024) Empowering sustainability: green human resource management as a catalyst in the circular economy. M U Iktisadi ve Idari Bilimler Dergisi. https://doi.org/10.14780/muiibd.1459745

Ben Abdelaziz F, Chen Y-T, Dey PK (2024) Supply chain resilience, organizational well-being, and sustainable performance: a comparison between the UK and France. J Clean Prod 444:141215. https://doi.org/10.1016/j.jclepro.2024.141215

Benchekroun HT, Benmamoun Z, Hachimi H (2019, July) Sustainable procurement in the public sector case study of Morocco. In: Proceedings of the International Conference on Industrial Engineering and Operations Management, Pilsen, Czech Republic, pp 23–26

Benmamoun Z, Hachimi H, Amine A (2017) Green logistics practices. In: 2017 International renewable and sustainable energy conference (IRSEC), pp 1–5. https://doi.org/10.1109/IRSEC.2017.8477578

Benmamoun Z, Fethallah W, Ahlaqqach M, Jebbor I, Benmamoun M, Elkhechafi M (2023a) Butterfly algorithm for sustainable lot size optimization. Sustainability 15(15):11761. https://doi.org/10.3390/su151511761

Benmamoun Z, Fethallah W, Bouazza S, Abdo AA, Serrou D, Benchekroun H (2023b) A framework for sustainability evaluation and improvement of radiology service. J Clean Prod 401:136796. https://doi.org/10.1016/j.jclepro.2023.136796

Berardi PC (2021) Supply chain collaboration for a circular economy—from transition to continuous improvement. J Clean Prod

Brun A, Karaosman H, Barresi T (2020) Supply chain collaboration for transparency. Sustainability 12(11):4429. https://doi.org/10.3390/su12114429

Calzolari T, Genovese A, Brint A (2021) The adoption of circular economy practices in supply chains—an assessment of European Multi-National Enterprises. J Clean Prod 312:127616. https://doi.org/10.1016/j.jclepro.2021.127616

Castillo VE, Mollenkopf DA, Bell JE, Bozdogan H (2018) Supply chain integrity: a key to sustainable supply chain management. J Bus Logist 39(1):38–56. https://doi.org/10.1111/jbl.12176

Chokshi H, Agrawal R, Samadhiya A, Kumar A (2023) IJMEMS. https://www.ijmems.in/article_detail.php?vid=8&issue_id=41&article_id=532

Colombo, B., Boffelli, A., Gaiardelli, P., Kalchschmidt, M., Madonna, A., & Sangalli, T. (2022). A multiple case study on collaboration for a circular economy: a focus on the Italian textile supply chain. In D. Y. Kim, G. Von Cieminski, & D. Romero (Éds.), Advances in production management systems. Smart manufacturing and logistics systems: turning ideas into action (Vol. 663, p. 408–415). Springer, Cham. https://doi.org/10.1007/978-3-031-16407-1_48

Didenko NI, Klochkov YS, Skripnuk DF (2018) Ecological criteria for comparing linear and circular economies. Resources 7(3):48. https://doi.org/10.3390/resources7030048

Dieckmann E, Sheldrick L, Tennant M, Myers R, Cheeseman C (2020) Analysis of barriers to transitioning from a linear to a circular economy for end of life materials: a case study for waste feathers. Sustain For 12(5):1725. https://doi.org/10.3390/su12051725

Eisenreich A, Füller J, Stuchtey M, Gimenez-Jimenez D (2022) Toward a circular value chain: impact of the circular economy on a company's value chain processes. J Clean Prod 378:134375. https://doi.org/10.1016/j.jclepro.2022.134375

Elma Sibonghanoy Groenewald (2024) Circular economy strategies in supply chain management: towards zero waste. Power Syst Technol 48(1):464–480. https://doi.org/10.52783/pst.291
Eyo-Udo NL, Odimarha AC, Ejairu E (2024a) Sustainable and ethical supply chain management: the role of HR in current practices and future directions. Magna Sci Adv Res Rev 10(2):181–196. https://doi.org/10.30574/msarr.2024.10.2.0058
Eyo-Udo NL, Odimarha AC, Kolade OO (2024b) Ethical supply chain management: balancing profit, social responsibility, and environmental stewardship. Int J Manag Entrep Res 6(4):1069–1077. https://doi.org/10.51594/ijmer.v6i4.985
Farooque M, Zhang A, Liu Y (2019a) Barriers to circular food supply chains in China. Supply Chain Manag 24(5):677–696. https://doi.org/10.1108/SCM-10-2018-0345
Farooque M, Zhang A, Thürer M, Qu T, Huisingh D (2019b) Circular supply chain management: a definition and structured literature review. J Clean Prod 228:882–900. https://doi.org/10.1016/j.jclepro.2019.04.303
Fatimah YA, Govindan K, Sasongko NA, Hasibuan ZA (2024) The critical success factors for sustainable resource management in circular economy: assessment of urban mining maturity level. J Clean Prod 469:143084. https://doi.org/10.1016/j.jclepro.2024.143084
Ghisellini P, Passaro R, Ulgiati S (2023) Environmental assessment of multiple "cleaner electricity mix" scenarios within just energy and circular economy transitions, in Italy and Europe. J Clean Prod 388:135891. https://doi.org/10.1016/j.jclepro.2023.135891
Gonzalez M (2014) Advancing human rights in supply chains through community engagement and investment. PM World J Adv Hum Rights Supply Chains Through 3
Graczyk-Kucharska M, Hojka K (2021) Conceptual model of human resource management for the efficient management of a circular economy. Eur Res Stud J XXIV(Special Issue 5):234–247. https://doi.org/10.35808/ersj/2715
Hartley K, Baldassarre B, Kirchherr J (2024) Circular economy as crisis response: a primer. J Clean Prod 434:140140. https://doi.org/10.1016/j.jclepro.2023.140140
Haseli G, Nazarian-Jashnabadi J, Shirazi B, Hajiaghaei-Keshteli M, Moslem S (2024) Sustainable strategies based on the social responsibility of the beverage industry companies for the circular supply chain. Eng Appl Artif Intell 133:108253. https://doi.org/10.1016/j.engappai.2024.108253
Hazen BT, Russo I, Confente I, Pellathy D (2021) Supply chain management for circular economy: conceptual framework and research agenda. Int J Logist Manag 32(2):510–537. https://doi.org/10.1108/IJLM-12-2019-0332
Huo B, Flynn BB, Zhao X (2017) Supply chain power configurations and their relationship with performance. J Supply Chain Manag 53(2):88–111. https://doi.org/10.1111/jscm.12139
Jaeger-Erben M, Jensen C, Hofmann F, Zwiers J (2021) There is no sustainable circular economy without a circular society. Resour Conserv Recycl 168:105476. https://doi.org/10.1016/j.resconrec.2021.105476
Jain S, Jain NK, Metri B (2018) Strategic framework towards measuring a circular supply chain management. BIJ 25(8):3238–3252. https://doi.org/10.1108/BIJ-11-2017-0304
Jebbor I, Benmamoun Z, Hachimi H (2023) Optimizing manufacturing cycles for advantage production: application in shipyards traditional industry. https://doi.org/10.20944/preprints202308.1398.v1
Jebbor I, Benmamoun Z, Hachmi H (2024) Revolutionizing cleaner production: the role of artificial intelligence in enhancing sustainability across industries. J Infrastruct Policy Dev 8(10):7455. https://doi.org/10.24294/jipd.v8i10.7455
Kadio AKC (2023a) Circular economy and supply chain change: the case of industries in a French Port. Eur J Logist Purch Supply Chain Manag 11(2):17–30. https://doi.org/10.37745/ejlpscm.2013/vol11n21730
Kadio AKC (2023b) The benefits of the circular economy for companies' circular supply chains in a French Port area. Eur J Logist Purch Supply Chain Manag 11(2):31–47. https://doi.org/10.37745/ejlpscm.2013/vol11n23147

Kazancoglu I, Sagnak M, Kumar Mangla S, Kazancoglu Y (2021) Circular economy and the policy: a framework for improving the corporate environmental management in supply chains. Bus Strateg Environ 30(1):590–608. https://doi.org/10.1002/bse.2641

Khan F, Ali Y (2022) Implementation of the circular supply chain management in the pharmaceutical industry. Environ Dev Sustain 24(12):13705–13731. https://doi.org/10.1007/s10668-021-02007-6

Kharola S, Ram M, Kumar Mangla S, Goyal N, Nautiyal OP, Pant D, Kazancoglu Y (2022) Exploring the green waste management problem in food supply chains: a circular economy context. J Clean Prod 351:131355. https://doi.org/10.1016/j.jclepro.2022.131355

Lahane S, Kant R (2021) Evaluating the circular supply chain implementation barriers using Pythagorean fuzzy AHP-DEMATEL approach. Clean Logist Supply Chain 2:100014. https://doi.org/10.1016/j.clscn.2021.100014

Lahane S, Kant R, Shankar R (2020) Circular supply chain management: a state-of-art review and future opportunities. J Clean Prod 258:120859. https://doi.org/10.1016/j.jclepro.2020.120859

Le TT (2022) How humane entrepreneurship fosters sustainable supply chain management for a circular economy moving towards sustainable corporate performance. J Clean Prod 368:133178. https://doi.org/10.1016/j.jclepro.2022.133178

Liu Z, Kringos N (2024) Transition from linear to circular economy in pavement engineering: a historical review. J Clean Prod 449:141809. https://doi.org/10.1016/j.jclepro.2024.141809

Lopes De Sousa Jabbour AB, Rojas Luiz JV, Rojas Luiz O, Jabbour CJC, Ndubisi NO, Caldeira De Oliveira JH, Junior FH (2019) Circular economy business models and operations management. J Clean Prod 235:1525–1539. https://doi.org/10.1016/j.jclepro.2019.06.349

Mangla SK, Luthra S, Mishra N, Singh A, Rana NP, Dora M, Dwivedi Y (2018) Barriers to effective circular supply chain management in a developing country context. Prod Plan Control 29(6):551–569. https://doi.org/10.1080/09537287.2018.1449265

Marino A, Pariso P (2016) From linear economy to circular economy: research agenda. Int J Res Econ Soc Sci 6(5):2249–7382

Matarneh S, Piprani AZ, Ellahi RM, Nguyen DN, Mai Le T, Nazir S (2024) Industry 4.0 technologies and circular economy synergies: enhancing corporate sustainability through sustainable supply chain integration and flexibility. Environ Technol Innov 35:103723. https://doi.org/10.1016/j.eti.2024.103723

Moktadir MA, Rahman T, Rahman MH, Ali SM, Paul SK (2018) Drivers to sustainable manufacturing practices and circular economy: a perspective of leather industries in Bangladesh. J Clean Prod 174:1366–1380. https://doi.org/10.1016/j.jclepro.2017.11.063

Montag L (2023) Circular economy and supply chains: definitions, conceptualizations, and research agenda of the circular supply chain framework. Circ Econ Sustain 3(1):35–75. https://doi.org/10.1007/s43615-022-00172-y

Prajapati D, Pratap S, Zhang M, Lakshay, Huang GQ (2022) Sustainable forward-reverse logistics for multi-product delivery and pickup in B2C E-commerce towards the circular economy. Int J Prod Econ 253:108606. https://doi.org/10.1016/j.ijpe.2022.108606

Quevedo FJ, Pradeep G (2020) The four anchors of brand appeal: a study into the gaps in the literature. J Humanit Soc Sci 3(2). https://doi.org/10.33140/JHSS.03.02.04

Raouf Y, Benmamoun Z, Hachimi H, Ikhlef J, Mehdi H, Mehdi A (2023) Towards a smart and sustainable industry: optimizing cycle time in the aeronautics company using lean six-sigma. Sigma. https://doi.org/10.1109/IRASET57153.2023.10153046

Remme D, Jackson J (2023) Green mission creep: the unintended consequences of circular economy strategies for electric vehicles. J Clean Prod 394:136346. https://doi.org/10.1016/j.jclepro.2023.136346

Roshan R, Balodi KC, Datta S, Kumar A, Upadhyay A (2024) Circular economy startups and digital entrepreneurial ecosystems. Bus Strateg Environ 33(5):4843–4860. https://doi.org/10.1002/bse.3727

Šapalaitė T, Kiaušienė I (2023) Assessment of financial condition in companies applying circular economy principles. Vilnius Univ Proc 37:104–111. https://doi.org/10.15388/VGISC.2023.15

Sapan Kumar S, Jhuma JS (2024) The transition in economic theory from linear to circular for the sustainability—a case study. Int J Multidiscip Res 6(4):26488. https://doi.org/10.36948/ijfmr.2024.v06i04.26488
Sariatli F (2017) Linear economy versus circular economy: a comparative and analyzer study for optimization of economy for sustainability. Visegrad J Bioecon Sustain Dev 6(1):31–34. https://doi.org/10.1515/vjbsd-2017-0005
Saroha M, Garg D, Luthra S (2020) Pressures in implementation of circular supply chain management for sustainability: an analysis from Indian industries perspective. Manag Environ Qual 31(5):1091–1110. https://doi.org/10.1108/MEQ-08-2019-0178
Shrivastava R (avec Jain K) (2024) Humanizing businesses for a better world of work, 1st ed. Emerald Publishing Limited
Singh R (2022) Circular economy in supply chain. J Prod Oper Manag Econ 23:35–45. https://doi.org/10.55529/jpome.23.35.45
Skalli D, Charkaoui A, Cherrafi A, Garza-Reyes JA, Antony J, Shokri A (2023) Analyzing the integrated effect of circular economy, lean six sigma, and industry 4.0 on sustainable manufacturing performance from a practice-based view perspective. Bus Strateg Environ 33:1208–1226. https://doi.org/10.1002/bse.3546
Skrypko T, Popadynets N, Yakhno T, Shulla R, Vlasenko T, Irtyshcheva I, Boiko Y (2021) Optimizing the polymer waste supply chains based on circular economy. Uncertain Supply Chain Manag 9(2):343–350. https://doi.org/10.5267/j.uscm.2021.2.008
Sørensen PB (2018) From the linear economy to the circular economy: a basic model. FinanzArchiv 74(1):71. https://doi.org/10.1628/001522118X15097191506475
Suchek N, Fernandes CI, Kraus S, Filser M, Sjögrén H (2021) Innovation and the circular economy: a systematic literature review. Bus Strateg Environ 30(8):3686–3702. https://doi.org/10.1002/bse.2834
Theeraworawit M, Suriyankietkaew S, Hallinger P (2022) Sustainable supply chain management in a circular economy: a bibliometric review. Sustainability 14(15):9304. https://doi.org/10.3390/su14159304
Upadhayay S, Alqassimi O (2019) Transition from linear to circular economy. Westcliff Int J Appl Res 2:62–74. https://doi.org/10.47670/wuwijar201822OASU
Villalba-Eguiluz U, Sahakian M, González-Jamett C, Etxezarreta E (2023) Social and solidarity economy insights for the circular economy: limited-profit and sufficiency. J Clean Prod 418:138050. https://doi.org/10.1016/j.jclepro.2023.138050
Zhou S, Smulders S (2021) Closing the loop in a circular economy: saving resources or suffocating innovations? Eur Econ Rev 139:103857. https://doi.org/10.1016/j.euroecorev.2021.103857

Chapter 2
The Circular Supply Chain Framework: Strategies, Concepts, and Advantages of Embracing Circular Economy in Supply Chains

K. Vanisri, A. Senthamizhselvi, Sunitha Sankaravarman, and I. Joseph Arc Leo

Abstract The rapid growth in technology has led to the manufacturing of many technological products and gadgets, making them overly abundant that an oversupply and shortage situation occurs in most aspects of the modern world. There is a country where much food goes to waste, while others have too many electronics; some countries do not have enough food and electronic supplies to sustain their needs. This created quite an imbalance and inefficiency. Collection and reuse or reprocessing of the waste products of different companies are also not satisfactory. This is the area where SCM can contribute significantly to developing the reusability and reprocessing potential of used products. Companies need to redesign supply chains in light of a circular economy that creates value out of waste by way of reusing and recycling products. However, the competencies of the employees required for such business models have remained largely unmanaged. In that respect, in this paper, a full taxonomy of skills that companies need to move towards a circular economy business model has been presented. Research into the contribution that supply chains can make to the circular economy is only just starting to take place and is in its early stages. As a result, the chapter illustrates the place of supply chains in a conceptual and broader framework that puts them at the center of it all. This conceptual framework now calls for purposed and detailed discussion on how

K. Vanisri (✉)
VEMU Institute of Technology, Tirupati, Andhra Pradesh, India

A. Senthamizhselvi
VIT-AP School of Business, VIT-AP University, Amaravati, Andhra Pradesh, India
e-mail: Senthamizhselvi.a@vitap.ac.in

S. Sankaravarman
Ranipettai Engineering College, Walaja, Ranipet, Tamil Nadu, India

I. Joseph Arc Leo
Department of English, Loyola College, Chennai, Tamil Nadu, India
e-mail: leojoseph@loyolacollege.edu

K. Mathiyazhagan et al. (eds.), *Humanizing Circular Supply Chain Management*, Circular Economy and Sustainability,
https://doi.org/10.1007/978-3-031-86562-6_2

exactly supply chain practices position themselves in enabling the effective implementation of circular-economy practices. The chapter also identifies areas of convergence that could foster cooperation among scholars and practitioners in the fields of Supply Chain Management, Circular Economy, and related disciplines.

Keywords Circular economy · Supply chain process · Benefits of circular supply chain process · Human resource management

2.1 Introduction

Business administration in the current age is getting closer to the very urgent concept of sustainability. This search has bent towards a high tendency, referring to the concept of Circular Economy, which helps reduce resource intake and less waste emission. The concept of the "Circular Economy means a systemic change in the operational structure of business—mating economic activity with environmental well-being, also calling for the development of new business models that drive better ecological, social, and economic results" (Montag 2023). From this perspective, circular economy is linked with the supply chain management practices at a deeper level, since both are dependent on the effectiveness of management in business operations. Indeed, global dialogue and approach to supply chain management are changing, and sustainability is one of the areas that companies are increasingly focusing on. Within this quest, the idea of the circular economy has followed through with new ways of doing business within the context of logistics and resource governance. Traditionally, companies have achieved sustainability by investing in a vertically integrated supply chain structure. Contrarily, considering the circular economy perspective, enterprises are building sustainable cross-industry networks that facilitate the establishment of interconnected supply chains guided by the principles of recycling, reducing, and reusing resources.

Companies and policymakers often view CE as the most aspirational pathway in placing economic growth on a trajectory aligned with sustainable development. Circular economy logistics emphasizes closing loops through the design of supply chain processes that limit waste, and promotes re-use, repair, and re-manufacturing to attain a regenerative system. Circular economy acceptance is driven within India by supply chain enterprises through a change in conventional methodology and the adoption of sustainable resolutions (Del Giudice et al. 2020). Moreover, diffusing the principles of CE in SC entails the acquiring, processing, and applying relevant information and knowledge to perform the operational changes by the firm effectively (González-Sánchez et al. 2020). AI, big data analytics, and the IoT are disruptive technologies that are changing decision-making in the supply chain manager's lens. Next chapters of the study will examine the implications in a supply chain of circular economy practices—especially the creation of human touch within these transactions.

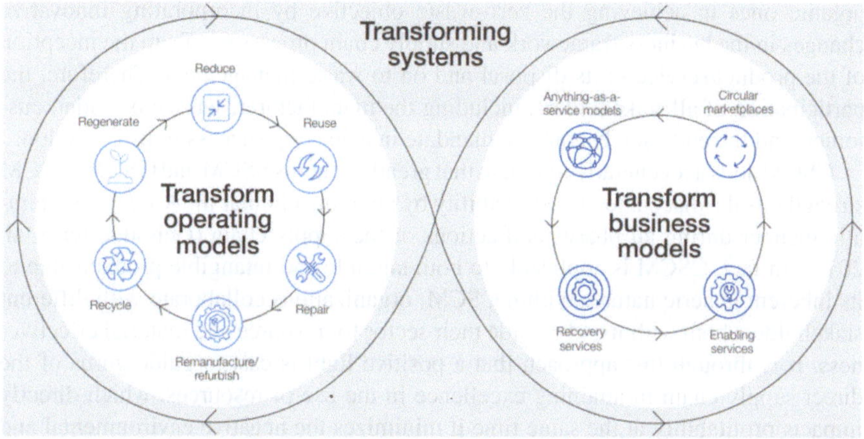

Fig. 2.1 Transforming industries into circular economies. (Source: World Economic Forum 2023; https://www.weforum.org/events/world-economic-forum-annual-meeting-2023)

This study will therefore provide, at the very least, two significant contributions to existing literature. First, it makes an investigation of the interrelationship that exists between the concepts of CE and SC in such a way that a further, still-growing understanding of circular economy supply chains and valuable lessons for all practitioners within industry are derived. In so doing, it guides the personification of the principles of CE within the supply chain. This is followed by consideration of the need to extend a research study on the relationship between the circular economy and the supply chain. First, it discusses how CE influences SCM and details the role of CE in SCM; second, it defines what competencies an employee must possess to manage successfully circular supply chains, based on successful practices; and lastly, with a discussion on Circular Economy Human Resource Management in the light of Circular Economy Supply Chains.

However, let us embark on a journey to comprehend the essence of circular economy logistics and its impact on enhancing the efficiency of supply chain management (Fig. 2.1).

2.2 Review of Literature

The concept-writing approaches the infusion of Circular Economy (CE) into Supply Chain Management (SCM) as simple: common, it denotes the term 'circular supply chain' (CSC) in the diverse scholarly research investigations of the topic (Calzolari et al. 2023; Matos et al. 2023; Nguyen et al. 2023). Circular supply chain management involves the incorporation of circular principles in the management of the supply chain and its associated business and natural ecosystems. This strategic approach works on the regeneration of scientific materials and the renewal of organic ones in achieving the zero-waste objective by incorporating innovative changes in the business framework and supply chain processes—from the inception of the product/service to its disposal and on to waste management. Therefore, the participation of all stakeholders, including the manufacturer, service provider, customer, and the end user, becomes a mandate in achieving success in its execution.

CSCM adds a regenerative element that greatly benefits SSCM and GSCM. CSCM strengthens the discourse of sustainability by applying circular thinking in a systematic manner during all phases and actions of the supply chain (Geissdoerfer et al. 2017). In fact, CSCM is applicable to both tangible and intangible products due to its inherent generic nature. Within CSCM, organizations collaborate with different stakeholders both within and outside their sector for resource and material effectiveness. It is through this approach that a positive light is cast to guide gurus of the direct supply chain in attaining excellence in the use of resources, which directly impacts profitability at the same time it minimizes the negative environmental and social/economic spillovers (Fig. 2.2).

Ideally, a circular supply chain is expected to account for no wastes at all due to its strategic process for renewing and regenerating the resources from the industrial and natural ecosystems in which it serves. The ideal of CSC encompasses two distinct classifications of resource flow: "primary resource flow and circular resource flows", as illustrated in Fig. 2.1.

Primary resource flows are associated with the simple flow of commodities in traditional linear and closed-loop supply chains. The circular resource flows, on the other hand, pertain to the flows associated with goods/materials/energy being "recycled, retained, reused, repaired, remanufactured, refurbished, recovered", etc. In the practical field, CSCM intends to achieve the goal of waste reduction by introducing innovations along the system to create values out of materials that have been stated as 'waste.' For example, in the construction field, there is incorporation of recycled

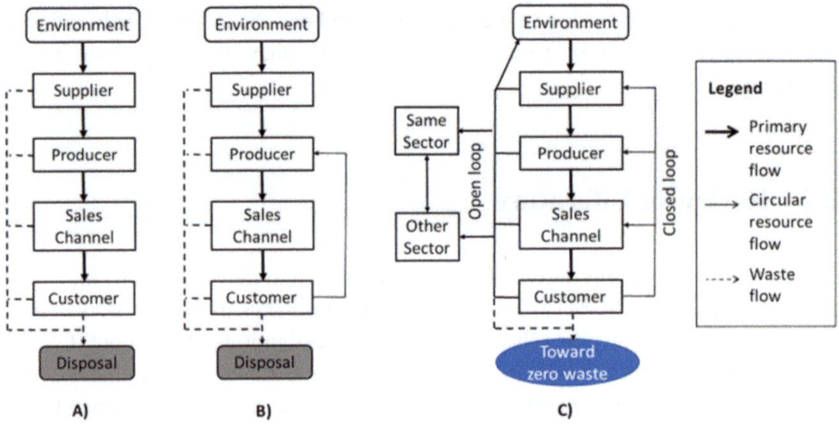

Fig. 2.2 Circular supply chain loops. (**a**) Linear supply chain. (**b**) Closed loop supply chain. (**c**) Circular supply chain. (Source: Farooque 2019)

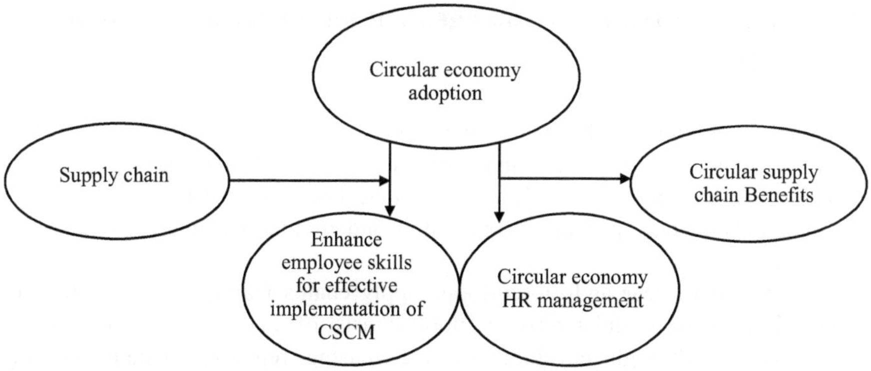

Fig. 2.3 Conceptual framework. (Source: Authors own compilation)

PET bottles; by filling them with lightweight concrete, the bottles become insulated walls for residential buildings (Scheel and Bello 2022).

A company may use the wasted textile material to produce insulation, for example, to be utilized by the construction sector, according to Nasiri et al. (2018). Similarly, in the case of a food supply chain, as Calzolari et al. pointed out in 2023, waste cooking oil is gathered and delivered to a refinery to be transformed into biodiesel. Source reduction initiatives are very critical, while the rest of the food waste is suitable for composting or anaerobic digestion to yield methane as renewable energy and fermentation, which is an excellent fertilizer in agriculture and horticulture. Guided by theoretical constructs of CSCM outlined earlier, we have prepared a groundwork literature review in this emerging area. We wish that this contribution serves significantly toward the development of CSCM and introduces another new perspective for researchers within the field of SCM, with noteworthy implications for management practices, policies, public health, and environmental well-being (Fig. 2.3).

The integration of supply chain management within a circular economy framework significantly develops employees' competencies by immersing them in sustainable practices, advanced technologies, critical analysis, problem-solving, teamwork, regulatory knowledge, and ethical considerations. This exposure not only expands their understanding but also inspires innovative thinking regarding the reuse and repurposing of materials and products. Employees receive training in cutting-edge technologies related to recycling, remanufacturing, and reverse logistics, thereby enhancing their technical expertise and ensuring they remain informed about industry developments. They are also encouraged to devise inventive solutions to logistical challenges, promoting a culture of ongoing improvement and innovation. Furthermore, they acquire valuable insights into regulatory and market dynamics, equipping them to foresee changes and adjust their strategies accordingly.

2.3 Human Resources Management for a Circular Economy Supply Chain

Inherently, a circular supply chain is to have no waste, based on the systematic manner of replenishing and revitalizing resources not only in the industrial and but also in natural ecosystems within which it operates. The CSC integrates two types of resource movements: "primary resource flow and circular resources," as illustrated in Fig. 2.1.

Conventional linear and closed-loop supply chains feature the direct flow of goods. In contrast, circular resource flows consist of material, good, and energy flows that pass through recycling, stacking, reusing, repairing, remanufacturing, refurbishment, and recovery, among other similar operations. Instilling practice, CSCM seeks to minimize waste by inventive application through the system to derive alternative value from the materials that were formerly labelled as "waste products". An example of such practice in using recycled PET bottles within construction has it whereby waste PET bottles are filled with lightweight concrete, then such bottles can be used as insulated walls within homes (Scheel and Bello 2022).

For example, a company would divert waste resulting from textiles, producing insulation material utilized in building industries (Nasiri et al. 2018), and excess cooking oil from a food chain would go through a refining process to produce biodiesel (Calzolari et al. 2023). The effort to minimize the origin of wastage needs to be dramatically increased, and the remaining food scraps are suitable for anaerobic digestion to produce methane as renewable energy and fermentation, which doubles as value-added fertilizer in agri-business and horticulture. Our research team developed the following theoretical framework of CSCM based on the seminal literature related to this emergent research area. We do hope that this work has significantly accelerated the development of CSCM and raised a fresh perspective for the scholars interested in the area of SCM into which a number of important implications exist for managerial practice, policy, public health, and environmental well-being.

2.4 Strategies for Integrating HR and CSCM

To effectively integrate HR and CSCM, organizations can adopt the following strategies:

2.4.1 *Cross-Functional Collaboration*

Encourage collaboration between HR, supply chain, and sustainability teams to align goals and share best practices:

Cross-Functional Teams: Lead cross-functional teams with respect to projects and initiatives on sustainability.

Integrated Planning: Involve human resources in all strategic planning sessions on supply chain sustainability.

2.4.2 *Comprehensive Training Programs*

Develop training programs on general principles of sustainability as well as practices on circular supply chain:

- **On boarding Programs**: Add sustainability and the principles of the circular economy into the on boarding programs of new staff.
- **Continuous Learning**: Provide regular training and professional development opportunities on CSCM.

2.4.3 *Employee Empowerment and Engagement*

Develop the means and opportunities involving employees in sustainability aspects:

- **Idea platforms**: Provide platforms where employees will share and discuss ideas on how to make chains of supply more sustainable.
- **Volunteering programs**: Employees are encouraged to be involved in community activities or voluntary environmental work.

2.4.4 *Transparent Communication and Reporting*

Keep the lines open towards stakeholders regarding progress, hurdles, and goals related to sustainability:

- **Regular Updates**: Keep updating about sustainability initiatives and successes continuously.
- **Employee Involvement**: Involve employees in the sustainability reporting process and obtain their feedback.

2.4.5 *Sustainability-Linked Incentives*

Link the incentives and rewards with sustainability goals to engage employees:

- **Performance Bonuses**: Provide performance bonuses if the fixed sustainability targets are achieved.

- **Recognition Programs**: Design recognition programs that involve the identification of the sustainability champions working in the organization.

The exemplar companies effectively demonstrate the implementation of CSCM, as illustrated in the following case studies.

Case Study 1: Interface
Interface, a global manufacturer of modular carpets, has integrated HR and CSCM to achieve its sustainability goals. The company's "Mission Zero" initiative aims to eliminate any negative impact on the environment by 2020. HR plays a key role by:

- **Employee Training:** Providing extensive training on sustainability and circular economy principles.
- **Engagement Programs:** Engaging employees through programs that encourage innovation and participation in sustainability projects.

Case Study 2: Unilever
Unilever has made significant strides in integrating HR and CSCM through its Sustainable Living Plan. Key HR initiatives include:

- **Leadership Development:** Training leaders on sustainability and circular supply chain management.
- **Employee Empowerment:** Empowering employees to contribute to sustainability goals through innovation and continuous improvement programs.

2.5 Role of Circular Economy in Supply Chain Management

Circular logistics to a circular economy is a new concept that diverts away from the current 'take-make-waste' model (Straub et al. 2023). It's a system designed by the eventual development of closed-loop systems where material resources circulate and remain in use as long as possible. The effect of supply chain companies in driving the transition to a circular economy is huge. These companies, as one of the most important stakeholders in global logistics, can drive sustainable practices with their influence through a few ways.

2.5.1 Lower Resource Use

Embracing a circular economy logistics scheme lowers resource use through resource and energy use promotion with high productivity and minimum wastage. Circular supply chains focus on maximising product life cycle through higher levels of recycling, refurbishment, and restoration, as opposed to dispense of products after their first use.

2.5.2 *Low Waste Production and Recycling*

Conventional linear supply chains are one of the biggest reasons which are generating waste and polluting the environment. The philosophy of circular economy logistics is not that waste would be generated and thrown into landfills; instead, such methods should be used that waste will be collected and reintegrated into some other products or application. The wastes of a product and valuable resources can be markedly reduced if reverse logistics, which is defined as a process of recovery, repair, and reintegrate of products in the supply chain, is used.

2.5.3 *Ethical Sourcing and Sustainable Practices*

Circular economy logistics calls for the ethical sourcing of materials and the widespread of sustainable practices in the entire supply chain. Ethical sourcing helps such organizations conserve their respective natural habitats and also in safeguarding weaker communities.

2.5.4 *Innovative Supply Chain Design*

Indian supply chain companies should be designed by enabling circular principles:—Reimagine product life cycles—Consider end-of-life solutions at the design phase: Closed-loop systems. An optimized design will reduce energy consumption, waste, and inefficiency in general.

2.5.5 *Strategic Partnerships and Collaborative Networks*

Collaboration is the key to circular economy success. Supply chain corporations are better placed to foster collaborative networks and partnerships. Working in close collaboration with the suppliers, manufacturing plants, customers, and recycling plants, these corporations are able to ensure the effective flow of materials and products through the reversed chains.

Efficient product collection, recovery, and refurbishment systems are basic and indispensable components of circular economy logistics. They enable supply chain companies to adopt reverse logistics processes for increasing product lifespan, material reuse, and waste reduction.

Supply chain firms can invest and lobby the installation of circular technologies, including advanced recycling methods and remanufacturing processes. These firms, through the incorporation of advanced technologies, can effectively close the loop, convert the waste material into valuable resources and reduce the dependency on virgin materials.

2.5.6 Stakeholder Education

Education and raising awareness remain vital for the uptake of circular economy. Indian supply Chains should educate their employees' customers' and partners' on circular economy logistics benefits—what their contribution could be and from their side in the drive for sustainability initiatives.

2.5.7 Matchless Benefits of Circular Economy Supply Chain

The transition to a circular economy in logistics has many benefits to the supply chain and the investment climate in general.

2.5.8 Unleash the Unlimited Opportunity of Efficiency of Resources and Cost Cuts

Professionals from logistics solutions providers will leverage gargantuan economies in the efficient use of resources and minimizes waste. There will be a considerable fall in operating costs for material used and apparent levels of energy efficiency will spike the profitability m.

2.5.9 Increased Resilience and Risk Management

Supply chains on a circular economy basis are more resilient to any form of supply chains' disturbances in the supply of raw materials. This is basically because their sourcing of raw materials is diversified, and reverse logistic mechanisms can be implemented. Risks that primarily cause supply chains to break and raw commodities to run out can be overcome.

2.5.10 Better Reputation and Customer Loyalty

Today's consumer environment requires sustainable and ethically made products. Using circular economy principles strengthens the brand's reputation and attracts customers who are aware of environmental impact; it finally results in long-term loyalty.

2.5.11 *Compliance and Market Access*

It therefore follows that with increasingly strict environmental policies globally, companies that have incorporated circular economy approaches in their supply chains are considerably better placed, by extension, to conform to the said regulations. Such conformity ensures markets that envisage procurement standards that tend towards sustainability.

The implication of circular economy logistics results in a tremendous drop in environmental impacts, including greenhouse gas emissions and resource depletion. Supply chain bodies contributing effort toward sustainable initiatives play a pivotal role in combating climate-driven impacts on supply chains and safeguarding ecosystems.

Thus, this proves to be an optimistic area for supply chain management service providers in driving and leading this change toward logistics under circular economy principles, with a view for a more sustainable future. Working collaboratively with these innovative supply chain firms will pave the way for a prosperous and environmentally conscious business landscape.

2.6 Skills Required by Employees for Efficient CSCM

The Bureau of Labour Statistics (BLS) ranks logistics expertise, IT and Database experience, Critical thinking skills, and Customer service talents as the most vital in supply chain employment. A few others are as follows.

2.6.1 *Seeing the Big Picture*

The ability to stand back and view operations from start to finish is important if practitioners are to be successful in supply chain environments. "The successful supply chain expert will be able to predict what can go wrong with everything from packaging to shipping and delivery," says Jon M. Quigley, author and principal of Value Transformation LLC. He opines that a mix of critical thinking skills and past experience can often enable the practitioner to foresee this. "Every single action has a consequence," says Christoph Seitz, president of CFR Rinkens, an International shipping firm. "You need to be ready to create a backup plan if needed to keep the supply chain running constantly. Planning for the unexpected is part of logistics."

2.6.2 Keep Calm on High Pressure

Supply chain working conditions are always quite rapid, and each process depends on how well the previous stage performs its duties. Quigley concurs that the circumstances may be a bit hectic.

"A manufacturing line that doesn't have the right material or incorrect material can be shut down," he says, adding that this may lead to a lot of time wasting.

2.6.3 Mature Problem-Solving Skills

Yeah, it sounds a little trite, Lauderdale acknowledges, but he still insists that robust problem-solving capabilities are a requirement for professional success in SCM.

"I've worked with some of the most book-smart people you could meet, and if solid problem-solving ability is absent, often so is their success in the field," Lauderdale asserts. "To deliver a [job], a great supply chain management professional will draw on a range of problem-solving tools, from analysis to business development knowledge to operational information to psychology, and more. You'll need to utilise, and sometimes cajole, technology, data, things and people to help your organisation deliver amazing outcomes."

2.6.4 Always Improving

"A hyper-focus in continuous improvement expertise is, and will continue to be, key when separating a company's supply chain from the rest of the pack," Lauderdale says. The position requires understanding of the supply chain, he says, but also the scope and the dynamics of your company's goals so you can help with the systems in place to reach them.

"Once you see what the finish line is supposed to look like, you'll be able to use information gathered from your various systems to monitor, analyse and adjust processes accordingly," Lauderdale says. "As the name suggests, continuous process improvement is a never-ending cycle."

2.6.5 Competent in Project Management

Michelle Gomez is an author, career counselor, and "Logistics and Supply Chain Specialist with nearly 20 years of experience". "It's key to plan out your day and tasks as much as possible," she says. "Supply chain management involves so many key details that must be gone through carefully and planned for accordingly."

2.6.6 Relationship Skills

Supply chain managers must be exceptionally good at collaboration, teamwork, and relationship building—all of which begin with communication skills. These are basic competencies for leading teams, cross-functional collaboration with other internal organizations, and, perhaps most importantly, building alliances and relationships with the companies that form a supply chain.

2.6.7 Information Systems and Technology Competency

As in all other corporate functions, information technologies are driving efficiencies, performance, and production in supply chain management. Analytics over data helps the management discover inefficiencies and risks in the supply chain (Maranesi and De Giovanni 2020). As a result of this dynamic rendering of prescriptive insights on top changes in circumstances to businesses through integrated analytics software and information systems, they can allow agility in anticipation, pivoting, and quick response to supply chain difficulties.

2.7 Benefits of a Circular Economy Supply Chain

2.7.1 Environmental Benefit

Resource efficiency, which is the key theme in non-linear production literature, particularly in connection with materials, has such typical features as reduced raw material consumption and reduced waste, carbon, greenhouse emissions. It was cited that products coming as a result of non-linear production methods have shown convincingly low carbon emitting levels for their life-cycle compared to the linear route. Closed-loop, industrial ecology, industrial symbiosis are found to exert potentially influential positive impulse on lowering energy consumption. However, these practices are not mentioned in reverse logistics and CE studies. Water quality impacts are hardly studied; the effects of water use are poorly researched in all sectors. The literature on environmental impacts is by and large normative and appealing and based on 'comparative mathematical' models of linear and circular production chains (Aragão and Jabbour 2017).

2.7.2 Economic Benefit

Literature indicates that shifting from non-linear to circular production has the potential to generate high financial profits, mainly through value creation. Value may be created by a decline in "costs, marginal costs, virgin material costs, waste disposal, and lower environmental taxes". Value creation can also be translated into revenue increases from inter-industry exchanges, for example, through selling the waste as the input of another industry or generating energy from waste. However, the literature also highlights that the production in functions through industrial symbiosis does demand higher fixed costs and a greater service intensity, whereas geographical proximity between firms involved in resource exchanges is more important for decreasing variable transport costs (Maji et al. 2023). Being involved in recycling, repair, reconditioning and remanufacturing, therefore, has the potential of improving profitability, as reported by. On the other hand, it is observed herein that only reconditioning and remanufacturing operations are found to boost sales. It also shows a possibility of drop in repaired product sales with strong pressures of stakeholders screening down to influence manufacturing profitability. Different research by, Paquin et al. 2015, demonstrates promising outcome for Eco-efficiency performance. However, the level of value created depends on variables like the firm experience level within industrial symbiosis, the generated quantity of waste, and also on the intensity of participation of the dedicated firms to waste.

2.7.3 Operational Benefit

It is operational benefit where the non-linear strategies of CE have relatively similar productivity and efficiency advantages. However, the approach towards recycling has placed CE short of value activity. Research into reverse logistics has established a number of benefits regarding product quality, lower inventory and lead times, incineration, and higher service levels; few adverse operational impacts exist. While research into CE, closed-loop systems, and reverse logistics does establish a unanimous view of increased uncertainty when operations are dependent upon external supplies of waste—thereby having the potential to reduce operational robustness or requiring higher inventory levels to counterbalance supply variability—the negative impacts seem to be limited (Banihashemi 2022). Also, the perception of customers towards remanufactured products is adverse and this would be overcome only by closed-loop strategies and market segmentation. This barrier though can be reduced as firms can develop circular products of advanced quality and more attractive to customers.

2.7.4 Social Benefit

Murray et al. (2015), demonstrate that within most conceptualizations and implementations of the circular economy, social sustainability has poorly been developed. We have identified only 12 papers which deal with the social dimension of sustainability, and six of these do it indirectly. For example, Murray et al. (2015) assume that 'a fully implemented CE will create investments, which inter alia advance social justice'. The most common themes in this regard are those on employment and community development. Three papers reflect on social welfare in closed-loop and reverse logistics. Wang et al. (2015), show that raising incentives and penalties for closed-loop supply chain managers benefits social welfare. Zhalechian et al. (2016), provide a model that assesses the impact of reverse logistics chains on job creation and community development. Pishvaee et al. (2014), find a chain in which supply chains with recycle operations are related to higher costs but also higher environmental and social benefits. Paquin et al. (2015) show positive results regarding ecodevelopment: increased employment and reduced carbon dioxide emissions. And also a qualitative assessment of social consequences through their examination of closed-loop strategies in the meat sector in Italy.

The significance of CSCM can be better understood by examining the following case studies on CSCM from the previous literature.

Case Study 1

This study uses a case study research methodology to analyze the role of HR and HRM in Corporate Environmental Responsibility (CE) organizations. The research design is based on case studies on CE, which are considered suitable for examining CE practices within organizations. The study uses non-probability sampling strategies, including convenience and purposive methods, to select six cases from the Ellen MacArthur Foundation (EMF) case study repository. The study aims to analyze HR/HRM in firm-level CE operations comprehensively. Although the analysis of case studies impacts the outcomes, the originality and significance of the research remain intact. The paper offers a snapshot of the current situation regarding HR and HRM functions in CE business instances.

The growing number of organizations adopting the circular economy for sustainable development has resulted in a lack of clarity regarding the involvement of individuals in these organizations and their management. This study aims to offer crucial empirical data and insights into human resources (HR) and the role of HR management in circular economic (CE) endeavors by examining six specific cases. Previous studies have predominantly concentrated on the design of products and processes, supply chain operations, marketing strategies, and sales management. Examining these case studies revealed broader social impacts such as the generation of employment opportunities, enhanced well-being, and transformations in organizational culture.

Case Study 2

Chiarini (2014) and Linder and Williander (2017), there is limited and inconclusive evidence to support the idea of long-lasting environmental effects. The analysis of trade-offs among many dimensions of environmental impact is not conducted systematically, even in mathematical models. Several case studies rely on a limited number of interviews, sometimes even only one, which raises concerns about the level of theoretical saturation attained. Only a small number of qualitative research have examined the use of double-coding, and the lack of this technique raises issues about the consistency of the findings. Regarding quantitative studies, two cross-sector surveys have found evidence supporting the favourable impact of reverse logistics on resource efficiency and environmental performance (Chiarini 2014; Khor et al. 2016).

Paquin et al. (2015) utilize secondary data obtained from 313 waste exchanges in the UK to demonstrate favourable outcomes in terms of the overall reduction of garbage sent to landfills and the decrease in emissions. In a longitudinal investigation conducted by Chiarini (2014), it was discovered that reverse logistics has a crucial role in enhancing environmental performance in manufacturing for large enterprises in Europe. However, the same does not hold for services. Khor et al. (2016) discovered that prolonging the lifespan of items through repair and reconditioning improves environmental performance. This finding was based on a study conducted on a small sample of 89 companies in Malaysia.

2.8 Conclusion

It is for this reason that the concept of CE has been in the lime light of the twenty-first century as an enabler of the quest for sustainability. Appreciating the implications of the business decisions on both the environment and the talents of employees, organizations are recognized as a main driver in society's transition to CE. The competitive environment in place also requires that an organization be innovative in their production process and review of how they make use of available resources and manage their waste. A circular production system should increase production with reduced input of resources and fuel energy with low emissions. The concept of circularity needs to be subjected to firms' supply chains because the function of a firm's supply chain significantly affects the overall performance of the firm. Therefore, scholars are increasingly becoming interested in adjusting supply chains to the principles of the circular economy. After a deep analysis of the existing literature, a theoretical framework has been built with all the factors that could give birth and shape the execution of circular supply chains. Although renovated strategies and the enabling infrastructure are already available to guide businesses, the role of skills in Circular Business Models remains relatively under addressed in the literature and in practice. Although some literature, for example, Coleman et al. (2022), Janssen (2022), and Sumter et al. (2021), adds valuable insights, a comprehensive understanding of the place of employee skills in the successful implementation of

CBM has not been thoroughly explored or assimilated into practice. The current research effort attempts to fill this gap by not only delimiting the necessary skills for circular start-ups but also providing a framework that improves the understanding of these skills, which enables organizations to identify and develop them.

Following is what is created when one marries Human Resources with "Circular Supply Chain Management": greater sustainability, operational efficiency, and higher employee engagement. By aligning HR strategies in support of the CSCM goals, companies would have a resourceful and motivated workforce committed to driving the agenda for sustainable practices. This multipurpose approach aids not only environmental and social objectives but also strengthens the organization's brand, reputation, and prospects for long-term success. This chapter thus intends to phenomenally revisit the critical role of HR in advancing CSCM and integrating these two critical areas. More specifically, a culture of sustainability will be fostered, continuous training provided, and incentives structured so that they support sustainability—a resilience and responsibility in supply chains facilitate business contributions toward a sustainable future.

References

Aragão CG, Jabbour CJC (2017) Green training for sustainable procurement? Insights from the Brazilian public sector. Ind Commer Train 49(1):48–54. https://doi.org/10.1108/ICT-07-2016-0043

Banihashemi TA (2022) Investigating the effect of disposition options of reverse logistics on sustainability performance in the Australian manufacturing industry

Calzolari T, Bimpizas-Pinis M, Genovese A, Brint A (2023) Understanding the relationship between institutional pressures, supply chain integration and the adoption of circular economy practices. J Clean Prod 432:139686. https://doi.org/10.1016/j.jclepro.2023.139686

Chiarini A (2014) Sustainable manufacturing-greening processes using specific lean production tools: an empirical observation from European motorcycle component manufacturers. J Clean Prod 85:226–233. https://doi.org/10.1016/j.jclepro.2014.07.080

Coleman E, Radix AE, Bouman WP, Brown GR, de Vries ALC, Deutsch MB, Ettner R, Fraser L, Goodman M, Green J, Hancock AB, Johnson TW, Karasic DH, Knudson GA, Leibowitz SF, Meyer-Bahlburg HFL, Monstrey SJ, Motmans J, Nahata L et al (2022) Standards of Care for the health of transgender and gender diverse people, version 8. Int J Transgender Health 23(S1):S1–S259. https://doi.org/10.1080/26895269.2022.2100644

Del Giudice M, Chierici R, Mazzucchelli A, Fiano F (2020) Supply chain management in the era of circular economy: the moderating effect of big data. Int J Logist Manag 32(2):337–356. https://doi.org/10.1108/IJLM-03-2020-0119

Farooque C (2019) Circular supply chain management: a definition and structured literature review. Abstract. J Clean Prod 228:882–900

Geissdoerfer M, Savaget P, Bocken NMP, Hultink EJ (2017) The circular economy—a new sustainability paradigm? J Clean Prod 143:757–768. https://doi.org/10.1016/j.jclepro.2016.12.048

González-Sánchez R, Settembre-Blundo D, Ferrari AM, García-Muiña FE (2020) Main dimensions in the building of the circular supply chain: a literature review. Sustainability (Switzerland) 12(6). https://doi.org/10.3390/su12062459

Janssen V (2022) Circular business model in an R&D service business: a roadmap from initiation to implementation. Master's Thesis, University of Twente

Khor CC, Do T, Jia H, Nakano M, George R, Abu-Amero K, Duvesh R, Chen LJ, Li Z, Nongpiur ME, Perera SA, Qiao C, Wong HT, Sakai H, De Melo MB, Lee MC, Chan AS, Azhany Y, Dao TLH et al (2016) Genome-wide association study identifies five new susceptibility loci for primary angle closure glaucoma. Nat Genet 48(5):556–562. https://doi.org/10.1038/ng.3540

Linder M, Williander M (2017) Circular business model innovation: inherent uncertainties. Bus Strateg Environ 26(2):182–196. https://doi.org/10.1002/bse.1906

Maji S, Yadav N, Gupta P (2023) LGBTQ+ in workplace: a systematic review and reconsideration. Equal Divers Incl 43:313–360. https://doi.org/10.1108/EDI-02-2022-0049

Maranesi C, De Giovanni P (2020) Modern circular economy: corporate strategy, supply chain, and industrial symbiosis. Sustainability (Switzerland) 12(22):1–25. https://doi.org/10.3390/su12229383

Matos J, Martins C, Simões CL, Simoes R (2023) Comparative analysis of micro level indicators for evaluating the progress towards a circular economy. Sustain Prod Consum 39:521–533. https://doi.org/10.1016/j.spc.2023.06.002

Montag L (2023) Circular economy and supply chains: definitions, conceptualizations, and research agenda of the circular supply chain framework. In: Circular economy and sustainability, vol 3(1). Springer, pp 35–75. https://doi.org/10.1007/s43615-022-00172-y

Murray KA, Preston N, Allen T, Zambrana-Torrelio C, Hosseini PR, Daszak P (2015) Global biogeography of human infectious diseases. Proc Natl Acad Sci U S A 112(41):12746–12751. https://doi.org/10.1073/pnas.1507442112

Nasiri M, Rantala T, Saunila M, Ukko J, Rantanen H (2018) Transition towards sustainable solutions: product, service, technology, and business model. Sustainability (Switzerland) 10(2). https://doi.org/10.3390/su10020358

Nguyen K, Akbari M, Quang HT, McDonald S, Hoang T-H, Yap TL, George M (2023) Navigating environmental challenges through supply chain quality management 4.0 in circular economy: a comprehensive review. Sustain For 15(24):16720. https://doi.org/10.3390/su152416720

Paquin RL, Busch T, Tilleman SG (2015) Creating economic and environmental value through industrial symbiosis. Long Range Plan 48(2):95–107. https://doi.org/10.1016/j.lrp.2013.11.002

Pishvaee MS, Razmi J, Torabi SA (2014) An accelerated benders decomposition algorithm for sustainable supply chain network design under uncertainty: a case study of medical needle and syringe supply chain. Transp Res E Logist Transp Rev 67:14–38. https://doi.org/10.1016/j.tre.2014.04.001

Scheel C, Bello B (2022) Transforming linear production chains into circular value extended systems. Sustainability (Switzerland) 14(7). https://doi.org/10.3390/su14073726

Straub L, Hartley K, Dyakonov I, Gupta H, van Vuuren D, Kirchherr J (2023) Employee skills for circular business model implementation: a taxonomy. J Clean Prod 410:137027. https://doi.org/10.1016/j.jclepro.2023.137027

Sumter D, de Koning J, Bakker C, Balkenende R (2021) Key competencies for design in a circular economy: exploring gaps in design knowledge and skills for a circular economy. Sustainability 13(2):776

Wang D, Gan C, Wu C, Wang D (2015) Ethical leadership and employee voice: employee self-efficacy and self-impact as mediators. Psychol Rep 116:751–767. https://doi.org/10.2466/01.07.PR0.116k29w9

World Economic Forum. World Economic Forum Annual Meeting 2023, Davos. 2023. https://www.weforum.org/events/world-economic-forum-annual-meeting-2023

Zhalechian M, Tavakkoli-Moghaddam R, Zahiri B, Mohammadi M (2016) Sustainable design of a closed-loop location-routing-inventory supply chain network under mixed uncertainty. Transp Res E Logist Transp Rev 89:182–214. https://doi.org/10.1016/j.tre.2016.02.011

Chapter 3
Training and Development for the Circular Supply Chain Workforce

Cristina Ciliberto and Mathiyazhan Kaliyan

Abstract The transition to a circular economy requires a completely new way of thinking about resources and their utilisation. Human resources are central to this, as people need to have different skills and attitudes to live according to the principles of the circular economy. This urgent need to close the skills gap is at the centre of our chapter on how we can prepare employees to drive sustainable change.

This chapter takes an interdisciplinary approach, highlighting the discrepancies between what is currently expected of employees and what the circular economy actually requires. It argues that training should not only be technical, but must also include systems thinking, collaboration and environmental awareness. It also outlines a broad framework for training that includes various methods, such as innovative training techniques, digital tools or continuous professional development programmes and more. Successful case studies and best practices from different companies are used to show how companies have been able to effectively upskill their employees. In addition, lessons are drawn for each industry and examples are given that can be adopted by others. It also highlights the collaboration between different stakeholders such as governments, educational institutions, industry, etc. to ensure the prioritization of circular skills development. The chapter concludes by envisioning a future where an experienced and circular economy-ready workforce drives sustainable economic development by building a resilient circular economy that produces zero waste. Companies can drive the transformation brought about by circular economy principles, opening the doors to a better future by equipping their employees with the necessary skills.

Keywords Human resource · Skill · Competencies · Circular economy · Upskilling · Reskilling · Sustainable development

C. Ciliberto (✉)
Department of Economics, University of Messina, Messina, Italy
e-mail: cristina.ciliberto@unime.it

M. Kaliyan
Thiagarajar School of Management, Madurai, Tamil Nadu, India

K. Mathiyazhagan et al. (eds.), *Humanizing Circular Supply Chain Management*, Circular Economy and Sustainability,
https://doi.org/10.1007/978-3-031-86562-6_3

3.1 Introduction

The concept of the circular economy, which considers a restorative and regenerative system that minimizes waste and maximizes resource efficiency, has emerged as a convincing approach to addressing urgent environmental problems and advancing sustainable development (Kirchherr et al. 2017). Comprehensive transformation across all industries is required to shift into this model with people being at the centre point.

When companies adopt circular economy principles like waste prevention, maintaining materials and products in use and restoring natural systems, training their workforce becomes paramount. The workforce's skills and mindset have been shaped by traditional linear models hence there is need for skillset development towards circularity (Straub et al. 2023).

Business processes from product design up to disposal should take on a circular approach. It is therefore strategically important for organizations to train their staff on these ideologies since success of transition lies in their hands alongside unlocking its full potential (Kirchherr and Piscicelli 2019). Entities must identify gaps in skills and knowledge followed by crafting training programs that will equip employees with necessary competencies needed for thriving within such an environment.

As the global economy seeks to move to a circular model, research is focused on the exact skills, far removed from the current labour force, required to effectively lead this paradigm shift. This skills gap constitute a major difficulty and is slowing down the circular economy shift in different sectors and job areas.

Nevertheless, the area where there is the most need for research is determining and monitoring the specific abilities that are needed across different industries with which to implement circular economy principles efficiently. This work is designed to bridge the gap by identifying the required skills in different sectors to move from linear to circular models systematically. The importance of this study lies in its ability to affix to the target skills and qualifications of those who can be successfully upskilled and re-skilled which, in turn, will assure that organizations and governments are able to develop a workforce that can lead the transformation of the circular economy.

Identifying the mismatch between current workforce skills and circular economy requirements are the first steps towards solving the problem. The skills of professionals have been shaped by traditional linear economic systems which often prioritize short-term efficiency and resource exploitation over long-term sustainability and circular economy principles (Kirchherr et al. 2017). Consequently, many employees do not have the know-how, attitude or hands on experience needed to handle the intricacies of closed-loop systems, resource recovery among others in a circular economy.

In order to come up with focused upskilling and reskilling programmes, it is important to evaluate sectoral and occupational skills shortages. For instance, workers in the manufacturing industry may require training on eco-design basics, selection of materials for circularity as well as sustainable production methods

(Gusmerotti et al. 2019). Professionals within waste management should be equipped with knowledge about advanced recycling technologies; material recovery processes; reverse logistics among others (Beducci et al. 2024).

It is essential to bridge the skills gap in the circular economy right now because it directly affects how well organizations and economies can make the most of the circular economy. If we fail to do this, we may find ourselves at a great disadvantage when competing with others and it could also mean that there is less innovation, efficiency in using resources or even the general changeover to sustainable ways of living (Kirchherr and Piscicelli 2019).

Thus, the elimination of the circular economy gap by conscious organizations and competent policymakers can significantly foster the talent and expertise of the staff, thus, these individuals will be even more capable to the catalysts of the change in the circular economy.

3.2 Literature Review

The complete transition to a circular economy may take some time. In addition, the ruling of the mindsets must be imposed that each of them be interconnected, namely, economics, ecology, and society, and therefore they have to be treated from a system-thinking approach. This originates from the idea that everyone is included and gets to know the cause-and-effect relationship and the interaction of different acts in various fields that result in a chain of subsequent effects in other things. Furthermore, the formation of a culture that fosters collaboration through the medium of cross-functional mindset is just as important, with the engagement of multiple disciplines in problem-solving process being the norm (Kraaijenhagen et al. 2016; Ul-Durar et al. 2023).

Whatever is bringing us nearer to the acquisition of the essential circular economy habit of using resources in a manner that is exact and sufficientness is the key, bringing about a sense of responsibility, stewardship, and conservation among the populations. The sense of responsibility should be part and parcel of the employees' minds that the resources they are using are finite, scarce, and their practices are causes of harming the planet and the future generations (Kirchherr and Piscicelli 2019). By promoting this thinking, companies will enhance their staff to be change catalysts that will create unparalleled progress, utilization, and implementation (de los Rios and Charnley 2017).

As the literature suggests, different skills are of utmost priority in the development of training programs in which the circular economy is the main issue. In the table that will follow, the main skill categories and specific skills of the circular economy and the authors are listed (Table 3.1).

Strengthening the relevant skills and attitudes in these areas through targeted training and development measures will enable companies to take full advantage of a circular economy; the employees they need to bring about sustainable change and be successful in the long term. Therefore, they need to create a workforce that can

Table 3.1 Circular economy skills

Skill category	Specific skills	References
	Knowledge related to material and recycled material	Chari et al. (2018), Chen (2022), de los Rios and Charnley (2017), Dokter et al. (2021), Thorley et al. (2021), Virtanen et al. (2017), Watkins et al. (2021)
Design skills	Design skills	Janssens et al. (2021), Leube and Walcher (2017)
	Material flow analysis	Avadanei et al. (2020)
	Eco-design	Cramer (2020), Pal et al. (2021), Rodriguez-Andara et al. (2018), Watkins et al. (2021), Wolf et al. (2022)
	Ability to apply big data to eco-design	Lopes de Sousa Jabbour et al. (2019)
Production skills	Skills on new green technologies	Declich et al. (2020), del Vecchio et al. (2021), Giannoccaro et al. (2021), Lopes de Sousa Jabbour et al. (2019)
	Circular manufacturing and cleaner production	Rizos et al. (2016), Rodriguez-Andara et al. (2018)
	Remanufacturing skills	del Vecchio et al. (2021), Lanz et al. (2019), Pal et al. (2021), Sumter et al. (2018), Vogt Duberg et al. (2020), Wolf et al. (2022)
Recovery skills	Cleaning	Casper and Sundin (2018)
	Repairing	Beyeler and Jaeger-Erben (2016), Phung (2019)
	Recycling and reuse	Oliveira Silva and Morais (2022), Wolf et al. (2022)
	Eco-labelling	Avadanei et al. (2020)
Communication skills	Circular economy communication	Minguez et al. (2021), Sumter et al. (2021)
Legal skills	Environmental legislation	Bassi and Guidolin (2024), Eisenreich et al. (2022)
Measuring skills	Life cycle assessment	Avadanei et al. (2020), Ballie and Woods (2018), Khan et al. (2020), Santa-Maria et al. (2022)
	Sustainable reporting and accounting	Bag and Pretorius (2020)

deliver sustainable change and be successful in the long term. "Critical thinking" and "creativity" are also circular skills that need to be developed in employees. These skills are not only needed for future jobs, but are also important for other sectors (Whalen et al. 2018a).

3.3 Methodology

The approach used and the research methods adopted to identify the required skills in this circular transitioning era have been summarized in Table 3.2.

Table 3.2 Methodology

Methodology stages	Procedure	Approach applied
Problem statement	Examination of the research topic	Research objective, discussion of the problem and research questions
Research design	Definition of the research approach	Qualitative, case studies
Literature review	Identification of key literature on circular economy skills	Understanding the topic Highlighting the impact of circular economy skills on green resource management
Case study analysis	Review of case studies and circular economy skills to improve corporate circularity	Recommendations from the case studies
Results and conclusion	Report on results	Role of circular economy skills in the corporate structure

In particular, it can be seen that a qualitative approach was used based on a literature review and subsequent analysis of few case studies in light of the theoretical findings.

This study was conducted through a literature review on the state of the art of circular economy skills in Green Resource Management. Specifically, the purpose of the literature review was to examine the scope of circular economy skills in the green transition. The literature review was conducted on the basis of the scheme proposed in the literature by authoritative scholars (Denyer and Tranfield 2009) and is structured in five basic steps:

(a) Identification of the research topic and elaboration of the research questions;
(b) Identification of the criteria;
(c) Selection and evaluation of the studies on the basis of the inclusion/exclusion criteria;
(d) Analysis of the studies on the basis of relevance to the research topic;
(e) Processing of the results.

Looking specifically at the different points, we have:

(a) Identification of the research topic and elaboration of the research questions

The research questions to be asked are as follows:

1. What are the circular economy skills required to achieve a greater degree of circularity in companies that train and implement them?
2. What is their impact on business sustainability?

(b) Identification of criteria

The Scopus and Google Scholar databases were used to search for articles. Articles were found using the advanced search function with the key terms described below, over the period 2014–2024. This timeframe was considered in view of the enormous volume of articles that would not have allowed a thorough skimming.

(c) Selection and evaluation of studies based on inclusion/exclusion criteria

Journal articles with the terms 'circular economy skills' and/or 'circular supply chain workforce' and/or 'sustainability' were selected. However, books, reports, dissertations, unpublished working papers, non-English papers and papers that cannot be downloaded were excluded in order to maintain the integrity and consistency of the results over the selected 10-year period.

(d) Analysis of the studies on the basis of relevance to the research topic

The content of the papers was examined to enable 'the researcher to analyse the textual information and systematically identify its properties'.

(e) Processing of results

The results of the analysis were used to deduce on the topic.

This thesis was also based on the analysis of some few case studies. According to authoritative scholars, case studies are defined as 'an in-depth study of a single unit (a relatively limited phenomenon) in which the scholar's aim is to elucidate the characteristics of a larger class of similar phenomena'. According to other authors, case studies have not always been regarded as an appropriate method for scientific research. Underlying this was the fact that case studies are situation-specific and provide a narrow basis for generalisation. However, learning from specific cases was later considered a strength rather than a weakness. Furthermore, in-depth case studies provide the means to understand the interplay of a phenomenon and a specific context. Based on this, case studies have become a common method to be used in many different scientific areas.

In particular, case studies are preferable 'when a problem is directed towards the analysis of a number of interdependent variables in complex structures'.

3.4 Managerial Implications

Upskilling means improving and upskilling workers' skills and knowledge to adapt them to the changing demands of their jobs or industries. In the context of the circular economy, upskilling measures aim to equip people with the right skills, perspectives and capabilities that would lead to a shift in sustainability and regenerative redevelopment of economic models (Ling 2022; World Economic Forum 2020). This differs from reskilling or retraining, where workers are prepared for new jobs related to new technologies or changes in the industry (Ling 2022), creating new roles or novel processes. Successful upskilling programs should include sophisticated training tailored to deepen the concept of circular economy, such as resource efficiency, waste management, product life cycle analysis, closed-loop systems and the principles of eco-design (Kirchherr et al. 2017; Whalen et al. 2018b). Using a systems approach allows workers to understand the intricacies of the economic, environmental and social aspects that are the drivers of the circular economy and thus better recognize the opportunities of the circular economy (Bocken et al. 2016). In addition, the use of digital tools and teaching with immersive methods, such as virtual reality simulations, applications of the augmented reality principle and gamification, are the techniques that can be used for students to achieve such knowledge transfer and retention. Furthermore, creating a development environment that emphasizes lifelong learning is even more important. Among the methods for the workforce to be trained are short-term activities like seminars, workshops, trainings with the online tutors and even mentorship that will help to update employees in the multidimensional framework of the circular economy. Some integrative ideas and approaches are the inclusion of a growth mindset as well as the building of interdependencies, cross-functional partnership, and information sharing within and across the different branches creating a descriptive embodiment; such practices can be the key to new circular economies (Kirchherr et al. 2017).

The change to a circular way of existence is a hard one that brings about profound changes in the way companies think, plan and act. Influential leadership is essential in spearheading this transformation and make the necessary changes at companies. Circular Leadership depicts a leadership methodology that gives main priority to the concepts of the circular economy, brings forth a culture of sustainability, and gives the workers the opportunity to adopt and participate in the circular development (Soni et al. 2023). It means that leaders have to be very committed to issues concerning the sustainability, to use the resources in the best way possible and to design the organization in a closed-loop way, thus creating core values and decision-making processes (Gusmerotti et al. 2019). Company executives perform the role of enablers in bringing about the circular economy transition within their organization. They are the ones to determine the policy and to pull on the vision and to gain the stakeholders' trust in the circular economy agenda (Kirchherr et al. 2017). Leaders should demystify the benefits of the circular economy, these include better resource management, lower costs, and getting it done faster than competitors, in order to win stakeholders' hearts (Gusmerotti et al. 2019). They should find and promote technologies that will enable products to be services, or initiatives that will enable industries to share resources. For example, transformational leaders need to rely on a style that motivates and liberates their employees in order to take initiative and promote circular economy initiatives. Transformational Leaders are models, who transmit the common purpose, who encourage intellectual stimulation, and who give person-centered support while also endorsing and affirming (Bass and Riggio 2006). They not only nurture their colleges to be open-minded but let them ask questions and come up with original concepts that further the goals of the circular economy. These leaders who implement the community of learners model have a somewhat different role in that they may encourage workers to engage in circularity and share their vision or even teach them new skills (Centobelli et al. 2020). Sustainable Circular Leadership or transformational leadership compels creating a work environment that stimulates and, at the same time, rewards the introduction of circular operating mechanisms and innovation. The first step could be to create appropriate systems and procedures that will reward and promote the employees who come up with the circular solutions implemented, for example, resource utilization strategies, waste reduction initiatives, and product redesign for circularity (Straub et al. 2023). This refers to performance-based bonuses, a system for employee recognition, as well as money set aside for innovation for programs that are linked to circularity. The environment that allows them to take calculated risks and experiment and also, share their expertise, while training and mentoring is provided, will lead to the development of the actionable approaches by employees. Organizations can apply the Circular Leadership method and by doing this make culture and operation that are compulsory for a smooth and perfect move to the circular economy and also form employees who are excited, and self-governing and able to follow the sustainable principles.

3.5 Case Study Examples

The real-life examples are a necessity to prove all the specific benefits that companies obtain from training their workforce but also that it is through these creative and innovative ways training that such could be done in many other fields. For example, one of the best cases is Interface, a global maker of modular carpet tiles. The company came to the inspiration of the circular economy long ago setting "Mission Zero"—environment impact elimination by 2020 as a must for them. Interface has worked on this incredibly bold objective through capacity building hence they launched the comprehensive training programs that discussed topics like sustainable design, life cycle analysis, and etc. (Kok et al. 2013). The industry employed simulation technology that takes employees into the virtual portions of the manufacturing plant to train them and get them to apply all the skills learned in real-time. Interface also made sure it promotes continuously learning through the spread of the word in the chain and collaboration of the functional teams.

Also, we have the instance of Philips, a Dutch company that is active in the health commodities and consumer electronics segments. Philips is one of the companies that has carried on the lightbulbs to the "Circular Lighting" and to have more customer satisfaction other than just the sales of the lightbulbs, and others related products (Bocken et al. 2016). Philips received positive feedback from both employees and external vendors about the usefulness of the new way of working and the lighting as a service strategy. Philips made the decision to run a thorough retraining program that integrates conventional lectures and seminars with interactive workshops and online learning sessions to facilitate the transition to the service-providing business model. In this regard, a new program called "The Circular Economy Academy" brought staff from various functions together to share new ideas and best practices (Bakker et al. 2020).

By analysing these case studies, one can grasp the essence of the message, and thus, the desire to pull additional factors and the readiness to retrain the workforce is generated by the corporations. Initially, organizations must propose a structure centered on theory as well as the application of the new concepts in the organization as well. Additionally, the company will succeed if it comes up with the tools for its human resources to undergo online learning, or doing real activities by introducing practical information—technology and making them participate in hands-on learning sessions. In addition, what is central to the reformation of the human organization is the creation of space for continuous learning and the cultivation of cross-functional and interactive co-operation that promotes the circular-minded and the interdisciplinary problem-solving, among the grouping. Indeed, it is the senior management's full commitment to and support for cultural and operational adjustments that will affect the successful transformation into a circular enterprise.

Organizations can contemporize the aforementioned outstanding skills and incorporate them into specific areas in their industries thereby mastering the process of transition to a circular economic model precisely.

3.6 Conclusion

The conversion to a circular economy which is worldwide and the result of our new understanding that compels us to adjust how we produce, use, and deal with resources is a challenging task. One of the major means to this is the implementation of a competent and practically knowledgeable workforce in the circular economy. This, in turn, implies the adoption of a mental attitude that is quick, decision-making that is consistent, and skills that are relevant to the eco-equitable trade. This part pinpoints one of the most important aspects of human capital development and the emergence of circular economies. Another way to achieve this is by training the staff and equipping them with requisite skills. Thereby they could be comfortable with the concept of circular economy. The companies could be a positive influence on their employees as the workers would turn out being proactive enough to be able to chronicize the opportunities of circular economy. Indirectly, their participation will boost the development of a company that is truly committed to the principles of circularity through their awareness to circularity, their ability to think in systems, and their skilfulness in resource management as they do their job. Vision relegation is the most important issue faced by company leaders due to the fact that it is necessary to convince the audience, and within a similar, external environment that allows the motivation to run the company through the layer of sustainability, responsibility, and accountability to circular processes to take place. Transformational leadership styles that incorporate sustainability deep into the core values of their organizations would be the method by which the leaders would be able to facilitate cultural transitions and operational changes that would be necessary for a corporate circular economy to thrive. Nevertheless, the education model is paramount in which the circular economy style can be learned by the future generation that cannot be solely based on the corporate organization. It is really a joint venture which couples such societal institutions as governments, the educational sector, and businesses. Thanks to setting out the fiscal backing and governance and through the user of partnerships between the government and private sector a catalyst for skill development will be created. It is very important that universities draw from best practice international and national in this area to provide students with the necessary knowledge and skills to become leaders in the circular economy. Equally important, enterprises are expected to spend more money on the professional development of their employees and to get directly involved with educational institutions in the materialization of circular economy leaders.

First, circular workforce development could stimulate economic growth if the correct and practical application of circular principles was governed. It is, therefore, anticipated that a sustainable economy will have already become a part of the future. Giving a brief example would be re-designing the production process to be more resource efficient, energy-saving by recycling waste, and creating the green economy—which expands the advanced science and machinery. It is our youths' future in which the human capital is the only way of empowerment to these challenges, besides that, young people's education should not be solely about clean energy but also should be integral part of the investigations into the mitigation of the growing climate changes. When dealing with the competition due to the waves of circular economy, it is the community's desire to focus human resources that is very important for the realization of this vision. Their mantles of the social citizenship are worn by these new green peasants who familiarize with eco-living through programs together with young leaders as well as regeneration projects. The young people grow together not only through their personal growth but also it grows the town into a green and developing one.

3.7 A Future Research Roadmap: Green Human Resource Management and Digital Innovation for Circular Workforce Development

Although we talked earlier about ways to promote circular thinking through workforce upskilling, there is another way forward, which is to incorporate green human resource management (GHRM) systems with modern technologies to achieve sustainable workforce development.

A dual approach in terms of technology and human behaviour, the workforce accepted as sustainability is a critical thing (Graves et al. 2019). The reason the trend of sustainable technology has grown significantly is that green technologies have developed at a rapid pace (Khan and Ulucak 2020). So many organizations around the world are not able to deal with the difficulties that are related to creating a more sustainable workforce (Ren and Jackson 2019).

A very common opinion today is that human resource management (HRM) is subject to changes such as biodiversity loss and climate change (Aust et al. 2020). According to the analysis of Boudreau et al. (2003), only when HRM is strategically in tune with organizational goals companies meet their goals. Apart from the financial crisis, a market meltdown that is largely triggered by speculative trading, HRM saves the day by strategic restructuring, which is a prevalent competitive reality occupying the business space. Subsequently, there has been a need for more research on the degrees of significance of HRM with regard to the encouragement of sustainable organizations.

'Green human resource management' (GHRM) (Renwick 2018) is a concept that has been introduced to tackle the ongoing conversation on the possibilities, constraints, and greening effects on the workforce to be able to form greener organizations. Green HRM has not only influenced other areas of management research but has also been shared in other HRM contexts. For example, the papers have delved into how GHRM can influence operations management (Daily et al. 2012), the adoption of environmental management systems (Jabbour and Santos 2008), and the potential impact that HRM may have on movement toward the contemporary environmental trends, for example, such as the circular economy (Jabbour et al. 2019). Essentially, interweaving 'green' topics into traditional HRM is a trending concept that the future of human resources will unpack.

Looking to the future, the use of GHRM in combination with future technologies seems to be an issue for the labour force, which will have many alternatives that are energy efficient and environmentally friendly. R&D should therefore have smart, industry-specific goals. In addition, data protection policy and the digital divide are the components that this addresses. HR, IT and sustainability departments need to be merged for operational business. Therefore, success and sustainability are determined by reporting and improving synergies such as resource utilisation and profitability. The power of the top companies to solve ecological concerns and problems is, for this reason, the most important condition, so they should be in sufficient competition in order to draw and keep the most qualified and long-term employees and ensure their green goals.

References

Aust I, Matthews B, Muller-Camen M (2020) Common good HRM: a paradigm shift in sustainable HRM? Hum Resour Manag Rev 30:100705

Avadanei M, Olaru S, Ionescu I, Ursache M, Ciobanu L, Alexa L, Luca A, Olmos M, Aslanidis T, Belakova D, Silva C (2020) ICT new tools for a sustainable textile and clothing industry. Industria Textila 71(5). https://doi.org/10.35530/IT.071.05.1726

Bag S, Pretorius JHC (2020) Relationships between industry 4.0, sustainable manufacturing and circular economy: proposal of a research framework. Int J Organ Anal. https://doi.org/10.1108/IJOA-04-2020-2120

Bakker C, den Hollander M, van Hinte E, Zijlstra Y (2020) Products that last: product design for circular business models. TU Delft Library

Ballie J, Woods M (2018) Circular by design: a model for engaging fashion/textile SMEs with strategies for designed reuse. In: Unmaking waste in production and consumption: towards the circular economy. https://doi.org/10.1108/978-1-78714-619-820181010

Bass BM, Riggio RE (2006) Transformational leadership, 2nd edn. Lawrence Erlbaum Associates Publishers

Bassi F, Guidolin M (2021) Resource efficiency and circular economy in European smes: investigating the role of green jobs and skills. Sustain For 13(21). https://doi.org/10.3390/su132112136

Beducci E, Acerbi F, Pinzone M, Taisch M (2024) Unleashing the role of skills and job profiles in circular manufacturing. J Clean Prod 449:141456. https://doi.org/10.1016/j.jclepro.2024.141456

Beyeler L, Jaeger-Erben M (2022) How to make more of less: characteristics of sufficiency in business practices. Front Sustain 3. https://doi.org/10.3389/frsus.2022.949710

Bocken NMP, de Pauw I, Bakker C, van der Grinten B (2016) Product design and business model strategies for a circular economy. J Ind Prod Eng 33(5):308–320. https://doi.org/10.1080/21681015.2016.1172124

Boudreau J, Hopp W, McClain JO, Thomas LJ (2003) On the interface between operations and human resources management. Manuf Serv Oper Manag 3(5):179–202

Casper R, Sundin E (2018) Addressing Today's challenges in automotive remanufacturing. J Remanuf 8(3). https://doi.org/10.1007/s13243-018-0047-9

Centobelli P, Cerchione R, Esposito E (2020) Environmental sustainability and energy-efficient supply chain management: a review of research trends and proposed guidelines. Energies 13(2):275. https://doi.org/10.3390/en13020275

Chari A, Niedenzu D, Despeisse M, Machado CG, Azevedo JD, Boavida-Dias R, Johansson B (2022) Dynamic capabilities for circular manufacturing supply chains—exploring the role of Industry 4.0 and resilience. Bus Strat Environ 31(5). https://doi.org/10.1002/bse.3040

Chen Y (2022) Advantages of 3D printing for circular economy and its influence on designers. Proc Des Soc 2:991–1000. https://doi.org/10.1017/pds.2022.101

Cramer JM (2020) The function of transition brokers in the regional governance of implementing circular economy - a comparative case study of six Dutch regions. Sustain For 12(12). https://doi.org/10.3390/su12125015

Daily BF, Bishop JW, Massoud JA (2012) The role of training and empowerment in environmental performance: a study of the Mexican maquiladora industry. Int J Oper Prod Manag 32(5):631–647

de los Rios IC, Charnley FJS (2017) Skills and capabilities for a sustainable and circular economy: the changing role of design. J Clean Prod 160:109–122. https://doi.org/10.1016/j.jclepro.2016.10.130

Declich A, Quinti G, Signore P (2020) SME's, energy efficiency, innovation: a reflection on materials and energy transition emerging from a research on SMEs and the practice of Energy Audit. Mater Tech 108:5–6. https://doi.org/10.1051/mattech/2020036

del Vecchio P, Secundo G, Mele G, Passiante G (2021) Sustainable entrepreneurship education for circular economy: emerging perspectives in Europe. Int J Entrep Behav Res 27(8). https://doi.org/10.1108/IJEBR-03-2021-0210

Denyer D, Tranfield D (2009) Producing a systematic review. In: Buchanan DA, Bryman A (eds) The Sage handbook of organizational research methods. Sage, pp 671–689

Dokter G, Thuvander L, Rahe U (2021) How circular is current design practice? Investigating perspectives across industrial design and architecture in the transition towards a circular economy. Sustain Prod Consum 26. https://doi.org/10.1016/j.spc.2020.12.032

Eisenreich A, Füller J, Stuchtey M, Gimenez-Jimenez D (2022) Toward a circular value chain: impact of the circular economy on a company's value chain processes. J Clean Prod 378. https://doi.org/10.1016/j.jclepro.2022.134375

Giannoccaro I, Ceccarelli G, Fraccascia L (2021) Features of the higher education for the circular economy: the case of Italy. Sustain For 13(20). https://doi.org/10.3390/su132011338

Graves LM, Sarkis J, Gold N (2019) Employee proenvironmental behavior in Russia: the roles of top management commitment, managerial leadership, and employee motives. Resour Conserv Recycl 140:54–64

Gusmerotti NM, Testa F, Corsini F, Pretner G, Iraldo F (2019) Drivers and approaches to the circular economy in manufacturing firms. J Clean Prod 230:314–327. https://doi.org/10.1016/j.jclepro.2019.05.044

Jabbour CJC, Santos FCA (2008) Relationships between human resource dimensions and environmental management in companies: proposal of a model. J Clean Prod 16(1):51–58

Jabbour CJC, Sarkis J, de Sousa Jabbour ABL, Renwick DWS, Singh SK, Grebinevych O et al (2019) Who is in charge? A review and a research agenda on the 'human side' of the circular economy. J Clean Prod

Janssens L, Kuppens T, van Schoubroeck S (2021) Competences of the professional of the future in the circular economy: evidence from the case of Limburg. Belgium J Clean Prod 281. https://doi.org/10.1016/j.jclepro.2020.125365

Khan D, Ulucak R (2020) How do environmental technologies affect green growth? Evidence from BRICS economies. Sci Total Environ 136504:136504

Khan O, Daddi T, Iraldo F (2020) Microfoundations of dynamic capabilities: insights from circular economy business cases. Bus Strat Environ 29(3). https://doi.org/10.1002/bse.2447

Kirchherr J, Piscicelli L (2019) Towards an education for the circular economy (ECE): five teaching principles and a case study. Resour Conserv Recycl 150:104406. https://doi.org/10.1016/j.resconrec.2019.104406

Kirchherr J, Reike D, Hekkert M (2017) Conceptualizing the circular economy: an analysis of 114 definitions. Resour Conserv Recycl 127:221–232. https://doi.org/10.1016/j.resconrec.2017.09.005

Kok L, Wurpel G, Ten Wolde A (2013) Unleashing the power of the circular economy. IMSA

Kraaijenhagen C, van Oppen C, Bocken N (2016) Circular business: collaborate and circulate. Circular Collaboration

Lanz M, Nylund H, Lehtonen T, Juuti T, Rattya K (2019) Circular economy in integrated product and production development education. Procedia Manuf 33. https://doi.org/10.1016/j.promfg.2019.04.058

Leube M, Walcher D (2017) Designing for the next (circular) economy. An appeal to renew the curricula of design schools. Des J 20(Suppl. 1). https://doi.org/10.1080/14606925.2017.1352999

Ling L (2022) Reskilling and upskilling the future-ready workforce for industry 4.0 and beyond. Inf Syst Front. https://doi.org/10.1007/s10796-022-10308-y

Minguez R, Lizundia E, Iturrondobeitia M, Akizu-Gardoki O, Saez-de-Camara E (2021) Education in circular economy: focusing on life cycle thinking at the university of the Basque country, Lecture notes in mechanical engineering. https://doi.org/10.1007/978-3-030-70566-4_57

Oliveira Silva WD, Morais DC (2022) Impacts and insights of circular business models' outsourcing decisions on textile and fashion waste management: a multicriteria decision model for sorting circular strategies. J Clean Prod 370. https://doi.org/10.1016/j.jclepro.2022.133551

Pal R, Samie Y, Chizaryfard A (2021) Demystifying process-level scalability challenges in fashion remanufacturing: an interdependence perspective. J Clean Prod 286. https://doi.org/10.1016/j.jclepro.2020.125498

Phung CG (2019) Implications of the circular economy and digital transition on skills and green jobs in the plastics industry. Field Actions Science Report. J Field Actions (Special Issue 19):100–107

Ren S, Jackson SE (2019) HRM institutional entrepreneurship for sustainable business organizations. Hum Resour Manag Rev 30:100691

Renwick DWS (ed) (2018) Contemporary developments in green human resource management research: towards sustainability in action? Routledge research in sustainability and business series. Routledge, Abingdon, pp 1–199

Rizos V, Behrens A, van der Gaast W, Hofman E, Ioannou A, Kafyeke T, Flamos A, Rinaldi R, Papadelis S, Hirschnitz-Garbers M, Topi C (2016) Implementation of circular economy business models by small and medium-sized enterprises (SMEs): barriers and enablers. Sustain For 8(11). https://doi.org/10.3390/su8111212

Rodriguez-Andara A, Río-Belver RM, Rodríguez-Salvador M, Lezama-Nicolás R (2018) Roadmapping towards sustainability proficiency in engineering education. Int J Sustain High Educ 19(2). https://doi.org/10.1108/IJSHE-06-2017-0079

Santa-Maria T, Vermeulen WJV, Baumgartner RJ (2022) How do incumbent firms innovate their business models for the circular economy? Identifying microfoundations of dynamic capabilities. Bus Strat Environ 31(4). https://doi.org/10.1002/bse.2956

Soni V, Gnekpe C, Roux M, Anand R, Vann Yaroson E, Banwet DK (2023) Adaptive distributed leadership and circular economy adoption by emerging SMEs. J Bus Res 156:113488. https://doi.org/10.1016/j.jbusres.2022.113488

Straub L, Hartley K, Dyakonov I, Gupta H, van Vuuren D, Kirchherr J (2023) Employee skills for circular business model implementation: a taxonomy. J Clean Prod 410:137027

Sumter D, Bakker C, Balkenende R (2018) The role of product design in creating circular business models: a case study on the lease and refurbishment of baby strollers. Sustain For 10(7). https://doi.org/10.3390/su10072415

Sumter D, de Koning J, Bakker C, Balkenende R (2021) Key competencies for design in a circular economy: exploring gaps in design knowledge and skills for a circular economy. Sustain For 13(2). https://doi.org/10.3390/su13020776

Thorley J, Garza-Reyes JA, Anosike A (2019) The circular economy impact on small to medium enterprises. WIT Trans Ecol Environ 231. https://doi.org/10.2495/WM180241

Ul-Durar S, Awan U, Varma A, Memon S, Mention A-L (2023) Integrating knowledge management and orientation dynamics for organization transition from eco-innovation to circular economy. J Knowl Manag 27(8):2217–2248. https://doi.org/10.1108/JKM-05-2022-0424

Virtanen M, Manskinen K, Eerola S (2017) Circular material library. An innovative tool to design circular economy. Des J 20(Suppl. 1). https://doi.org/10.1080/14606925.2017.1352685

Vogt Duberg J, Johansson G, Sundin E, Kurilova-Palisaitiene J (2020) Prerequisite factors for original equipment manufacturer remanufacturing. J Clean Prod 270. https://doi.org/10.1016/j.jclepro.2020.122309

Watkins M, Casamayor JL, Ramirez M, Moreno M, Faludi J, Pigosso DCA (2021) Sustainable product design education: current practice. She Ji 7(4). https://doi.org/10.1016/j.sheji.2021.11.003

Whalen KA, Milios L, Nussholz J (2018a) Bridging the gap: barriers and potential for scaling reuse practices in the Swedish ICT sector. Resour Conserv Recycl 135:123–131. https://doi.org/10.1016/j.resconrec.2017.07.029

Whalen K, Berlin C, Ekberg J, Barletta I, Hammersberg P (2018b) 'All they do is win': lessons learned from use of a serious game for circular economy education. Resour Conserv Recycl 135:335–345. https://doi.org/10.1016/j.resconrec.2017.06.021

Wolf M, Ketenci A, Weyand A, Weigold M, Ramsauer C (2022) Learning factories and sustainable engineering—competencies for students and industrial workforce. IEEE Eng Manag Rev 50(3):115–122. https://doi.org/10.1109/EMR.2022.3195452

World Economic Forum (2020) The future of jobs report 2020. http://www3.weforum.org/docs/WEF_Future_of_Jobs_2020.pdf

Chapter 4
Human-Centric Skills for Efficient Circular Supply Chain Management in the Hospital Industry

Hepzibah Gnana Sinthiya. R

Abstract The Hospital industry is experiencing a digital transition with an emphasis on sustainability, supply chain, and regulatory compliance. This study explores the impact of Compliance with Healthcare Regulations, organisational commitment to sustainability, supply chain collaboration, technological information particularly human-centric competence in supply chain management on 'circular supply chain performance' (CSCP). The questions of the research were explored from lens of 'Resource Dependency Theory, Stakeholder Theory, Theory of planned behaviour, Resource based view theory, Diffusion of Innovations Theory'. Quantitative approach was adopted, and the required data was collected from 150 managers working in 200 bed multi-speciality hospitals. The gathered data was examined by utilising 'Partial Least Squares Structural Equation Modelling' (PLS-SEM) to test hypotheses and relationships among variables. It was found that Compliance with Healthcare Regulations, organisational commitment to sustainability does not have direct significance on CSCP. But Supply Chain Collaboration (SCC), technological information and Human-centric competence in circular supply chain management has a significant factor in shaping Circular supply chain performance. Further it was revealed that SCC, technological information has strong indirect influence on CSCP through the mediation of human-centric competence in circular supply chain management. Hospital industries can enhance further by implementing of compliance with regulations, training the various department in the hospital and supporting the management's commitment to the concept of sustainability, engaging the suppliers and nurturing the partnership, advancing the adoption of technology and supporting the hospital staff to acquire adequate knowledge, skill and to achieve humanised circular Supply Chain Management.

Keywords Humanising circular supply chain · Human competence · Organisational commitment · Compliance with healthcare standards · Sustainability

---

H. G. Sinthiya. R (✉)
Department of Commerce, Voorhees College, Vellore, Tamil Nadu, India

K. Mathiyazhagan et al. (eds.), *Humanizing Circular Supply Chain Management*, Circular Economy and Sustainability,
https://doi.org/10.1007/978-3-031-86562-6_4

4.1 Introduction

Today, companies everywhere want their distribution networks to increase the sustainability economically, environmentally, and socially (Duque-Uribe et al. 2019). Sustainability has become a hot topic, especially in supply chain management. Traditionally, production follows a linear model of taking resources, making products, consuming them, and then disposing of them. Consequently, in recent days healthcare services also have environmental, social, and economical effects (Hutchins and Sutherland 2008). Hospital logistics is defined as the responsible utilization of resources while at the-same time minimizing the amounts of harm that is done to the natural environment and also ensuring that health supply chains are socially responsible and economically feasible (Pereno and Eriksson 2020). Thus, the healthcare industry is a global with extended supply chains and multiple links that incorporate activities such as outsourcing, which often entails employing the low-waged labour and providing dangerous working conditions for them, predominantly in the developing countries (Roseland 2000). In general, this study supports the idea that healthcare organizations need to promote human rights and, at the same time, satisfy economic needs which is said a new concept known as degrowth. The Challenges involved are the increased costs for medical services and the growing population need for such services while opportunities include the development of valuable strategies that enable hospitals to save money, work more effectively, and consequently, demonstrate higher financial results (Meara et al. 2015). The identified healthcare sector plays great role in the society, its primary purpose is to enhance the state of human's health (Pereno and Eriksson 2020). But it also is a major thrust to environmental degradation since it involves the creation of large amounts of waste, energy consumption, and emission of greenhouse effects (Nosratabadi et al. 2019). Earlier literature evidence shows that hospitals and clinics generate wastewater, solid waste, hazardous medical waste and Single use plastics which are very dangerous to the environment and the people. Many health care organizations practice sustainable measures such as waste minimization, reuse, and energy conservation, thereby minimizing the health care supply chain's impact on climate change (Goyal et al. 2019). 'Sustainable Supply Chain Management' (SSCM) in health care is defined as process of sourcing, acquiring, converting, and providing information while maintaining focus on the utilisation of natural resources, people, and organisations responsibly. (Nosratabadi et al. 2019). Technologies such as the blockchain or RFID can help in creating more supply chain transparency that help with recall issues and avoid certain risks. Working in partnership is vital for collaboration with the stakeholders (Dutta et al. 2020). Thus, ethical sourcing plays a major role to achieve collaboration, which means choosing suppliers not only by their compliance with social responsibility, their treatment of workers, and their attitude towards the environment. In any health care organization ethical purchasing serves to protect patients by promoting the quality of procured products. In another way, an organization can reduce cases such as child labour as well as harm to the environment through scrutiny of the suppliers for different standards (Völker and Kistemann 2013). Healthcare organizations should require improvement, which means they should allow their employees to look for the betterment or for new technologies and working procedures or strategies that could be adopted (Gutachter et al. 2018). The application of continuous improvement in the learning organizations enables them to achieve greater satisfaction by making changes that are needed in the current market relative to the sustainable development goals (Musse et al. 2018). SSCM can be summarized as a socially, ethically, and environmentally responsible management of supply chains which delivers business value for sustainable stakeholder returns through effective SC activities in areas of value delivery, costs, risks. Integrating these components assists the healthcare organization to address the environment concern and social as well as economical accountability necessary to retain sustainability and deliver quality health care (Kanokphanvanich et al. 2023).

Although there has been an increase in attention given to management of CSC (Circular supply chain) in the hospital industry, very few research were conducted to establish the competency profile of employees for implementation of humanising circular supply chain management. Previous research is prevalent in general sustainability activities or applied technologies without considering the bilateral relationship of Employee competence (EC) with regards to other variables like Compliance with Healthcare Regulations (CHR) (McNab et al. 2016; Sharma and Luthra 2023), Organizational Commitment to Sustainability (OCS) (Epstein et al. 2017), Supply Chain Collaboration (SCC) (Sarkis and Zhu 2018) and Technological Innovation (TI) (Gupta and Barua 2017; Lieder and Rashid 2016). Furthermore, the overall performance of circular supply chains (CSCP) have been extensively discussed, but the roles and employees' skills (Bocken et al. 2016; Zahid et al. 2019) needed to operate within and contribute to these domains have received limited attention. Additionally, it is crucial to bridge this gap in order to create training interventions and policies that will build the efficiency of CSCM in hospitals. Thus, the study postulates the following research questions: (1) What core competencies are needed in employees to achieve humanised circular supply chain management in the hospital industry given the factors of Compliance with healthcare regulations (CHR), Organizational Commitment to Sustainability (OCS), Supply Chain Collaboration (SCC), Technological Innovation (TI), Employee Competence in Circular Supply (ECCS)? (2) Does CHR, OCS, SCC, IT directly influences CSCM? (3) To reveal the mediating relationship of ECCS between CHR, OCS, SCC, IT and CSCM.

The study analysed various theories associated with the variables which resulted in the proposed framework to find answers to the research questions mentioned above. PLS-SEM analysis is used to test the framework and hypothesis reliability and validity. From the derived results the findings were given which provides valuable insights to achieve effective circular supply chain management in hospital industries.

4.2 Literature Review

Sustainable practices refer to the incorporation of ecological, social, and economic policies and methods in different undertakings. This include using of renewable energy, recycling and reusing of products, use of energy efficient technologies and sustainable procurement of supplies (Emmanuel 2014). Additionally, it is also referring to the development of communities, issues of labour and employment, and the encouragement of local businesses. Organizations that opt for sustainable development avoid causing harm to the environment and are also equally protective of society in the process they ensure the greatest financial returns in the long run to support the betterment of the world (Shelton et al. 2018).

4.2.1 Theories and Its Integration

Theories associated with the variables were considered from previous literatures, to get an in-depth view on the considered factors which further paved way to build a theoretical framework. Particularly, Resource Dependency Theory (RDT), Stakeholder Theory (ST), Resource based view theory (RBV), Diffusion of Innovations Theory (DIT), Theory of planned behaviour (TPB) were incorporated to construct a theorical model. The Resource Dependency Theory assumes that organizations meet requirements to get resources and avoid sanctions in relation to regulations and standards (Biermann and Harsch 2017). This way, stakeholders who require certain compliance maintain proper functioning and resource availability in organizations. Away from it, the ST postulates that organizations exist to serve all stakeholders, including the customers, employees, investors, and the society. Another form of stakeholders includes shareholders, and commitment to sustainable has the organization's commitment to these people (Mahajan et al. 2023). According to the Social Capital Theory, both material and non-material resources that are embodied in and originate from the structure of relations among supply chain partners include relationships, networks, and trusting attitudes that enable cooperation and exchange of information. However, trust, reciprocity and other aspects related to social capital constitute the ground for developing effective cooperation between entities in order to provide the maximum improvement for all the partners and the common long-term goals (Zaini 2018). From RBV it can be proposed that sustained competitive advantage lies in the ability to acquire and deploy resources that are valuable, rare, hard to imitate and for which there are no good substitutes. Thus, financial investment in circular initiatives can build such resources and thereby generate competitive advantage. According to Diffusion of Innovation Theory, it describes the process and at what pace, new ideas or technology is adopted. Like all other technologies sustainable technologies are adopted via a diffusion process that involves early adopters that determine the level of acceptance by the general society (Bertram 2016). According to TPB "attitude, subjective norms, and perceived control" are the determinants for behaviour. The actual behaviour in the application of sustainable actions depends on the perception that the organisation has towards the notion of sustainability, the pressure it receives from the market players, and the perceived effort of implementing sustainable practices (Hassan et al. 2016).

The integration of RDT, ST, TPB, RBV and DIT theories can offer a theoretical framework for developing of the skills for efficient usage of single and multiple SCM for the employees working in a hospital value chain. This entails knowledge in asset acquisition and use, people management, and change implementation. Training solutions should therefore deal with general organizational factors such as activities that aim to facilitate the effective use of resources, communication, and change. These could include seminars, on-going education and computerized support to the teams which will in turn enhance green practices and performance at the workplace. With this integration, hospitals will be in a position to develop a workforce to address circular supply chains hence promoting sustainability in the healthcare system.

Humanizing supply chain management is a crucial process to be implemented in the hospital industry during this digital era (Nosratabadi et al. 2019). As most hospitals have adopted recent technologies to simplify, accelerate, and improve efficiency in their operations, this has resulted in underutilizing the skills of their employee's (Sudusinghe and Seuring 2022). This study identifies the variables that enhance human-centric circular supply chain management in the hospital industry, which are discussed in detail below.

4.2.2 Employee Competence in CSCM (ECCSCM)

ECCSCM is important for successful implementation of circular SCM since employee competence is necessary for circular SCM. This involves making the employees aware and capable of undertaking developments, implementations and management of chain of supplies which are oriented to reuse, Recycling and reduction of wastage (Foltynowicz et al. 2024). This competence encompasses knowledge on the circular economy concepts like product life cycle, resource optimization, and sustainable procurement. Thus, increasing employee competence in ECCSCM will help achieve organisational objectives related to improving corporate performance, decreasing organisation's negative effect on the environment, and stimulating new developments in the direction of sustainable SCM (Tseng et al. 2024).

4.2.3 Circular Supply Chain Performance (CSCP)

It concentrates on suggesting end-to-end, closed loop supply chain approach within which resources are composed, returned, and renewed as few as possible times as possible. Some of the key elements of CSCP include enhancing the product durability and its returnability by means of efficient reverse logistics, improving resource productivity, and the ability to reuse products. If companies quantify and increase the value of their CSCP, the overall sustainability in the economy will impact positively by embracing more sustainable production and quantifiable sustainable economy thus given organizations the ability to withstand tough economic times and thus can be very sustainable (Sudusinghe and Seuring 2022).

4.2.4 Compliance with Healthcare Regulations (CHR)

Regulations and standards remain a mandatory function being implemented in healthcare. Being an element of healthcare logistics policy that encompasses various requirements which ensure the safety and quality of products as well as the efficiency of services (Tsytsyna et al. 2024). In health sector, there are requirements on quality standards like ISO 9001 that focuses on general quality management system and ISO 13485 that focuses on medical devices to provide assurance of product excellence within and across the supply chain. Moreover, it is a must to abide by the regulations from 'FDA (Food And Drug Administration), EMA (European Medicines Agency)' and other organizations that regulate pharmaceutical and biotech products, elements that should be included on product packaging and labeling, temperature control procedures, origin tracking, and record keeping on different activities (Rowan 2024). Business environment regulations are also significant for imposing specific guidelines for the disposal of medical waste, limitation of emission levels of vehicles, and promotion of environmentally friendly practices in carrying out business. Compliance risk management is goes in hand with risk management, which means that it unveils compliance risks to prevent organization from suffering penalties, reputational loss or disruption of operations (Kakemam and Liang 2023). When quality and standards compliance features are prioritized, together with the compliance with the regulations, the healthcare logistics can have the guarantees for the calibre of delivered materials, equipment, patient protection and the gained trust of the other participants of supply chain (Bertram 2016). Ergo, the study put forth following hypothesis:

H1: CHR has direct significant relationship with CSCP.
H2: CHR has direct significant relationship with ECCSCM.

4.2.5 Organizational Commitment to Sustainability (OCS)

The Organizational Commitment to Sustainability (OCS) is an organizational management concept that entails the practice of such sustainability in an organizational structure and framework. OCS management by foreign players ensure that the facilities manage to minimize carbon emissions, conserve the resources and trim down the wastes (Zarim et al. 2017). Another component of OCS is Facility & Economic Sustainability, which considers cost and business performance efficiency of sustainable solutions, cost and resource savings and entrepreneurialism. OCS also entails engagement with the employees, customers, investors, and other associates with regards to sustainability strategies and management of sustainable strategies (Hasim et al. 2018). Goh and Marimuthu (2016) in his study indicates how organisations that subscribe to OCS are making periodic changes in their practices and approach in order to embrace sustainability and owe to the centre of contemporary issues and trends. By using OCS, it is possible to launch long-term changes in the development of businesses and guarantee stakeholders that they will be beneficial for them in the framework of systemic and sustainable development when making decisions in the framework of uncertain world trade environment. As a result, the study proposes the following hypothesis:

H3: OCS has direct significant relationship with CSCP.
H4: OCS has direct significant relationship with ECCSCM.

4.2.6 'Supply Chain Collaboration' (SCC)

SCC is essential in this present highly competitive business climate as the business organizations identify the importance of linking their distributive partners to collaborate in achieving common organizational goals such as reduced costs, increased productivity and satisfaction of the customers (Syahrir et al. 2018). There is always need for performance measurement to determine the efficiency of the collaborative agreements and where there is need for enhancement. Co-operations and information sharing a vital aspect in SCC since it has the ability to cause better forecasts, savings, and quality services (Bvuchete et al. 2021). If the working relations between the parties involved in business processes have to be improved and overall business objectives have to be met, it is crucial to build a solid foundation for measuring the performance level in SCC (Bvuchete et al. 2020). Thus, SCC should be integrated with such strategic priorities as goals and objectives of collaboration, as well as with the emphasis on culture-oriented performance indicators for sustainable improvement and benefit optimization (Ramanathan 2014). Therefore, it is pivotal that companies rely on strategic partnerships and apply various measurement approaches to boost the performance of supply chain and assure long-term business success in the context of the contemporary business environment.

H5: SCC has direct influence on CSCP.
H6: SCC has direct influence on ECCSCM.

4.2.7 Technological Innovation (TI)

It is a known fact that through the lens of technology, societies around the globe are actively evolving over time (Sinthiya 2024). Ranging from medical discoveries that change the lives of many to recognizing new ways of communicating or moving from place to another, innovation provides a means through which new frontiers are opened and human efficiency is boosted (Moshood et al. 2022). Utilising recent technology is always powerful in transforming the economies, enhancing the productivity of the business and opening new horizons for development. But with all these technologies come issues like unemployment, and moral problems to solve, besides the need for better controls to be placed to reduce on the misuse of these technologies (Mele et al. 2014). The roles of the government are to encourage cooperation in different sectors such as research, industries as well as the government to ensure that various problems are solved. Also, it is necessary to apply such components as sustainability and ethical issues to ensure that environmentally friendly approaches are adopted, and ethical practices are maintained. If technology is managed wisely over these and other concerns and possibilities, then it can still remain humanity's driving force in making the world a better place and a safer place for generations to come (Omachonu and Einspruch 2010) (Fig. 4.1).

H7: TI has direct positive influence on CSCP.
H8: TI has direct positive influence on ECCSCM.
H9: ECCSCM has a positive effect on CSCP.

To test mediating effect among the variable the following hypothesis are proposed:

H10: CHR has positive effect on CSCP when ECCSCM mediates.
H11: OCS has significant effect on CSCP when ECCSCM mediates.
H12: SCC has significant impact on CSCP when ECCSCM mediates.
H13: TI has significant impact on CSCP when ECCSCM mediates.

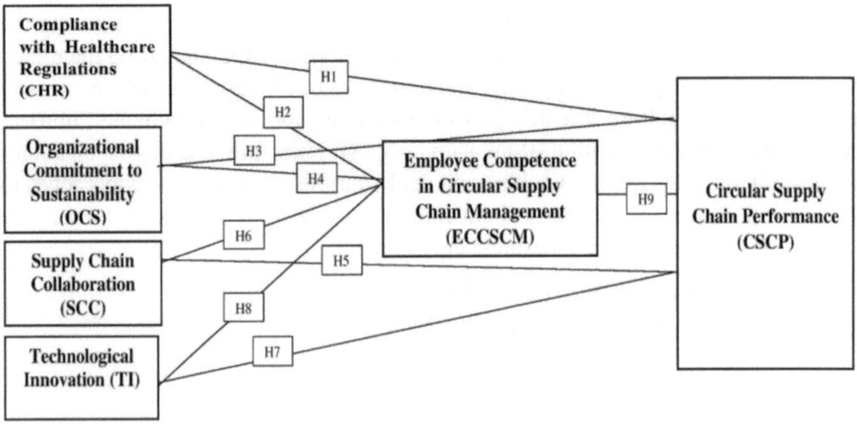

Fig. 4.1 Theoretical framework

4.3 Methodology

Quantitative methodology is used in this study to seek answers to the research queries. In data acquisition, quantitative analysis is effective in ensuring that the information obtained is credible and unbiased and also useful in getting insights into the tendencies in the data (Michel et al. 2011). A Framed questionnaire is used for data collection, from the employees employed in the hospitals who are in the designation of manager level. Firstly, the database of the multi-speciality hospitals having more than 200 beds in Tamil Nadu were collected from the website of "government of Tamil Nadu chief minister health insurance scheme" and the consent of the respondents were asked through the telephone. Initially 300 hospitals were contacted to participate in the survey. Nearly 210 hospital management agreed to participate in the survey. To conduct pilot test the researcher used 25 respondents and made necessary changes in the questionnaire to make it more reliable. The structured questionnaire was sent to 210 mangers of the Multi-speciality hospitals who agreed to participate. Nearly 157 questionnaires were received in return, where the researcher scrutinised by eliminating unfilled questionnaire and rounded off to 150 questionnaires finally. The proposed framework of the study is analysed using Smart PLS 4-SEM. The SEM enables to estimate the complicated models with numerous constructs, numerous indicators, and many structural paths that did not require empirical distributional assumptions (Hair et al. 2019). Firstly, reliability and validity of proposed framework was assessed using measurement model. Secondly, to assess the link between the constructs structural model is used.

4.3.1 Measurement

From various previous scholarly literatures, the scales were adapted which were further used to develop a structured questionnaire. To measure the variables Likert scales (5-point) from 1 (Strongly Disagree) to 5 (Strongly Agree) were used. The independent variables Compliance with Healthcare Regulations scales were derived from Guerra-Bretaña and Flórez-Rendón (2018), McNab et al. (2016), Reiman and Pietikäinen (2012), Sharma and Luthra (2023), Testa et al. (2011), and Zobel (2013). Organizational Commitment to Sustainability (OCS) scales were taken from Adams and Frost (2008), Epstein et al. (2017), and Geissdoerfer et al. (2017). Supply Chain Collaboration (SCC) scales were taken from Genovese et al. (2017), Liu and Bai (2014), Sarkis and Zhu (2018), and Tate et al. (2010), For Technological Innovation (TI) (Gupta and Barua 2017; Hanisch et al. 2015; Lieder and Rashid 2016). The dependent variable Circular Supply Chain Performance (CSCP) scales were taken from Genovese et al. (2017), Liu and Bai (2014), Sarkis and Zhu (2018), and Tate et al. (2010). The mediating variable Employee competence on circular supply chain management (CCSCM) scales were taken from Bocken et al. (2016), Zahid et al. (2019), and Zhong et al. (2016).

4.3.2 Data Analysis

4.3.2.1 Assessing Measurement Model (Fig. 4.2)

The construct's reliability is analysed using the statistical measure Cronbach's Alpha (CA). CA is widely used, which distinctly measure the internal consistency. The construct is said to be highly reliable when the CA values are more than 0.70 (Hair et al. 2019). As per Hair et al. (2019) a construct is said to be valid when it satisfies the following three criteria. Firstly, all the measurement factors loadings should be more than 0.50 to be considered significant. Secondly, the AVE (Average Variance Extracted) of every single construct should be greater than 0.50. Finally, values of composite reliability (CR) must be larger than 0.70 (Hair et al. 2019). Evidence presented in the Table 4.1 postulate that the construct having strong convergent validity as it satisfies all the three criteria mentioned above.

4.3.2.2 Discriminant Validity

Henseler et al. (2015) proposed HTMT (Heterotrait–Monotrait) ratio which is used to test discriminant validity of the measurement model. The HTMT ratio should be below 0.9. which is said to be threshold value. Table 4.2 shows that most of the correlations lie below 0.90 except for OCH-CHR (0.904). The variables were retained in the model as it is very close to 0.90.

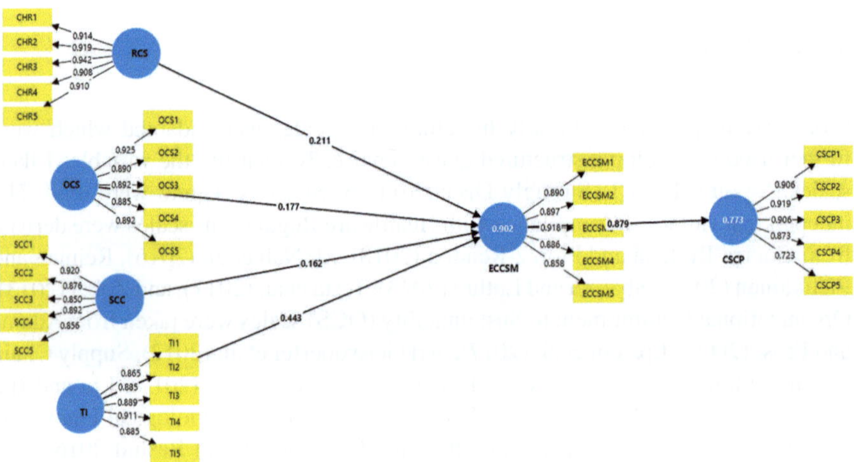

Fig. 4.2 "Measurement model assessment-Depicting Path Coefficients, Factor Loadings, and Construct Relationships". (Source: PLS SEM)

Table 4.1 'Determination of convergent validity'

	Factor loading	CA	CR	AVE
Compliance with healthcare regulations (CHR)				
CHR1	0.914	0.945	0.949	0.822
CHR2	0.919			
CHR3	0.942			
CHR4	0.908			
CHR5	0.910			
Organisational commitment to sustainability (OCS)				
OCS1	0.925	0.944	0.944	0.817
OCS2	0.890			
OCS3	0.829			
OCS4	0.885			
OCS5	0.892			
Supply chain collaboration (SCC)				
SCC1	0.920	0.941	0.944	0.809
SCC2	0.876			
SCC3	0.850			
SCC4	0.892			
SCC5	0.856			
Technological innovation (TI)				
TI1	0.865	0.941	0.941	0.809
TI2	0.885			
TI3	0.889			
TI4	0.911			
TI5	0.885			
Employee competence in circular supply chain management (ECCSCM)				
ECCSCM1	0.890	0.942	0.945	0.813
ECCSCM2	0.897			
ECCSCM3	0.918			
ECCSCM4	0.886			
ECCSCM5	0.858			
Circular supply chain performance (CSCP)				
CSCP1	0.906	0.960	0.961	0.836
CSCP2	0.919			
CSCP3	0.906			
CSCP4	0.873			
CSCP5	0.723			

(Source: PLS SEM)

Table 4.2 HTMT

	Heterotrait-Monotrait ratio (HTMT)
CSCP ↔ CHR	0.840
ECCSCM ↔ CHR	0.823
ECCSCM ↔ CSCP	0.758
OCS ↔ CHR	0.904
OCS ↔ CSCP	0.816
OCS ↔ ECCSCM	0.811
SCC ↔ CHR	0.810
SCC ↔ CSCP	0.872
SCC ↔ ECCSCM	0.860
SCC ↔ OCS	0.883
TI ↔ CHR	0.821
TI ↔ CSCP	0.829
TI ↔ ECCSCM	0.891
TI ↔ OCS	0.840
TI ↔ SCC	0.806

(Source: PLS SEM)

Table 4.3 Inner values

	VIF
CHR → ECCSCM	1.152
ECCSCM → CSCP	1.000
OCS → ECCSCM	1.353
SCC → ECCSCM	2.226
TI → ECCSCM	2.157

(Source: Smart PLS output)

4.3.3 Structural Model Evaluation

4.3.3.1 Collinearity Assessment

Strong correlations between two indicators can provide methodological and interpretational challenges to the structural model, which is known as collinearity and multicollinearity (Hair et al. 2019). Therefore, the issue of collinearity is solved using variance inflation factor test. The values of VIF below 5 rules out the issue of collinearity (Hair et al. 2019). Table 4.3 value below satisfies the thumb rule which shows that there is no problem in the collinearity (Table 4.4).

Table 4.4 Outer values

CHR1	2.931
CHR2	2.048
CHR3	3.930
CHR4	3.651
CHR5	3.287
OCS1	4.210
OCS2	4.293
OCS3	2.713
OCS4	3.803
OCS5	4.207
SCC1	4.095
SCC2	3.180
SCC3	3.163
SCC4	4.973
SCC5	4.064
TI1	3.175
TI2	4.426
TI3	3.926
TI4	4.090
TI5	3.963
ECCSCM1	3.148
ECCSCM2	2.393
ECCSCM3	1.397
ECCSCM4	1.380
ECCSCM5	2.055
CSCP1	1.348
CSCP2	2.407
CSCP3	1.890
CSCP4	3.259
CSCP5	1.432

(Source: Smart PLS output)

4.3.3.2 Assessing Structural Model (Fig. 4.3)

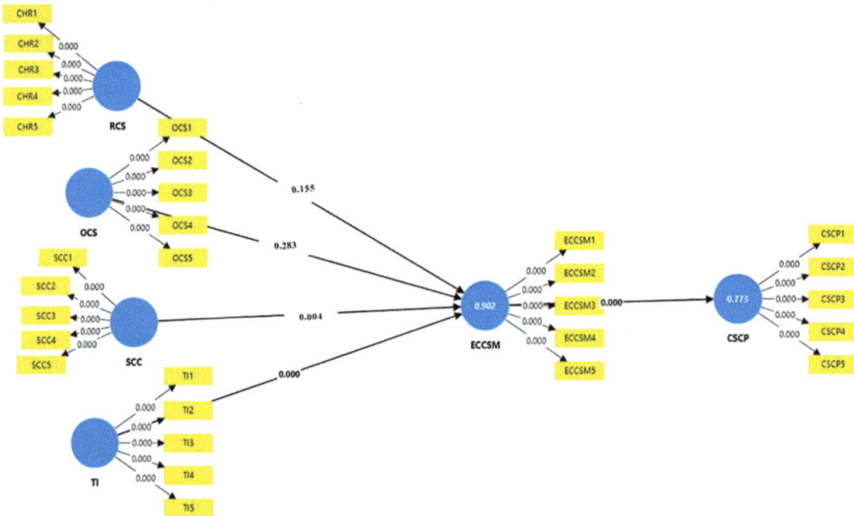

Fig. 4.3 "Structural model-Depicting Path Coefficients and Factor Loadings for Constructs". (Source: PLS SEM)

4.3.4 Coefficient of Determination (R2)

To measure the efficacy of the structural model and to analyse predictive capacity, the R square values are used. The increase in R square values indicates the higher prediction accuracy (Hair et al. 2019). Table 4.5 shows R^2 value of construct Circular supply chain performance indicates 73% of variance and for Employee Competence in Circular Supply Chain Management indicates 90% of variance, explained by the model's independent or latent variables Compliance with Healthcare Regulations, organisational commitment to sustainability, supply chain collaboration and technological innovation indicating strong predictive accuracy. Ringle et al. (2023) suggested Q^2 test which is also known as the blindfolded Stone-Geisser test to analyze the predictive power of the framework. Table 4.5 shows the Q^2 values of all the dependent variables which are above 0. This shows the robust predictive power of the model.

The SRMR (Standardized Root Mean Square Residual) and NFI (Normalized Fit Index) is used to measure model fit. According to Hair et al. (2019) a model is said to be good fit when the SRMR values lies below 0.08 and NFI values closer to 0.9. From Table 4.6 it is evident that the proposed model fits well as it satisfies the thumb rule.

To analyze the significance of the framed hypothesis bootstrapping technique was used (Hair et al. 2019). Table 4.7 shows path outcome of PLS SEM model. The p values below the threshold limit of 0.05 were considered significant.

Table 4.5 R-square and Q-square values

	R-square	Q^2 (=1 − SSE/SSO)
CSCP	0.773	0.523
ECCSCM	0.902	0.742

(Source: Smart PLS output)

Table 4.6 Model fit

SRMR	0.072
NFI	0.794

(Source: PLS SEM)

Table 4.7 Path outcome-hypothesis testing

Hypothesis		β	Standard deviation	T statistics	P values	Supported
H1	CHR → CSCP	0.137	0.096	1.427	0.154	No
H2	CHR → ECCSCM	0.151	0.106	1.423	0.155	No
H3	OCS → CSCP	0.125	0.118	1.065	0.287	No
H4	OCS → ECCSCM	0.137	0.128	1.073	0.283	No
H5	SCC → CSCP	0.219	0.075	2.914	0.004	Yes
H6	SCC → ECCSCM	0.240	0.083	2.887	0.004	Yes
H7	TI → CSCP	0.434	0.049	8.765	0.000	Yes
H8	TI → ECCSCM	0.476	0.051	9.259	0.000	Yes
H9	ECCSCM → CSCP	0.912	0.029	31.973	0.000	Yes
H10	CHR → ECCSCM → CSCP	0.137	0.096	1.427	0.154	No
H11	OCS → ECCSCM → CSCP	0.125	0.118	1.065	0.287	No
H12	SCC → ECCSCM → CSCP	0.219	0.075	2.914	0.004	Yes
H13	TI → ECCSCM → CSCP	0.434	0.049	8.765	0.000	Yes

(Source: Smart PLS output)

The above table depicts, H1 is not accepted as value of p is greater than threshold value of 0.05 (SD = 0.096, β = 0.137, p = 0.154), therefore CHR does not have strong influence on CSCP. CHR also does not influence ECCSM since the value of p is higher than 0.05 (SD = 0.106, β = 0.151, p = 0.155) therefore, H2 is not accepted. The p value of OCS on CSCP is more than 0.05 (SD = 0.118, β = 0.125 p = 0.287), which emphasis that there is no strong positive impact between OCS and CSCP. Hence H3 is also not accepted. OCS does not have positive influence on ECCSM as the value of p is greater than 0.05 (SD = 0.128, β = 0.137, p = 0.283). This depicts that H4 is rejected. H5 is accepted as the p values is less than 0.05 (SD = 0.075, β = 0.219, p = 0.004), which states that SCC has significant influence on CSCP. The significance of SCC and ECCSM is highly positive, since the p value is less than 0.05 (SD = 0.083, β = 0.083, p = 0.004). Therefore, H6 is accepted. H7 is accepted since a strong positive relationship between TI and CSCP is found, as the value of p is less than 0.05 (SD = 0.049, β = 0.434 p = 0.000). The relationship between TI and ECCSM is significant since the p value is less than 0.05 (SD = 0.051,

$\beta = 0.476$, $p = 0.000$). Therefore, H8 is accepted. There is a positive relationship between ECCSM and CSCP as the p value lies below 0.05 (SD = 0.029, $\beta = 0.912$, $p = 0.000$). Hence H9 is accepted. H10 is rejected as the p value is less than 0.05 (SD = 0.096, $\beta = 0.137$, $p = 0.154$). Therefore, it is clear that there is no strong positive correlation between CHR and CSCP when ECCSM mediates. Similarly, H11 is also not accepted as P value is greater than threshold value of 0.05 (SD = 0.118, $\beta = 0.125$, $p = 0.287$), therefore OCS does not have strong influence on CSCP when ECCSM mediates. SCC influence CSCP when ECCSM mediates since the p value is less than 0.05 (SD = 0.075, $\beta = 0.219$, $p = 0.004$). Therefore, H12 is accepted. The p value of TI on CSCP when ECCSM mediates is less than 0.05 (SD = 0.049, $\beta = 0.434$, $p = 0.000$), which shows that there is a strong positive relationship between TI on CSCP when ECCSM mediates. Hence H13 is accepted.

4.4 Discussion

It is evident from the study that the research has endeavoured to identify pertinent factors within the circular supply chain management in healthcare sector context. It discusses the issues such as Compliance with Healthcare Regulations, commitment to sustainability, collaboration with the supply chain, technological advancement, and competency of employees as factors that will help in the internalization of sustainable practices and improvement of the supply chain in hospitals. Hypotheses were formulated to determine the interconnection of the variables. The finding rejects H1, H2, H3, H4 hypotheses where CHR have no influence on the CSCP. It also negates the scramble stating that CHR has negligible influence on ECCSM as suggested by literature. The empirical analysis results showed a negative link with OCS and both CSCP and ECCSM. These findings are supported by Gruchmann et al. (2018) and Pulido (2016) in their study while attempting to analyse "Industry 4.0 technologies in circular economy". Their findings revealed that RCA and OCS do not have significant relationship on CSCP and ECCSM. The results of H5 and H6, are that SCC has a positive effect on CSCP and ECCSM, which corresponds to prior research studies that highlight the centrality of SCC in CSCP and ECCSM. The given result corroborates the findings of other studies that indicate that SCC is critical to CSCP and ECCSM by Upadhyay et al. (2017). The result of research hypotheses H7 and H8 shows that TI is having significant positive relation with CSCP and ECCSM and H9 confirms that ECCSM has positive relation with CSCP. This paper's finding supports the earlier works which have affirmed that TI has a direct and positive relationship both with ECCSM and CSCP (Wang et al. 2020) and also support the previous research that ECSCM extent positively affects better CSCP (Sodhi et al. 2012). The study also aims at establishing the correlation of the independent variables of CHR and OCS with the dependent variable, namely the CSCP with ECCSM as the mediating variable. Thus, according to the results of the current study which negates hypotheses H10 and H11, there is a clear signal that the correlation between CHR and OCS with the CSCP when ECCSM as mediates diminishes relatively. The above results validates by prior research stating that SCP is a function of an organisations compliance with a specific compliance measure and the degree of commitment the organisation has towards supply chain management which also includes ECCSM (Di Maria et al. 2022). In the context of CSCP as dependent variable and ECCSM as the mediator for the independent variables SCC and TI, the results of this study affirm the hypothesis H12, H13 are valid, this confirms that, SCC and TI have a positive indirect link with CSCP enhanced by ECCSM. This result aligns with the literature that positions and maintains the centrality of supply chain collaboration (Flynn et al. 2021) and technological innovation (Wang et al. 2020) plays a great part on the supply chain performance if ECCSCM acts as a mediator (Lu and Ramamurthy 2011). Therefore, it has been suggested that CSCM can be adopted in the healthcare sector to improve overall supply chain efficacy, requirement compliance, organizational engagement, and employees' expertise.

4.4.1 Theoretical Implications

The study highlights the need to employ theories such as RDT, ST, TPB, RBV, and DIT to improve more sustainability practices and a better supply chain healthcare organization. Some of the skill training and development areas that have emerged from need analysis include human competencies for managing resources sustainably, ways of conducting effective communication and change management as means of enhancing the ability to implement humanised circular supply chain practices within the hospital value chain and hence improving performance. Therefore, the study reveals that there is an insignificant correlation between CHR and CSCP as well as the ECCSM. This tends to imply that regulatory structures may not have that much impact on the multiple aspects of the health sector as was once believed probably resulting from the special nature of this particular sector Gruchmann et al. (2018). Pulido (2016) states that the development of competence needs training and development interventions. The findings also provide evidence to show that while the extent of organisational commitment to sustainability may not be high in SC organisations, as it is negatively related to supply chain performance, which implies that commitment to sustainability does not necessarily translate to higher organisational performance in the terms of healthcare industry's supply chain (Lu and Ramamurthy 2011). The relative paucity of clear change points to the fact that such sustainability commitments must be properly supported by developed training and startups to increase the competence of employees. The works of prior researchers support this notion showing that there is a positive significant impact between SCC on CSCP and on ECCSM (Upadhyay et al. 2017). Joint initiatives enable the exchange of information and trainings since organizational prerequisites based on humanising circular supply chain management competence of employees are essential (Sinthiya 2024). It is revealed by the research that all technological innovations have resulted in increase in SCP which underlines the virtue of employing enhanced technologies to bolster performance of distribution network (Wang et al. 2020). Particularly the positive points underlined the importance of developing training schemes that prepare the staff for new technologies. The analysis illustrates that there is a strong correlation between ECCSM with CSCP which supports the literature finding that sustained the effect that employee competence is directly related to humanising supply chain performance conducted by Flynn et al. (2021). The outcome of this research also highlights relevance of context-specific evaluation activities, cross-functional cooperation, IT integration, as well as employee competence regarding the further advancement of humanising supply chain performance.

4.4.2 Practical Implications

Findings of this research postulate certain practical implications for Human-centric CSCM in the Hospital Industry. Firstly, Hospitals ought to be in a position to observe healthcare regulations and standards when disposing of wastes in an environmentally friendly manner (Piser 2019). For instance, Balrampur Hospital, Lucknow, India implement recycling of wastes in the hospital to lessen the effects on the environment and boost sustainability (Gupta and Boojh 2006). secondly, Healthcare facilities can enhance the implementation of sustainability through encouraging the employees to participate in the green activities and supporting the sustainability objectives (Lee and Lee 2022). For instance, Hospital 'All India Institute of Medical Sciences (AIIMS), New Delhi', organises awareness creation programmes on energy conservation among the staff members on the way to adopt environmentally friendly behaviour at workplace (Economic times 2022). Most of the circular supply chains involve suppliers and other stakeholders in the realization of sustainability (Kazancoglu et al. 2021). In the case of Apollo Hospitals, one strategy in the next 3–5 years is that the hospital can source products from local suppliers who have a sustainable sourcing policy. For example, selecting a medical equipment supplier that deals in refurbished equipment assists in cutting on waste and promoting the circular economy (Unique 2016). The implementation of RFID or blockchain is one of the many innovations suppliers could adopt for supply chain visibility and efficiency (Unhelkar et al. 2022). Here, Fortis Hospital adopt the concept of RFID tracking system, which would help the hospital track movement of medical and surgical requisites and maintain proper check and balance so that wastage could be minimized. Also, the same technology can be employed in the tracking of drugs to guarantee its originality and proper sources (Icegen Computing 2024). Providing more focus on human-centric CSCM also ensures the skills of the employees in chain management (Farooque et al. 2019). To support staff involvement in the application of sustainable waste management practices in Apollo Hospital, trainings are offered to staff members concerning best practices in reduction and recycling of wastes. For instance, conducting awareness sessions with the healthcare staff on proper handling and disposal of medical waste results in proper recycling and disposal of such wastes (Limited 2021). With these practical implications, the sustainability concepts can be incorporated into the supply chain management systems successfully with aim to avoid the insignificant impact on the environment and enhance operational results in the hospital sector.

4.5 Conclusion

To conclude, this paper focuses on the importance of putting into practice the concept of sustainable human-centric circular supply chain practices in the healthcare facilities. This paper outlines that by advocating for implementation of compliance with regulations in the respective hospital, by training the various department in the hospital and supporting the management's commitment to the concept of sustainability, engaging the suppliers and nurturing the partnership, advancing the adoption of technology and supporting the hospital staff to acquire adequate knowledge and skill in circular Supply Chain Management, the hospitals will be in a position to create the right foundation for a human-centric sustainable and resilient healthcare supply chain. Thus, the implementation of these approaches in the management of healthcare facilities will reduce insignificant effect on the environment and costs which paves way to improve patients' outcomes and the future sustainability of the healthcare industry sphere.

4.5.1 Limitations and Future Scope

The study concentrates only on the healthcare industry thus the findings and the results cannot be applied to other studies. The study has employed only a sample size of 150 that may cause bias of the results. Some external contextual factors that might influence circular supply chain were excluded from the research as the analysis was restricted to certain variables related to sustainability implementation in healthcare organisations. Further, the identification of potentials for the application of the new scientific know-how such as artificial intelligence and Internet of things in circular supply chain and sustainable development in health care sector can be explored. Advocating for policies and regulations in the field of healthcare supply chains that would improve behaviours of the circular economy activities can be investigated.

References

Adams CA, Frost GR (2008) Integrating sustainability reporting into management practices. Account Forum 32(4):288–302. https://doi.org/10.1016/j.accfor.2008.05.002

Bertram M (2016) The strategic role of software customization: managing customization-enabled software product development, pp 1–304. https://doi.org/10.1007/978-3-658-14858-4

Biermann R, Harsch M (2017) Palgrave handbook of inter-organizational relations in world politics, pp 135–155. https://doi.org/10.1057/978-1-137-36039-7
Bocken NMP, de Pauw I, Bakker C, van der Grinten B (2016) Product design and business model strategies for a circular economy. J Ind Prod Eng 33(5):308–320. https://doi.org/10.1080/21681015.2016.1172124
Bvuchete M, Grobbelaar SS, van Eeden J (2020) Best practices for demand-driven supply chain management in public healthcare sector: a systematic literature review. S Afr J Ind Eng 31(2):11–27. https://doi.org/10.7166/31-2-2006
Bvuchete M, Grobbelaar SS, van Eeden J (2021) A network maturity mapping tool for demand-driven supply chain management: a case for the public healthcare sector. Sustainability (Switzerland) 13(21):1–29. https://doi.org/10.3390/su132111988
Di Maria E, De Marchi V, Galeazzo A (2022) Industry 4.0 technologies and circular economy: the mediating role of supply chain integration. Bus Strateg Environ 31(2):619–632
Duque-Uribe V, Sarache W, Gutiérrez EV (2019) Sustainable supply chain management practices and sustainable performance in hospitals: a systematic review and integrative framework. Sustainability (Switzerland) 11(21). https://doi.org/10.3390/su11215949
Dutta P, Choi TM, Somani S, Butala R (2020) Blockchain technology in supply chain operations: applications, challenges and research opportunities. Transp Res E Logist Transp Rev 142:102067. https://doi.org/10.1016/j.tre.2020.102067
Economic Times (2022) Ushering Green & Digital Hospitals: enabling energy saving through ICT. Economic Times
Emmanuel AN (2014) Challenges of implementing sustainable health care delivery in Nigeria under environmental uncertainty. J Hosp Admin 3(6):113. https://doi.org/10.5430/jha.v3n6p113
Epstein MJ, Buhovac AR, Elkington J, Leonard HBD (2017) Making sustainability work: best practices in managing and measuring corporate social, environmental and economic impacts, pp 1–305. https://doi.org/10.4324/9781351276443
Farooque M, Zhang A, Thürer M, Qu T, Huisingh D (2019) Circular supply chain management: a definition and structured literature review. J Clean Prod 228:882–900. https://doi.org/10.1016/j.jclepro.2019.04.303
Flynn PJ, Bliese PD, Korsgaard MA, Cannon C (2021) Tracking the process of resilience: how emotional stability and experience influence exhaustion and commitment trajectories. Group Org Manag 46(4):692–736
Foltynowicz Z, Łupicka-Fietz A, Jeszka AM, Kowalczyk D (2024) Evolution of social competencies in sustainable supply chains. Sustainability (Switzerland) 16(6). https://doi.org/10.3390/su16062581
Geissdoerfer M, Savaget P, Bocken NMP, Hultink EJ (2017) The circular economy—a new sustainability paradigm? J Clean Prod 143:757–768. https://doi.org/10.1016/j.jclepro.2016.12.048
Genovese A, Acquaye AA, Figueroa A, Koh SCL (2017) Sustainable supply chain management and the transition towards a circular economy: evidence and some applications. Omega (United Kingdom) 66:344–357
Goh CY, Marimuthu M (2016) The path towards healthcare sustainability: the role of organisational commitment. Procedia Soc Behav Sci 224:587–592. https://doi.org/10.1016/j.sbspro.2016.05.445
Goyal S, Routroy S, Singhal A (2019) Analyzing environment sustainability enablers using fuzzy DEMATEL for an Indian steel manufacturing company. J Eng Design Technol 17(2):300–329. https://doi.org/10.1108/JEDT-02-2018-0033
Gruchmann T, Melkonyan A, Krumme K (2018) Logistics business transformation for sustainability: assessing the role of the lead sustainability service provider (6PL). Logistics 2(4):25
Guerra-Bretaña RM, Flórez-Rendón AL (2018) Impact of regulations on innovation in the field of medical devices. Res Biomed Eng 34(4):356–367. https://doi.org/10.1590/2446-4740.180054

Gupta H, Barua MK (2017) Supplier selection among SMEs on the basis of their green innovation ability using BWM and fuzzy TOPSIS. J Clean Prod 152:242–258. https://doi.org/10.1016/j.jclepro.2017.03.125

Gupta S, Boojh R (2006) Report: biomedical waste management practices at Balrampur Hospital, Lucknow, India. Waste Manag Res 24(6):584–591. https://doi.org/10.1177/0734242X06068342

Gutachter SK, Dierkes P, Scheersoi A (2018) Vom Fachbereich Biowissenschaften der Johann Wolfgang Goethe—Universität als Dissertation angenommen

Hair JF, Risher JJ, Sarstedt M, Ringle CM (2019) When to use and how to report the results of PLS-SEM. Eur Bus Rev (1):31, 2–24. https://doi.org/10.1108/EBR-11-2018-0203

Hanisch C, Loellhoeffel T, Diekmann J, Markley KJ, Haselrieder W, Kwade A (2015) Recycling of lithium-ion batteries: a novel method to separate coating and foil of electrodes. J Clean Prod 108:301–311. https://doi.org/10.1016/j.jclepro.2015.08.026

Hasim MS, Hashim AE, Ariff NRM, Sapeciay Z, Abdullah AS (2018) Commitment to sustainability: a content analysis of website for university organisations. IOP Conf Ser Earth Environ Sci 117(1):012046. https://doi.org/10.1088/1755-1315/117/1/012046

Hassan LM, Shiu E, Parry S (2016) Addressing the cross-country applicability of the theory of planned behaviour (TPB): a structured review of multi-country TPB studies. J Consum Behav 15(1):72–86. https://doi.org/10.1002/cb.1536

Henseler J, Ringle CM, Sarstedt M (2015) A new criterion for assessing discriminant validity in variance-based structural equation modeling. J Acad Mark Sci 43(1):115–135. https://doi.org/10.1007/S11747-014-0403-8

Hutchins MJ, Sutherland JW (2008) An exploration of measures of social sustainability and their application to supply chain decisions. J Clean Prod 16(15):1688–1698. https://doi.org/10.1016/j.jclepro.2008.06.001

Icegen Computing (2024) Orderly management—patient tracking solution for Fortis Hospitals—RTLS/RFID based. https://icegen.net/rfid-based-patient-tracking-system-for-fortis-hospitals/

Kakemam E, Liang Z (2023) Guidance for management competency identification and development in the health context: a systematic scoping review. BMC Health Serv Res 23(1):1–13. https://doi.org/10.1186/s12913-023-09404-9

Kanokphanvanich C, Rattanawong W, Vongmanee V (2023) A new model for a sustainable healthcare supply chain prioritizes patient safety: using the fuzzy Delphi method to identify healthcare workers' perspectives. Sustainability (Switzerland) 15(9). https://doi.org/10.3390/su15097123

Kazancoglu I, Sagnak M, Kumar Mangla S, Kazancoglu Y (2021) Circular economy and the policy: a framework for improving the corporate environmental management in supply chains. Bus Strateg Environ 30(1):590–608

Lee SM, Lee D (2022) Developing green healthcare activities in the total quality management framework. Int J Environ Res Public Health 19(11):6504

Lieder M, Rashid A (2016) Towards circular economy implementation: a comprehensive review in context of manufacturing industry. J Clean Prod 115:36–51. https://doi.org/10.1016/j.jclepro.2015.12.042

Limited AI (2021) Business responsibility and sustainability reporting 2021—full report, pp 28–69

Liu Y, Bai Y (2014) An exploration of firms' awareness and behavior of developing circular economy: an empirical research in China. Resour Conserv Recycl 87:145–152. https://doi.org/10.1016/j.resconrec.2014.04.002

Lu Y, Ramamurthy K (2011) Understanding the link between information technology capability and organizational agility: an empirical examination. MIS Q 35:931–954

Mahajan R, Lim WM, Sareen M, Kumar S, Panwar R (2023) Stakeholder theory. J Bus Res 166:114104. https://doi.org/10.1016/j.jbusres.2023.114104

McNab D, Bowie P, Morrison J, Ross A (2016) Understanding patient safety performance and educational needs using the "Safety-II" approach for complex systems. Educ Prim Care 27(6):443–450. https://doi.org/10.1080/14739879.2016.1246068

Meara JG, Leather AJM, Hagander L, Alkire BC, Alonso N, Ameh EA, Bickler SW, Conteh L, Dare AJ, Davies J, Mérisier ED, El-Halabi S, Farmer PE, Gawande A, Gillies R, Greenberg

SLM, Grimes CE, Gruen RL, Ismail EA et al (2015) Global surgery 2030: evidence and solutions for achieving health, welfare, and economic development. Lancet 386(9993):569–624. https://doi.org/10.1016/S0140-6736(15)60160-X

Mele V, Compagni A, Cavazza M (2014) Governing through evidence: a study of technological innovation in health care. J Public Adm Res Theory 24(4):843–877. https://doi.org/10.1093/jopart/mut016

Michel JB, Kui Shen Y, Presser Aiden A, Veres A, Gray MK, Pickett JP, Hoiberg D, Clancy D, Norvig P, Orwant J, Pinker S, Nowak MA, Aiden EL (2011) Quantitative analysis of culture using millions of digitized books. Science 331(6014):176–182. https://doi.org/10.1126/science.1199644

Moshood TD, Sorooshian S, Nawanir G, Okfalisa S (2022) Efficiency of medical technology in measuring service quality in the Nigerian healthcare sector. Int J Afr Nurs Sci 16:100397. https://doi.org/10.1016/j.ijans.2022.100397

Musse JO, Homrich AS, de Mello R, Carvalho MM (2018) Applying backcasting and system dynamics towards sustainable development: the housing planning case for low-income citizens in Brazil. J Clean Prod 193:97–114. https://doi.org/10.1016/j.jclepro.2018.04.219

Nosratabadi S, Mosavi A, Shamshirband S, Zavadskas EK, Rakotonirainy A, Chau KW (2019) Sustainable business models: a review. Sustainability (Switzerland) 11(6):1–30. https://doi.org/10.3390/su11061663

Omachonu VK, Einspruch NG (2010) Innovation in healthcare delivery systems: a conceptual framework. Innov J 15(1):1–20

Pereno A, Eriksson D (2020) A multi-stakeholder perspective on sustainable healthcare: from 2030 onwards. Futures 122:102605. https://doi.org/10.1016/j.futures.2020.102605

Piser A (2019) How do hospitals dispose of their healthcare waste? Daniels. https://www.danielshealth.com/knowledge-center/hospital-healthcare-waste

Pulido NA (2016) Effect of standard post-harvest interventions on the survival and regrowth of antibiotic-resistant bacteria on fresh produce. Virginia Tech

Ramanathan U (2014) Performance of supply chain collaboration—a simulation study. Expert Syst Appl 41(1):210–220. https://doi.org/10.1016/j.eswa.2013.07.022

Reiman T, Pietikäinen E (2012) Leading indicators of system safety—monitoring and driving the organizational safety potential. Saf Sci 50(10):1993–2000. https://doi.org/10.1016/j.ssci.2011.07.015

Ringle CM, Sarstedt M, Sinkovics N, Sinkovics RR (2023) A perspective on using partial least squares structural equation modelling in data articles. Data Brief 48:109074. https://doi.org/10.1016/j.dib.2023.109074

Roseland M (2000) Sustainable community development: integrating environmental, economic, and social objectives. Prog Plan 54(2):73–132. https://doi.org/10.1016/S0305-9006(00)00003-9

Rowan NJ (2024) Digital technologies to unlock safe and sustainable opportunities for medical device and healthcare sectors with a focus on the combined use of digital twin and extended reality applications: a review. Sci Total Environ 926:171672. https://doi.org/10.1016/j.scitotenv.2024.171672

Sarkis J, Zhu Q (2018) Environmental sustainability and production: taking the road less travelled. Int J Prod Res 56(1–2):743–759. https://doi.org/10.1080/00207543.2017.1365182

Sharma A, Luthra G (2023) Implementing a risk-based approach to quality management system ISO-13485 processes in compliance with EUMDR 2017/745 for medical device industry. J Pharm Res Int 35(13):8–19. https://doi.org/10.9734/jpri/2023/v35i137365

Shelton RC, Cooper BR, Stirman SW (2018) The sustainability of evidence-based interventions and practices in public health and health care. Annu Rev Public Health 39:55–76. https://doi.org/10.1146/annurev-publhealth-040617-014731

Sinthiya RHG (2024) A study on reliability of news shared in social media platforms (with reference to whatsapp users). 1(1)

Sodhi MS, Son B, Tang CS (2012) Researchers' perspectives on supply chain risk management. Prod Oper Manag 21(1):1–13

Sudusinghe JI, Seuring S (2022) Supply chain collaboration and sustainability performance in circular economy: a systematic literature review. Int J Prod Econ 245:108402. https://doi.org/10.1016/j.ijpe.2021.108402

Syahrir I, Suparno S, Vanany I (2018) Strategic management for logistics and supply chain operation in healthcare. IPTEK J Proc Ser 0(3):10. https://doi.org/10.12962/j23546026.y2018i3.3699

Tate WL, Ellram LM, Kirchoff JF (2010) Corporate social responsibility reports: a thematic analysis related to supply chain management. J Supply Chain Manag 46(1):19–44. https://doi.org/10.1111/j.1745-493X.2009.03184.x

Testa F, Iraldo F, Frey M (2011) The effect of environmental regulation on firms' competitive performance: the case of the building & construction sector in some EU regions. J Environ Manag 92(9):2136–2144. https://doi.org/10.1016/j.jenvman.2011.03.039

Tseng ML, Tran TPT, Fujii M, Lim MK, Negash YT (2024) Modelling hierarchical circular supply chain management enablers in the seafood processing industry in Vietnam under uncertainties. Int J Log Res Appl 27(1):30–58. https://doi.org/10.1080/13675567.2021.1965105

Tsytsyna E, Lintukangas K, Virolainen VM (2024) Business ecosystem concept in supply chain management research. Int J Proc Manag 20(2):240–266

Unhelkar B, Joshi S, Sharma M, Prakash S, Mani AK, Prasad M (2022) Enhancing supply chain performance using RFID technology and decision support systems in the industry 4.0—a systematic literature review. Int J Inf Manag Insights 2(2):100084. https://doi.org/10.1016/j.jjimei.2022.100084

Unique A (2016) Our journey of touching billion lives. 1–23

Upadhyay A, Khan MW, Chaurasia PK (2017) The impact of crowdsourcing in software engineering: software testing perspective. Int J Adv Res Comput Sci 8(7):250–252

Völker S, Kistemann T (2013) "I'm always entirely happy when I'm here!" Urban blue enhancing human health and well-being in Cologne and Düsseldorf, Germany. Soc Sci Med 78(1):113–124. https://doi.org/10.1016/j.socscimed.2012.09.047

Wang M, Wu Y, Chen B, Evans M (2020) Blockchain and supply chain management: a new paradigm for supply chain integration and collaboration. Oper Supply Chain Manag 14(1):111–122

Zahid T, Agha MH, Schmidt T (2019) Investigation of surrogate measures of robustness for project scheduling problems. Comput Ind Eng 129:220–227. https://doi.org/10.1016/j.cie.2019.01.041

Zaini M (2018) Analysis of port costs and loading-unloading costs on the revenue of PT. Abdi Nusantara Indonesia Line, Gresik Branch

Zarim ZA, Mohamad O, Rahman MS, Zaki HO, Sergio RP, Haladay DJ (2017) The role of organisational commitment, leadership style, strategic human resources practices and job satisfaction towards sustainable tourism industry: comparative study in the UAE and Malaysia. In: Springer proceedings in business and economics, December, pp 255–268. https://doi.org/10.1007/978-3-319-43434-6_21

Zhong RY, Newman ST, Huang GQ, Lan S (2016) Big data for supply chain management in the service and manufacturing sectors: challenges, opportunities, and future perspectives. Comput Ind Eng 101:572–591. https://doi.org/10.1016/j.cie.2016.07.013

Zobel T (2013) ISO 14001 certification in manufacturing firms: a tool for those in need or an indication of greenness? J Clean Prod 43:37–44. https://doi.org/10.1016/j.jclepro.2012.12.014

Chapter 5
From Fast Fashion to Sustainable Cycles: Exploring the Psychological and Social Drivers of Garment Recycling

Bhumika Achhnani, Pulok Ranjan Mohanta, and Sunali Bindra

Abstract The fast fashion industry has enhanced the economic growth by revolutionizing the garments market, but its deteriorating impact on the environment is lately found to be a matter of serious concern. While several researchers are working on multitude of approaches to address the underlying issues, garments recycling is being considered to have a significant impact. However, to ensure a promising result it is imperative that we understand the consumer's perspectives. This chapter examines the psychological factors that influence people's decision to garment recycling. The chapter explores different aspects of recycling behaviours such as attitudes, subjective norms, perceived behavioural control, and self-identity. It draws upon well-established frameworks such as the Theory of Planned Behaviour and Value-Belief-Norm theory. Further, keeping circular economy in the context, the text explores the economic feasibility of garment recycling, emphasising the importance of incentives and creative business models to foster sustainable practices. The impact of recycling behaviours is evaluated by examining social trends such as heightened environmental consciousness and the social media's influence. The chapter also explores the impact of social comparisons and personal branding on individuals' choices to recycle garments. Subsequently, the important barriers that impede garment recycling, and some solution approaches are discussed in detail. Towards the end the chapter delineates the future avenues of research. This chapter serves as a reference material for the decision makers, policy makers, industrialists, and researchers to create successful approaches that encourage garment recycling and support the development of a fashion industry that is circular and more sustainable.

B. Achhnani (✉) · P. R. Mohanta
Marwadi University, Rajkot, Gujarat, India

S. Bindra
Symbiosis Centre for Management and Human Resource Development, Symbiosis International (Deemed) University, Pune, Maharashtra, India

K. Mathiyazhagan et al. (eds.), *Humanizing Circular Supply Chain Management*, Circular Economy and Sustainability,
https://doi.org/10.1007/978-3-031-86562-6_5

Keywords Garment recycling · Circular fashion · Consumer psychology · Sustainable consumption · Sustainability · Circular economy

5.1 Introduction

Fast fashion is widely described as a business mode that has an agile response to the customer's demand for the latest fashion trends and makes them available for reasonable prices (Choi 2014). This business model has significantly influenced the processes of garment manufacturing, consumption, use, and disposal for both individuals and corporations (Niinimäki et al. 2020). Due to the increased demand, greater availability, low prices, and a culture that encourages discarding items after minimal use, in recent decades, the manufacturing and consumption of garments have experienced a substantial rise (Ozdamar-Ertekin 2017). Additionally, it has led to an abundance of generation of waste, leading to a huge environmental problem (Remy et al. 2016). Annually, the global production of textile waste amounts to 92 million tonnes (Woolridge et al. 2006). If significant alterations do not occur to the processes of manufacturing alongside the utilisation and disposal of garments, it is projected that the worldwide textile waste will reach to 134 million tonnes by 2030 (Igini 2024). The severity of the environmental hazard created by the fashion industry is depicted in Fig. 5.1.

Annually 100 billion garments manufactured worldwide.

The global fashion industry contributes 10% of total greenhouse gas emissions.

Producing a single cotton garment requires about 2,700 litres of water.

The garment industry accounts for 20% of global industrial wastewater contamination.

Individuals are purchasing 60% more garments and using them for just 50% of the duration

Synthetic fibres like polyester and acrylic account for almost 60% of worldwide garments purchases.

57% of the discarded clothes goes into landfills. Incineration accounts for 25%.

Synthetic fibres require 80 to 800 times more time to disintegrate compared to natural fibres.

Fig. 5.1 Environmental impact from fashion industry. (Source: Peters et al. 2021; Igini 2024)

This chapter intends to examine the psychological, social, economic, and policy factors that influence consumer behavior towards garment recycling, focusing on attitudes, social comparisons, circular economy models, and the barriers hindering adoption in the fast fashion industry. The chapter focuses on answering the following questions:

1. What psychological factors, including attitudes, subjective norms, and perceived behavioral control, influence consumers' decision to recycle garments in the fashion industry?
2. How do social trends, personal branding, and social media impact consumers' choices to engage in sustainable garment recycling?
3. What are the key barriers to garment recycling, and how can circular economy models and policy frameworks like Extended Producer Responsibility (EPR) address them to promote sustainability?

5.2 Literature Review

The environmental concerns arising out of these facts stimulated researchers and policy makers to ponder over sustainable consumption practices such as Recycling. However, gaining insight into the psychological determinants that impact customers' garment recycling behaviours is essential for fostering a sustainable approach to fashion consumption (Chaturvedi et al. 2020; Laitala 2014). Studies have discovered various important psychological factors that impacts people's choices to recycle their garments (Eppinger 2022). These factors include their attitudes, subjective norms, perceived behavioural control, and self-identity (Sun et al. 2024). In the context mentioned above, the Theory of Planned Behaviour (TPB) (Ajzen 1991) and the Value-Belief-Norm (VBN) theory (Stern et al. 1999) are useful frameworks for comprehending the psychological aspects that impact intentions and behaviours related to recycling.

5.2.1 Theory of Planned Behaviour (TPB)

Sun et al. (2024) noted that a person's behavior in clothes recycling is determined by his/her attitude towards that action, his/her perception of social norms regarding it and his/her perceived behavioral control i.e. intention to perform it depending on the Theory of Planned Behavior (TPB). According to Hu et al. (2022) individual who are more optimistic and have a greater belief in receiving support from others on recycling and are confident about their ability to recycle effectively are more likely to take part in textile recycling schemes.

5.2.2 *Value-Belief-Norm (VBN) Theory*

According to Ghazali et al. (2019) the importance of personal values and beliefs stressing biospheric meanings, related most significantly to environmental conservation is emphasized in the VBN hypothesis, for example—clothes recycling. According to this theory, individuals' self—identity i.e. their perception of themselves as someone interested in environmental issues, greatly contributes to forecasting their intentions to recycle clothes (Hong et al. 2024).

5.2.3 *The Combination of TPB and VBN*

Self-identification is how we see ourselves on the inside, and it has enduring properties. Ashforth and Mael (1989) were the first to introduce it into TPB, and this inclusion can be traced to their identity theory work. In order for one to gain identity as an individual according to this theory s/he has to play various roles based on his or her personality hence making it a habit to maintain certain actions that are supported by his or her own believe systems (Stets and Burke 2003). Self-identity is supposed to make sure that there is a link between how someone thinks and behaves thus forming some specific target points (Whitmarsh and O'Neill 2010). The greater the relevance of an identification, the greater the probability the corresponding behaviours were adopted by an individual (Swann Jr. et al. 2009). The use of these frameworks allows scientists to have better understanding of the emotional factors motivating people towards their participation in clothes recycling. Moreover, this knowledge provides a basis from which effective policies can be created with the aim of promoting the desired actions which in turn play a role in moving the mode of fashion operation closer to full circle.

Customers' psychology governs their clothes recycling patterns. It is very crucial to know all about these dimensions if we want to help to change for more sustainable fashion consumption patterns and lessen on the damaging environmental impacts produced by the fast fashion sector (Williams 2022). By taking a more intentional approach in decision-making processes and adopting eco-friendly attitudes towards life, we can all contribute significantly towards shaping sustainable future prospects within the fashion industry. This chapter examines the psychological determinants of people's activity in the recycling of clothes. It does so by bringing together insights from different areas including circular economy, economic viability, societal trends, social comparisons, personal branding as well as psychological motivations.

5.3 The Circular Economy and the Process of Garment Recycling

Various regions and countries such as China, Africa, the European Union, as well as the United States have quickly adopted the circular economy to become a key principle in their industrial and environmental policies (Camacho-Otero et al. 2018). It is also increasingly becoming popular with companies and local authorities as demonstrated by the collaborators of Ellen MacArthur Foundation in 2017 (A New Textiles Economy: Redesigning Fashion's Future n.d.). Supporters of this model argue that it helps in conserving resources, reducing waste, reducing emissions, and energy losses through long life design, maintenance at right time, repair, reuse, remanufacture, refurbish, and recycle to create a restorative system with less material and energy loops (Kirchherr et al. 2023). The purpose of a circular economy is to create a system where material flows are slow and cyclical in nature, moving from a consumer-driven society to a society guided by usage. This aims to dissociate resource use and environmental impact from economic growth. In the same way, circular economy models have designs that help to reduce costs, generate higher revenues and at the same time manage risks, yet they present chances for finance industry to be involved in steering societies towards sustainability (Barros et al. 2021).

The typical linear economy conforms to a 'take-make-waste' philosophy, unlike the circular economy approach (Kalmykova et al. 2018). In the linear economy, raw materials are obtained from the surroundings, processes into products, used up and thrown out as garbage. There are several reasons why this approach cannot work, these include (What Is the Linear Economy? 2023):

- Valuable resources and the environment are highly stressed by the continuous mining of finite naturally occurring resources.
- It leads to air pollution and degradation of the environment through constant dumping of tons of waste on landfills or burning them.
- In the long run, the model fails to capture the impacts of using up resources and piling up waste.

Alternatively, the purpose of the circular economy is to make wastes and pollution disappear through elongating the life spans of product and material. By adopting some strategies like reusing, repairing, remanufacturing, and recycling, these aims can be reached. Organisations and individuals would be able to reduce their dependence on new unused raw materials, limit the amount of trash generated and (re) establish more environmentally sustainable economic systems if they follow a circular approach. The circular economy bears promise for creative innovation, job creation, and environment preservation (McGinty n.d.).

The fashion business has typically followed a straight-line economy model where resources were obtained, commodities manufactured and eventually discarded when there was no use for them anymore. A circular economy, however, stands out as a better option for the fashion and garment industry due to lower perplexity and higher burstiness (Ki et al. 2020). This is achieved through strategies such as product durability with recycling capacity, wardrobe leasing, second-hand sales of clothes, and advancement of textile recycling technologies (Shirvanimoghaddam et al. 2020). By implementing principles of the circular economy, the fashion industry is capable of reducing waste, conserving natural resources and minimizing its huge ecological footprint. The fast fashion industry has been widely carped for its enormous ecological footprint due to waste attributed to it in production, consumption, and disposal of clothes. In light of these concerns about fashion's environmental impact, the concept of a circular economy has emerged as a potential blueprint for transforming the sector (Dan and Østergaard 2021).

Textile/clothes recycling forms an integral part of the circular fashion system, since it prevents throwing away valuable textile materials into dumpsites or burning them. For example, Patagonia's "Worn Wear" initiative allows consumers to exchange their used Patagonia gear for store acclaim in addition to making repairs and selling them back at a discounted price (Used Patagonia® Clothing and Gear | Worn Wear n.d.). The process of garment recycling includes several important stages: (1) collecting and sorting used clothes, (2) applying different recycling methods to them, and (3) making new items from them using remanufacturing. There are various ways of collecting used clothes: drop-off points, roadside collection programs, or merchant take-back programs. The sorting process is based on the materials' fabrics composition should they be cottons, wools, or polyesters leading to their condition and categorization for reuse or recycling (Sandvik and Stubbs 2019). Those in good condition that can still be worn are often taken to charity groups or thrift shops where they are sold cheaply and reused by others. However, where they cannot be reused any more, these garments end up at recycling plants for either mechanical, chemical, or thermal processing so that their derivatives can be utilised in making new products (Jacometti 2019) The main aim here is to dismantle such textile materials in order to recover main elements from them which can later serve as raw materials in manufacturing new goods thereby closing a loop and promoting circular fashion systems (Jacometti 2019). Nonetheless, there are some stumbling blocks in textile (garment) recycling sphere including concerns related to (1) quality of recycled product, (2) pollution during recycling process, (3) collection and transportation management, (4) financial sustainability and (5) consumer acceptance of recycled items. To enhance more environmentally friendly and circular garment industry it is important to overcome these problems while utilizing advantages presented by technology advancement along with administrative support and cooperation among various stakeholders (Mizrachi and Tal 2022). In this way, the fashion enterprises depending on these principles can minimize waste production; save natural resources and minimize their extensive environmental degradation thus leading to sustainable future (Fig. 5.2).

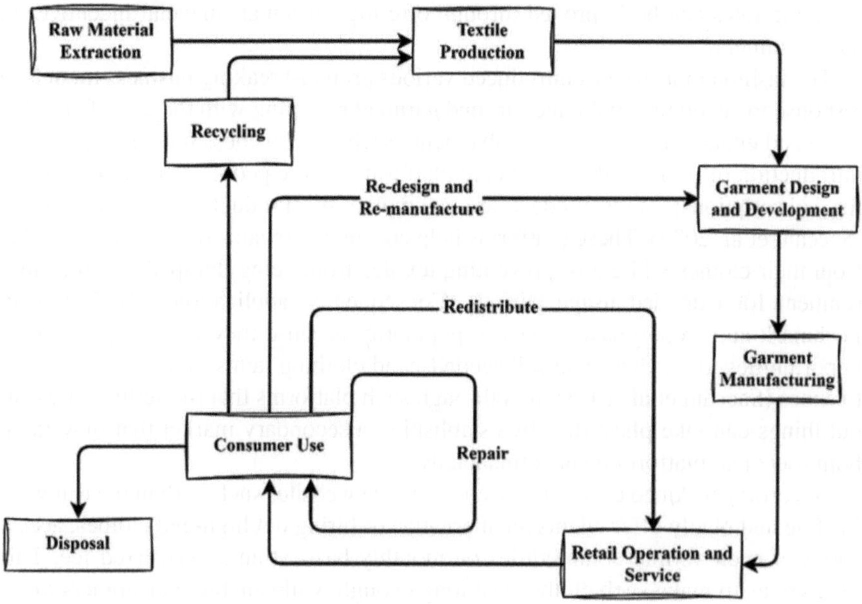

Fig. 5.2 Schematic representation of the garment recycling process. (Created by authors)

5.4 The Economic Viability and Policy Implications of Garment Recycling

The recycling of discarded clothes is an important means through which people can contribute to greener and more sustainable fashion business model. For both consumers and companies, however, generating widespread use of old clothing as a raw material for new products requires overcoming one major challenge, which is how to make this economically viable. Customers are more likely to take their worn-out garments back if they are given financial reasons to do so (Sandvik and Stubbs 2019). Regarding consumer perspectives, a survey conducted by Dan and Østergaard (2021) examined the various factors affecting consumers' psychological disposition towards garment recycling. It was shown that clients who perceived financial benefits such as earning from reselling worn-out garments or obtaining discounts off their next purchase were more likely to recycle garments. According to Saccani et al.'s recommendation based on an empirical study in the UK, respondents who considered potential monetary gains from recycling were more likely to engage in it with respect to this other specific issue involving earning money (Saccani et al. 2023). Additionally, a WRAP poll conducted in the UK showed that 45% of respondents listed 'making money' as a major reason for giving away or selling clothes they did not need (Gomes et al. 2022). This work thus suggests that garment recycling rates can be improved through offering particular financial incentives to the consumers.

The fashion industry has introduced various ground-breaking business models in response to economic challenges around garment recycling with the aim of creating value and enhancing customer involvement. Such new business models include the introduction of resale and recommerce platforms, where people can rent or sign up for subscriptions; other models are exampled by Product-as-a-Service offers (Saccani et al. 2023). These platforms help consumers regain some monetary value from their clothes while also preventing textiles from being dumped into the environment for extended usage periods. Consequently, applications like ThredUp; Poshmark and Depop have grown in popularity because they give shoppers alike opportunities to purchase and sell second-hand clothing items ranging from outfits to shoes (Saccani et al. 2023). It is through such platforms that the selling of worn-out things can take place thereby establishing a secondary market that may favor both users and platform owners financially.

According to Amed et al. (2019), e-commerce websites such as Rent the Runway; Le Tote and Nuuly offer clients an alternative of hiring out high-end clothes, accessories or even normal commodities on monthly basis at an agreed fixed fee. This idea serves to make such clothes last long enough while enabling customers make more fashion choices at reduced prices compared with traditional purchasing behaviors. In this aspect, Mud Jeans Company based in Netherlands has adopted the "Lease a Jeans" initiative which enables people to rent a pair of their favorite denims for a month and exchange them when they get bored (Sustainability Report | Mud Jeans n.d.). Besides, referring to this as an affordable method of getting top quality environmentally friendly clothing; it is said that it also serves to make them last longer. The economic performance of garment recycling strongly relies on governmental support as well as policy intervention. Taking responsibility for the management of their textiles at the end of the textile's life would lead to Extended Producer Responsibility (EPR) programs aimed at obligating manufacturers and importers in the disposal and recycling of garments and textile products (Extended Producer Responsibility and Economic Instruments n.d.). These schemes provide financial incentives for producers to create more easily recyclable goods and invest in recycling facilities' improvement. France requires that brands dealing with clothing, linen as well as shoes among others should bear the costs of collecting, sorting and recycling their products under the EPR program (King 2024). Thus, this law has resulted into increased investment in textile recycling while encouraging manufacturers to consider recyclability when designing new items. Governments could also come up with regulations prohibiting the disposal of such items in landfills whereas promoting recycling through other means. For example, various nations and jurisdictions have banned certain types of textile waste from being dumped into disposal sites forcing firms and individuals to look for other ways like recycling (Ghadimi et al. 2019).

5.5 The Impact of Social Trends on Garment Recycling

Clothes recycling initiatives' efficiency is greatly influenced by contemporary fashion trends and consumer behaviours. Indeed, customer awareness and concern about the ecology of fashion industry have tremendously heightened in the recent past prompting rise in consumers' devotion sustainability (Vladimirova et al. 2023). It is consumers' increased realization on the wastage; pollutants release as well as resource exhaustion in linear garment manufacture-consumption-disposal cycle that compels them to recycle clothing more so as to lessen their own carbon footprint while facilitating sustainable development. Studies show that environmentally conscious people tend to recycle their clothes more frequently' (Mandarić et al. 2022). Consumers are buying more environmentally friendly clothes. This behavior pattern leads to greater numbers of people going for sustainable brands that respect ethics in production process in environmentally conscious economies through their designs. Circular economy aims at steering clothes recycling in favour of sustainable fashion firms. Social media has been instrumental in promoting green awareness and sustainability trend in fashion industry. It involves the use of online platforms such as Facebook, Twitter, Instagram among others to share information about recycling garments as well as environmental damages caused by fashion industry. Online communities have sprung up on social media sites that exchange or resell used textile products, thereby promoting the circular system while reducing textile waste generation through reuse (Vladimirova et al. 2023). There have been changes in legislation and regulations regarding recycling of garments due to society's shift towards eco-friendly habits as well as desirability for sustainable clothes. Business entities are increasingly recycling their garments to adhere to the law or simply because it became part of their marketing strategy. To handle environmental footprints in the fashion industry, Garments recycling is supported and incentivized by the European Union's Circular Economy Action Plan (CEAP) as well as the Strategy for Sustainable Textiles (OECD 2008). The legislation and laws in the EU on textiles as well as market forces that drive consumer buying habits have made the recycling of clothes more and more appealing (Circular Economy Action Plan n.d.).

5.6 The Impact of Social Comparisons and Personal Branding on Garments Recycling

Peers' behaviour and views have an impact on what other people do and their compliance with social expectations. Someone's decision to recycle clothes might most likely be impacted by social comparison. If other people conserve the environment, then it can encourage one to follow suit (Gen Z Cares About Sustainability More Than Anyone Else—and Is Starting to Make Others Feel the Same Way 2022). Based on the average person's grid of friends who have practiced recycling clothes, there is a higher chance that any individual will recycle their own clothes to avoid being left out. This would mean that people recycle because others are also recycling, and it makes considering our own values or goals possible (Yan et al. 2021). People can appear more sustainable while gaining social acceptance through clothing recycling. Personal branding has become increasingly significant because individuals want to project themselves. Wearing clothes can also be a way of declaring personal values, interests, and even class. People who recycle their trendy garments show that they care for the earth and also think about the society in terms of waste management. According to Van Der Werff et al. (2013), recycling intentions can be precisely forecasted by self-identity and perceived behavioral control. Self-identity refers to one's environmental consciousness whereas perceived behavioral control means confidence in recycling. Having a powerful sense of environmental self-identity enhances the significance of garment recycling and corresponds with personal convictions. They will also tend to assume personal responsibility to reduce their environment impact and advance sustainability. Garments recycling intention is also influenced by perceived behavioural control among individuals. If individuals perceive that they possess knowledge, skills and resources for garment recycling, they are more likely to do so.

5.7 Psychological Motivations for Garments Recycling

It is important to know the underlying psychological reasons that motivate people to recycle clothing because this will encourage wider acceptance of sustainable practices in the fashion sector. Recycling practices are often influenced by both internal and external factors which can best be understood through established behavioral theories (Kollmuss and Agyeman 2002). An essential inner motive for garment recycling is environmental awareness and concern. Recycling is commonly done among individuals who are conscious about the environmental impacts of the textile industry and feel morally obligated to cut their ecological footprints. Furthermore, engaging in pro-environmental activities often makes people feel good about themselves; this sense of personal satisfaction serves as a strong motivator for some toward issues of sustainability and fair trade (Yang et al. 2004). There are also external factors which majorly affect clothing recycling behaviors. For example, someone may want to recycle his/her clothes so as to show how eco-friendly they are and earn respect from those surrounding them, with the aim of attracting attention or approval from other people. One more external drive is receiving material benefits like cash. Consequently, it is common for most people to recycle their old garments if there would be something they would get in return (Stancu et al. 2020).

Köhler et al. (2019) show that The Theory of Planned Behaviour (TPB) and the Value-Belief-Norm (VBN) theory offer organized methods to evaluate such psychological catalysts. TPB holds that an individual's willingness to involve with textile recycling hinges on their attitude, subjective norm and perceived behavioural control. Meanwhile, VBN emphasizes personal ethics as well as convictions that alter eco-positive acts including donating cloth. Through employing these models, scholars have gained useful information regarding what drives consumer engagement towards sustainable clothing trends.

5.8 Implications

5.8.1 Policy Implications

- Regulatory Support for Recycling Initiatives: This study suggests governments to develop or strengthen Extended Producer Responsibility (EPR) programs, mandating brands to handle final garment disposal. This could improve fashion industry recycling and sustainability.
- Incentive-based Policies: The study suggests government or local authorities to provide financial incentives (e.g., tax breaks or refunds) for garment recycling by consumers and corporations. This would foster circular economy.

5.8.2 Business Implications

- Adoption of Circular Fashion Models: Fashion companies can use the insights to embrace more sustainable models including resale platforms, garment leasing, and take-back programs to attract environmentally conscious customers and reduce waste. Projects like Patagonia's "Worn Wear" can demonstrate success.
- Business Redesign: Companies may redesign their production methods to make clothing with longer lifespans and easier recycling. Marketing sustainability as a fundamental value can help boost personal branding and social trends.

5.8.3 Environmental Implications

- Reduced Textile Waste: Promoting garment recycling could reduce textile waste in landfills and fast fashion's environmental impact, including greenhouse gas emissions and resource conservation.
- Helping Achieve the United Nations' Sustainable Development Goals: This study has the potential to bolster larger international initiatives aimed at achieving the SDGs, especially Goal 12 (Responsible Production and Consumption) and Goal 13 (Climate Action).

5.9 Obstacles to Garments Recycling

Despite boasting diverse advantages including waste minimization, resource preservation, as well as reducing the effect of the fashion industry to the environment, garment recycling faces many hurdles (Riemens et al. 2021). The challenges are brought forth by the intersection of several factors such as lack of consumer knowledge on recycling importance and available options, difficulty in securing necessary materials and infrastructure for garment recycling process, as well as misconceptions about recycling that make it appear cumbersome, hence the general belief in its difficulty (Wójcik-Karpacz and Stankowska 2023). Many people remain unaware of how their actions affect the environment especially with respect to the fashion industry while recycling remains an option towards its mitigation. In addition, recycling can be an occupying process in terms of resources making it necessary for particular machines and places which might not be easily reachable too. Some buyers view sorting, washing and preparing their clothes for recycled use as too time-consuming or less convenient according to Testa et al. (2017).

Educating the consumers, expanding infrastructure and giving incentives are key if these obstacles are to be overcome hence increasing the rate at which textiles are recycled. Campaigns through education serve to enlighten people about the importance of textile recycling & reusing clothes through which they can have a collection of options (Stancu et al. 2020). Establishment of appropriate textile recycling infrastructure will include setting up easily accessible collection points as well as investment in modern recycling technology and facilities. Monetary incentives or decrease in pricing for recycled clothes would also work as incentives towards this end (Vladimirova et al. 2023). This would increase the efficiency in recycling apparel by utilizing earthy fibers promoting easy disassembly and return/donation mechanisms for those clothes one does not need. If addressed and executed respectively by stakeholders in this sector including the government then industrialists will increase their recycling percentages significantly contributing to reducing fashion industry environmental stressors through promotion of self-sustaining economies and promoting durability.

5.10 Conclusion and the Way Forward

Given the environmental challenges posed by the traditional "take-make-waste" model, there is an increasing need for a more balanced and circular approach in the garment industry. Previous studies have shed important light on the drivers of garment recycling behaviors but there is still insufficient understanding that necessitates more exploration. It is important that future research concentrate on the current knowledge gaps concerning garment recycling behaviors. This involves identifying the psychological determinants which influence consumer attitudes and intentions across different demographic groups and cultures. An in-depth understanding of age, gender, socio-economic status (SES) as well as associated cultural norms versus environmentalism and inherent values as psychological drivers can assist researchers and policy makers in formulating more precise and effective ways to promote sustainable garments consumption & recycling.

At the same time, scholars should undertake further studies about how peer pressure and social comparison make people engage in recycling old clothes. This can be helpful in developing suitable communication messages which work for both online communities and social media. Apart from understanding why people recycle their clothes; it is suggested that authorities develop workable legislation that can even provide incentive schemes among other things aimed at promoting sustainability The implementation for instance entails passing policies that advocate Extended Producer Responsibility (EPR) where garment manufacturers must take back their textile products when they cannot be used anymore. Through the adoption of extended producer responsibility (EPR) programmes, manufacturers would bear financial burden related to disposal costs. Both approaches serve as powerful stimuli for fashion designs which are stronger, sustainable and more recyclable. As such, it's prudent to introduce initiatives like offering monetary rewards; subsidised pricing or even tax relief for recycling used clothes by individuals. Social incentives are known to be effective in changing consumer behaviour basically by reaching them through monetary means though they assist in evolving internal standards of cleanliness and environmental preservation. Rather than relying solely on governmental actions or incentives one alternative way to promote economic sustainability through garment recycling is the use of innovative business models in this context. For instance, second hand exchange or sale networks, repair services, alteration centres as well as rental shops will help create the closed-loop system for responsible utilization of garments. These systems contribute towards increased clothing durability, extra money sources and limited waste disposal in the space using them. However, it is only possible if fashion industry stakeholders and policymakers come together to facilitate their establishment through collaboration and coordination Traditional impediments to widespread garment recycling practices could be overcome if they were to combine their experiences, funds, and contacts with the goal of addressing common obstacles.

References

A New Textiles Economy: Redesigning Fashion's Future (n.d.) Ellen MacArthur Foundation

Ajzen I (1991) The theory of planned behavior. Organ Behav Hum Decis Process 50(2):179–211

Amed I, Berg A, Brantberg L, Hedrich S (2019) The state of fashion 2019: a year of awakening. The Business of Fashion

Ashforth BE, Mael F (1989) Social identity theory and the organization. Acad Manag Rev 14(1):20–39

Barros MV, Salvador R, Piekarski CM, de Francisco AC (2021) Circular economy as a driver to sustainable businesses. J Clean Prod 299:126848

Camacho-Otero J, Boks C, Pettersen IN (2018) Consumption in the circular economy: a literature review. Sustainability 10(8):2758
Chaturvedi H, Mathur S, Mandal A (2020) Recycling behavior of Indian consumers: evidence from textile industry. J Consum Stud 24(2):135–154
Choi T (2014) Fashion retail supply chain management: a systems optimization approach. Int J Prod Econ 147:448–459
Circular economy action plan. (n.d.). Environment. https://environment.ec.europa.eu/strategy/circular-economyaction-plan_enDan H, Østergaard M (2021) Circular fashion economy: trends and consumer attitudes. Sustain Text J 22(1):75–89
Eppinger S (2022) Consumer attitudes and behaviors towards garment recycling. J Sustain Recycl 33(4):489–502
Extended Producer Responsibility and Economic Instruments (n.d.) Organisation for Economic Co-operation and Development (OECD)
Gen Z cares about sustainability more than anyone Else—and is starting to make others feel the same way (2022) Business Insider
Ghadimi P, Heavey C, Dolgui A (2019) Eco-efficient supply chain strategies. J Clean Prod 233:1033–1047
Ghazali SM, Hui TK, Sayekti MP (2019) Value-belief-norm theory and recycling behavior: a study of Indonesian consumers. Asian J Sustain 28(3):221–231
Gomes F, Oliveira M, Saccani N (2022) Exploring the impact of financial incentives on recycling behaviour. J Consum Econ 35(2):137–149
Hong Y, Lee S, Jung H (2024) Self-identity and its role in sustainable consumer behavior. J Environ Psychol 58:82–90
Hu Z, Tang L, Sun Y (2022) Motivational factors in clothing recycling. J Text Recycl 19(3):299–317
Igini G (2024) Textile waste and environmental sustainability. J Environ Conserv 54(1):12–26
Jacometti V (2019) The challenges of textile recycling. Waste Manag J 39(2):1015–1023
Kalmykova Y, Sadagopan M, Rosado L (2018) Circular economy: critical review of progress, gaps and research needs. J Clean Prod 177:219–232
Ki CWC, Cuevas LM, Chong SM, & Lim H (2020) Influencer marketing: Social media influencers as human brands attaching to followers and yielding positive marketing results by fulfilling needs. Journal of retailing and consumer services, 55, 102133.
King A (2024) Regulatory frameworks for sustainable textiles. Environ Policy Gov 20(2):99–112
Kirchherr J, Reike D, Hekkert M (2023) The circular economy: barriers and enablers. J Ind Ecol 27(3):89–100
Köhler C, Hidalgo-Crespo E, Ferreira G (2019) Recycling habits: a psychological perspective. J Environ Psychol 63:24–35
Kollmuss A, Agyeman J (2002) Mind the gap: why do people act environmentally and what are the barriers to pro-environmental behavior? Environ Educ Res 8(3):239–260
Laitala K (2014) Consumers' clothing disposal behaviour–a synthesis of research results. International Journal of Consumer Studies, 38(5), 444–457.
Mandarić M, Ivanovic B, Arsic L (2022) Trends in sustainable fashion consumption: a cross-country study. Sustainability 14(5):2879
McGinty A (n.d.) Sustainable fashion and the circular economy. Eco-friendly Fash J
Mizrachi L, Tal A (2022) Textile waste management: emerging policies for a circular economy. J Environ Manag 323:118029
Mud Jeans (n.d.) Sustainability Report | MUD Jeans. https://mudjeans.eu/pages/sustainability
Niinimäki K, Peters G, Dahlbo H, Perry P, Rissanen T (2020) The environmental price of fast fashion. Nat Rev Earth Environ 1(3):189–200
ORGANISATION FOR ECONOMIC CO-OPERATION AND DEVELOPMENT, & Fielder, A. (2008). Promoting sustainable consumption [Report]. https://www.oecd.org/greengrowth/40317373.pdf
Ozdamar-Ertekin Z (2017) Sustainable fashion consumption: evidence from Turkey. Int J Consum Stud 41(3):453–462

Peters G, Niinimäki K, Dahlbo H (2021) Environmental impacts of clothing: life-cycle assessment. J Environ Manag 281:111847

Remy N, Speelman E, Swartz S (2016) Style that's sustainable: a new fast-fashion formula. McKinsey & Company

Riemens B, Schmidt G, Korpys C (2021) Barriers to sustainable fashion. J Fash Text 19(4):1023–1035

Saccani N, Gomes F, Oliveira M (2023) Economic incentives in textile recycling: a systematic review. J Ind Econ 35(2):199–218

Sandvik IM, & Stubbs W (2019) Circular fashion supply chain through textile-to-textile recycling. Journal of Fashion Marketing and Management: An International Journal, 23(3), 366–381.

Shirvanimoghaddam, Kamyar, Motamed, Bahareh, Ramakrishna, Seeram, & Naebe, Minoo. (2020). Towards sustainability in textile and fashion: A review of sustainable fibers and recycling techniques.

Stancu V, Vadillo RG, Jansson J (2020) Pro-environmental behavior in fashion consumption: a quantitative approach. J Clean Prod 255:120144

Stets JE, Burke PJ (2003) A sociological approach to self and identity. In: Handbook of self and identity, pp 128–152

Stern PC, Dietz T, Abel T, Guagnano GA, & Kalof L (1999) A value-belief-norm theory of support for social movements: The case of environmentalism. Human ecology review, 81–97.

Sun Y, Tang L, Hu Z (2024) Attitudes towards recycling in the textile industry. J Consum Psychol 17(1):35–48

Swann WB Jr, Gómez A, Dovidio JF (2009) Identity fusion: the interplay of personal and social identities in extreme group behavior. J Pers Soc Psychol 96(5):995–1011

Testa F, Pretner G, Iraldo F (2017) The role of consumer convenience in garment recycling. Waste Manag J 70:89–101

Used Patagonia® Clothing & Gear | Worn Wear (n.d.) Patagonia worn wear. https://wornwear.patagonia.com

Van Der Werff E, Steg L, Keizer K (2013) The value of self-identity in pro-environmental behavior. J Environ Psychol 34:55–63

Vladimirova D, Evans S, Kerrigan F (2023) Sustainable fashion trends in the digital age. J Consum Cult 23(1):89–105

Whitmarsh L, O'Neill S (2010) The green identity: the role of identity in pro-environmental behaviour. J Environ Psychol 30(3):305–314

What is the linear economy? (2023, February 10). https://www.ellenmacarthurfoundation.org/what-is-the-lineareconomy

Williams A (2022) Consumers and sustainability: changing behaviors in fast fashion. Sustain Pract 15(2):101–115

Wójcik-Karpacz A, Stankowska M (2023) Barriers to textile recycling. J Text Waste Manag 40(1):63–78

Woolridge AC, Ward GD, Phillips PS, Collins M, Gandy S (2006) Life cycle assessment for reuse/recycling of donated waste textiles compared to use of virgin material: an UK energy saving perspective. Resour Conserv Recycl 46(1):94–103

Yan X, Rashid A, Nguyen M (2021) Sustainable consumption patterns and garment recycling. J Fash Mark 23(1):45–64

Yang D, Tadeu SR, Sousa JC (2004) Environmental awareness and behavior: a study of fashion consumers. J Consum Mark 18(4):101–114

Chapter 6
Issues and Challenges in Humanizing Circular Supply Chain Management

Soumi Das Gupta, Anuradha Yadav, and Nancy Kumari

Abstract This research uses a structured literature review process to investigate the complexity and challenges of humanizing circular supply chain management (CSCM). By summarizing previous research, we identify major themes, difficulties, and gaps in this field. Our research focuses on natural, ethical, and organizational factors that are critical in integrating human-centred methods into CSCM. The report identifies major problems such as effective stakeholder participation, ethical quandaries, technical adaptability, and policy consequences. The paper focuses on a multidisciplinary approach to direct these concerns and presents a framework for incorporating human-centric practices into circular supply chains. This article adds to the existing knowledge of sustainable supply chain management by providing a thorough analysis of the challenges and opportunities for humanizing CSCM.

Keywords Circular supply chain management (CSCM) · Humanizing supply chains · Sustainable supply chain management · Technological integration · Challenges

6.1 Introduction

In the age of Industry 4.0, effective supply chain management ensures that items are made or purchased in adequate numbers. They are delivered to the correct locations at the appropriate times, increasing customer satisfaction and creating a competitive

S. D. Gupta
IILM University, Gurugram, India

A. Yadav (✉)
DPG Degree College, Gurugram, India

N. Kumari
SRM Institute of Science and Technology (SRM IST), Delhi-NCR Campus, Ghaziabad, India

K. Mathiyazhagan et al. (eds.), *Humanizing Circular Supply Chain Management*, Circular Economy and Sustainability,
https://doi.org/10.1007/978-3-031-86562-6_6

advantage (Zelbst et al. 2023). It also involves collaboration among suppliers, manufacturers, and retailers, leveraging and integrating technologies like automation, data analytics, cloud computing and blockchain to improve transparency and responsiveness (Vechain 2023; OriginTrail 2023). Then there are also various technologies such as IoT (Internet of Things) technologies and Industry 4.0 which has been majorly made for managing supply chain and various manufacturing or operating services in the business (Shayganmehr et al. 2021). It will not only help businesses to operate more smoothly even when they are receiving a huge number of orders but also for the customers who are receiving it (Ahamed et al. 2022). This is because if the operations are managed or controlled properly through the technologies then the business will be able to give authentic information as to what is the status of the product order i.e., the tracking details of the product (Shang et al. 2022). Technological adaptation like humanizing technology and upgradation also helps in reducing the carbon emissions that the machines working in the supply chain systems generate (Casciani and D'Itria 2024; Davies et al. 2024). These advanced technologies which have been created will also benefit the supply chain systems to handle any kind of external as well as internal events (Mollenkopf et al. 2024). So, it is crucial to streamline such processes and adapt to market changes, manage risks, and drive sustainable growth. If these strategies are adopted to the point required then it can also help in competing with the other companies in an efficient way or it can give a competitive edge among the competitors in the market (Kumar et al. 2021).

For example, during the COVID-19 pandemic, many companies' operations were disrupted one of them being supply chain management processes because people were not able to work in the warehouses or the manufacturing industry as they were required to be shut down for the timing and also because majorly the sales of the products that they were selling were also stopped as per the guidelines by the government due to which the businesses also could not gather the used products for procuring and remanufacturing. Only the essential items were allowed which also resulted in losses of a lot of companies.

The theoretical frameworks used in this research include sustainability, engagement of different stakeholders and human centred design. It extends the scope of circular economy beyond technological and natural resource management solutions to emphasize on the shareholder engagement and on the humanitarian issue. Drawing upon themes derived from ethical theory and organisational behavior the paper explores the challenges associated with implementing the human-centred activities into supply chain. Specifically, it employs scholarly perspectives from a range of disciplines in order to generate a framework that would de-militarize Circular Supply Chain Management (CSCM).

Further sections include literature review about the study, research methodology, analysis and discussion and conclusion of the research.

6.2 Literature Review

A lot of technological advancements in the supply chain also could not happen due to this external event (Apte and Viswanathan 2002). But as the companies believed in updating their supply chain systems as it came it became easier for these companies to revive their upcoming operations. The table below mentions the major external events with examples that may affect supply chain management (Mollenkopf et al. 2024).

Several factors which are external and affect supply chains. Some of these events include natural calamities like; forest fires, and floods, which makes it difficult to transport raw materials, and manufactures goods and services. Pandemics such as COVID-19 disrupt people's working capacity and affect the transportation of products. It considers operational changes that include recessions and inflation, changes that affect costs on an organization. Disregard of authorities and territorial conflicts obviously affect the business pace disrupting the transport of products.

Along with technological adoption, it is also the obligation of the companies or the businesses to manage internal and external events to reduce the suffering of those who work in the supply chain systems. This can be done by fulfilling all the requirements that they have while managing such events (Jenkins and Henley 2014). It can be solved through planning, implementation and control of the material flow and goods together with the related information throughout a system from the point of supply until the point of use. There are now many more organisations participating in supply chains than there were a long time ago, but this hasn't stopped the improvement of humanitarian response by meeting the needs of individuals affected by internal or external events with more goods and services (Pateman et al. 2013).

After managing all the internal and external events based on technology, adoption of the technologies, procuring raw materials, packing and delivering, operating the complete supply chain process smoothly and handling external events such as covid-19 and also addressing the requirements of the employees working in such systems there is a major challenge that is arising is of procuring sustainable raw materials or previously used raw materials from the external environment (Rasi et al. 2023). Nowadays due to climate change and customers knowing the harmful effects of the materials used in the products purchased there is an increased demand for sustainable products and to make them there is a requirement for sustainable raw materials or materials that can be refurbished. This used raw materials can be procured from the company's customer itself or by any other company's customer materials that they are not using anymore (Marshall et al. 2015). This concept might also be known as circular economy where there is consumption of the products manufactured, usage of the product, reuse as well as repair, collections of those products and recycling of the same. In this way, the lifecycle of a product is extended and implies reducing waste to minimum for the wellbeing of the environment (Giudice et al. 2020) (Fig. 6.1).

Several Indian companies are conscious and have become leaders in the use of recycled materials to make new goods. These businesses are prime examples of the circular economy's tenets, which emphasise minimising waste and constantly

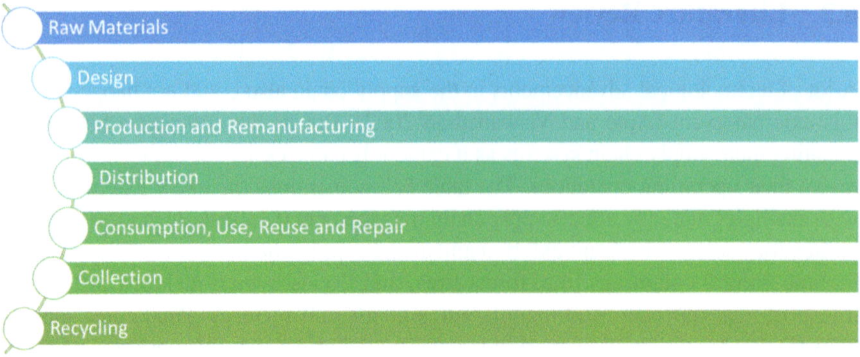

Fig. 6.1 Circular economy steps followed in supply chain management

repurposing resources to add value. ITC Limited's Waste Paper Recycling Unit marketed under Paperkraft is illustrative of sustainability as it recycles waste paper, reduces landfill and Ministers of Environment of Thy, and conserves water and forest (Marshall et al. 2015). On its part, Tata Group that is well-known for social accountability and creativity exhibits environmental accountability through Tata Steel. The organization utilizes steel scrap in the manufacturing of new products thus cutting on energy bills and emissions of carbon, sparing the natural resources in their many franchises. Mahindra Group's electric vehicle arm, Mahindra Reva, showcases sustainability in car manufacturing. They implement recycling programs for parts and batteries, emphasizing recyclable materials in vehicle production. This strategy promotes sustainable manufacturing and addresses end-of-life car waste effectively.

6.3 Research Methodology

The study uses a systematic literature review (SLR) to assess the difficulties of implementing humanization in Circular Supply Chain Management (CSCM). It entails the formulation of research questions, conducting a literature review and the use of inclusion and exclusion criteria. Information is harvested thematically to establish the key issues and Miller and De Matta (2008) deficits. To refine the proposed human-centred practices within CSCM the following framework is developed: Quality assessment and Peer review.

This research makes use of systematic review, which employs the SCOPUS database. After fine-tuning the criteria for selection, 50 articles were selected for the final analysis using the keywords such as "Circular Supply Chain Management" OR "ccm" or "dehumanizing supply chains" AND "Sustainable" OR "Supply Chain Management" AND "Challenges". The following is the breakdown of the papers selection process: Fig. 6.2 brings out the paper selection process towards the final review. The final stage of the search and selection of articles was done systematically from articles that have been published between the year 2015–2024.

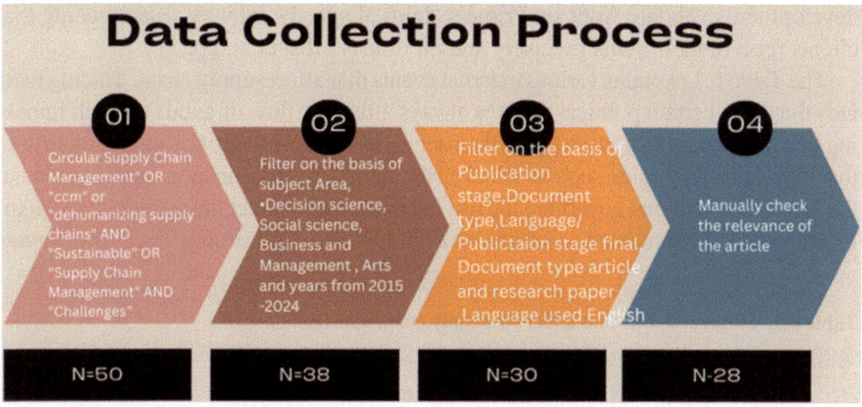

Fig. 6.2 Data collection process

6.4 Systematic Review

Several companies follow the major technologies to ensure that the products manufactured or sourced are delivered in the accurate quantities to the accurate locations, and at the accurate time, enhancing customer satisfaction and competition, and involving collaboration among suppliers, manufacturers, and retailers, leveraging technologies like automation, data analytics, cloud computing and blockchain to improve transparency and responsiveness (Akbari 2023). The major companies following these practices are Amazon and Flipkart. Two of the biggest e-commerce companies in India, Flipkart and Amazon, have created strong and advanced supply chain management systems to satisfy the wide range of demands of their enormous client bases. In order to guarantee maximum reach and efficiency, Flipkart has established a vast network of warehouses, fulfilment centres, and last-mile delivery partners. These locations are carefully chosen (Maheshwari et al. 2020). Artificial intelligence (AI) and machine learning, which are utilised for demand forecasting using big data, inventory management, and route optimisation, enhance the company's supply chain. Ekart, Flipkart's internal logistics division, is essential to overseeing delivery operations and making sure that goods get to consumers' locations on time and with reliability (Rothengatter and Petronella 2022). The table below mentions the major technologies used in supply chain management.

Amazon has established a similarly intricate and effective supply network in India by utilising its experience in international logistics. The business runs sizable fulfilment facilities, local sorting facilities, and a network of outside carriers and delivery partners. Proprietary technology, such as the utilisation of Kiva robots (Kiva robots simplify warehouse operations by moving product shelves near human workers, resulting in much shorter picking times and higher productivity. These robots greatly contribute to Amazon's ability to deliver goods quickly and reliably by speeding up order processing, maximising space utilisation, and reducing labour expenses.) in warehouses to simplify operations and advanced data analytics to optimise supply chain processes, further improving Amazon's supply chain. These developments validate Amazon Prime's claim of speedy delivery, guaranteeing that clients receive their items precisely and on time (Zelbst et al. 2023).

The Table 6.1 explains various external events that affect supply chain. Trading partners that have certain political motives always affect the flow of goods through imposing or adjusting tariffs and engaging themselves in trade conflicts. It is important to note that certain technologies such as Blockchain and AI would enhance supply chains in the future. Regulations governing the use of environment affect the techniques of manufacture and implementation of sustainable systems. Concerning disruptive threats

Table 6.1 External events that affect the supply chain management

S.no.	External events	Examples of external events	Source
1.	Natural disasters	Natural disasters of various kinds, including forest fires, hurricanes, floods, and earthquakes, are potentially disruptive events able to hinder direct delivery and transportation and/or affect manufacturing	Mangla and Luthra (2020), Petrudi et al. (2020)
2.	Pandemics	Extended health crises meaning such as the current Covid-19 pandemics may lead to labour shortages, shutdowns and restrictions of movement of goods	Craighead et al. (2020), Rinaldi et al. (2022)
3.	Economic shifts	The operations and cost drivers of supply chain can be affected by factors such as recessions, inflation, and changes in customers' buying behaviours	Closs et al. (2014), De Angelis et al. (2018)
4.	Political instability	Conflicts such as war, civil strife, and instabilities in governments are some of the enemies that slow down supply chains and accessibility to markets	Jadallah and Bhatti (2020), Kharsun et al. (2022)
5.	Trade policies	The aspects related to price and flow of goods across borders are subject to changes in tariffs, trade conflicts and other regulatory actions.	Blanchard et al. (2016), Heise et al. (2015)
6.	Technological advances	Innovations such as the use of Blockchain; AI and automation has the capabilities of positively transforming or altering the nature of supply chain	Apte and Viswanathan (2002), Singh (2003)
7.	Environmental regulations	There could be concerns on the environmental legislation affecting supply chain as well as production methods due to the changes in stringency of environmental legislation and sustainability practices	Chen and Sheu (2009), Yang and Sheu (2011)
8.	Cybersecurity threats	Operations can be interrupted due to cyberattacks and data breaches due to which supply chain information systems can be compromised	Lamba et al. (2017), Boiko et al. (2019)
9.	Transportation issues	There are few factors that might result in delays in the shipment of the products such as strikes, fluctuations in fuel prices and infrastructure problems or an increase in logistical costs	Sainathuni et al. (2014), Mason et al. (2003)
10.	Globalization trends	Due to the changes in the international trade trends and to the opening up of new markets the partnerships and supply chain tactics can be altered	Kumar (2016), Mattsson (2003)

cyber-attacks is defined as threats that disrupt operations. Labour strikes and also problems with the transport systems lead to delay in the shipment of goods. In the last place, globalization trends may also impact on partner and supply chain relations because of shifts in the patterns of trade and new market opportunities.

Both businesses prioritise a customer-centric strategy and are always coming up with new ideas to increase delivery times, dependability, and overall client happiness. They use a variety of tactics, including automated systems for improved inventory control, real-time tracking for transparency, and same-day and next-day delivery alternatives (Bag et al. 2021). Moreover, Flipkart and Amazon both dramatically expand their operations during busy times like the holidays, hiring more staff, building temporary warehouses, and improving their technological prowess to manage the spike in orders effectively. Flipkart and Amazon have been able to hold onto their top positions in the fiercely competitive Indian e-commerce market because to their strategic efforts in supply chain management. The Table 6.2 below shows the major technologies which can be used in the supply chain systems.

Table 6.2 Major technologies used in supply chain management

S.no.	Technologies	Definition	Role in supply chain management	Source
1	Artificial Intelligence (AI)	Artificial Intelligence has been made to learn and think the way a human does. It includes different technologies such as machine learning, natural language processing, and robots. This allows the systems to carry out functions including sensing, language translation and decision-making	By using artificial intelligence for automating operations, optimising inventory levels and forecasting demand may result in increase in efficiency and accuracy	Sharma et al. (2022), Younis et al. (2022)
2	Internet of Things (IoT)	The linked devices that communicate and share data via the internet is known as Internet of Things. These gadgets, ranges from to exchange and gather data to enhance productivity, industrial machinery to home appliances, automation and decision-making	The functions in supply chain management such as inventory control, streamlining of the logistics, real-time tracking and monitoring of commodities are transformed with the help of Internet of Things (IoT)	Rejeb et al. (2020), Birkel and Hartmann (2020)
3	Cloud Computing	The computing services on the internet such as servers, databases, storage, analytics, networking, and software are provided by cloud computing. It helps users to access and save data remotely instead of storing the data on local servers or personal devices	Optimization of supply chains with enhanced visibility, real-time data sharing, scalable resources, automation and advanced data analytics is done by cloud computing	Ivanov et al. (2022), Attaran (2020)

(continued)

Table 6.2 (continued)

S.no.	Technologies	Definition	Role in supply chain management	Source
4	Big Data	The large amount of speed, diversity of data produced by several resources, including social media, transactions, sensors, and more is termed as big data. In order to gain valuable information and facilitate decision-making, it includes huge data sets that conventional data-processing software is unable to manage	The supply chain companies can enhance inventory management, demand forecasting, operational efficiency and logistics by leveraging big data analytics	Maheshwari et al. (2020), Giudice et al. (2020)
5	Machine Learning	Machine learning deals with creating statistical models and algorithms that let computer analyse, learn, and make decisions based on the data. It is a branch of Artificial Intelligence	In supply chain the analyses of data for trends, optimization of logistics, forecasting demand, inventory and detecting anomalies through machine learning is resulting in improvement of efficiency, adaptability and cost effectiveness	Tirkolaee et al. (2021), Ni et al. (2019)
6	Blockchain	The distributed, decentralised ledger technology that maintains security, transparency, and immutability by recording transactions across several computers refers to blockchain. It helps make secure transactions and data sharing as safe and transparent as possible	The blockchain technology helps the supply chain with secure, traceable and transparent decentralized ledger and by also ensuring authenticity from raw materials to end consumers	Moosavi et al. (2021), Queiroz et al. (2019)

All the above technologies mentioned in the table also saves the costs of the company when it comes to the supply chain of the business. A lot of companies are adopting or have started to adopt these technologies which resulted in reducing the cost of the business, increasing efficiency, procuring more customers as the efficiency and reliability increased, and delivering the product through the supply chain systems as well as generating profits. Here are some more examples of a few companies adopting the technologies and benefitting from it in return (Rothengatter and Petronella 2022).

Now, the process of humanising circular supply chain management also entails incorporating ethical and human-centred principles into the design of a supply chain that emphasises sustainability by reusing, recycling, and repurposing materials to reduce waste and increase resource efficiency (Davies et al. 2024). By balancing economic goals with the welfare of people and communities along the supply chain, this all-encompassing strategy aims to promote a more socially conscious and inclusive company model (Saroha et al. 2021). Fair employment standards are the cornerstone of the humanising circular supply chain management approach. It is crucial to guarantee that every person in the supply chain receives respect, just compensation, and secure working conditions (Menon 2012). This dedication goes all the way from the first phases of gathering raw materials to the last steps of recycling and product repurposing. Companies may assist avoid exploitation and promote a more equal working environment by supporting fair labour practices. This can ultimately result in higher levels of worker satisfaction and productivity (Agarwal et al. 2020).

This human-centred approach also requires transparency and ethical sources. Giving stakeholders and customers access to the complete supply chain process enables them to comprehend how goods are created, utilised, and ultimately recycled. By guaranteeing that materials are sourced ethically and that recycling procedures are sustainable and do not harm vulnerable groups, this transparency contributes to the development of trust and accountability (Bag and Dhamija 2022). Additionally, it gives customers the power to choose products wisely and supports businesses that place a high value on morality and environmental responsibility (Pagell et al. 2024). Humanising circular supply chain management also requires active community engagement. It is essential to include these communities in the supply chain process and make sure that the activities help them rather than hurt them (Soundararajan et al. 2021). This can involve investing in community development projects, fostering local economies, and generating employment possibilities. By doing this, businesses can promote social progress and economic development in the communities in which they operate, which will help create a society that is more just and equal (Agarwal et al. 2020).

The most important part in humanizing the supply chain is also the employees working there. It is very important to give the employees working there with good working conditions, providing them with all the sorts of training required such as how to source and transport the necessary raw materials and how to handle the machines and many more in the supply chain systems and addressing their needs properly (De Camargo Fiorini et al. 2021). If they would be satisfied and trained in a proper manner then the supply chain works without any hindrance in this case. Also, proper monitoring of the supply chain systems will also happen (Soundararajan et al. 2021). The Table 6.3 below mentions the major factors that effect humanizing circular supply chain management.

In this method, consumer participation and education are crucial. By informing customers about the value of sustainability and their part in the circular supply chain can promote appropriate recycling and responsible consumption. To close the loop and make sure that materials are used for as long as feasible, businesses can support take-back programmes and other activities that let customers return used goods for recycling or other uses (Saroha et al. 2021). An essential component of humanising the circular supply chain and circular supply chain management is environmental supervision. It is essential to make sure that supply chain procedures are both environmentally friendly and sustainable. This entails cutting waste, encouraging the

Table 6.3 Major factors that effect in humanizing supply chain management

S.no.	Major factors	Examples of the factors	Source
1.	Stakeholder Collaboration	Stakeholder collaboration refers to the involvement of parties involved in the decision-making processes, such as staff members, suppliers, clients and local communities to guarantee that different viewpoints and requirements are taken into account	Co and Barro (2009), Baah et al. (2021)
2.	Employee Well-being	Employee well-being refers to the possibilities for safe working conditions, employee growth and fair labour standards at the top of the supply chain	Menon (2012), Mollenkopf et al. (2020)
3.	Customer Involvement	Customer involvement refers to motivating sustainable consumption practices, recycling, and product reuse in order to inform and involve customers in circular economy	Kumar et al. (2014), Sahay (2003)
4.	Transparent Communication	Transparent communication refers to open communication in all lines and being honest with all parties regarding supply chain procedures, environmental effects, and sustainability objectives	Cerullo et al. (2016), Brun et al. (2020)
5.	Ethical Sourcing	Ethical sourcing refers to the resources and products that are procured ethically, taking into account the well-being of the local people, human rights and the environment	Roberts (2003), Chen and Slotnick (2015)
6.	Local Community Engagement	Local Community Engagement refers to working together with neighbourhood groups to reduce detrimental effects on the environment and society, generate economic opportunities and promote local companies	Gonzalēz (2014), Jenkins and Henley (2014)
7.	Inclusive Design	Inclusive design refers to creating systems and products with the capabilities and needs of all people-including members of underrepresented groups and those with disabilities-in mind	Dong et al. (2004), Krishnan et al. (2019)
8.	Cultural Sensitivity	Cultural sensitivity refers to valuing and taking into consideration the regional traditions and cultural differences in supply chain operations to promote goodwill and community support	Gupta and Gupta (2019), Marshall et al. (2015)
9.	Sustainability Education	Sustainability education refers to training and educating staff members, vendors and customers on eco-friendly procedures and the value of the circular economy	Salinas-Navarro et al. (2022), Kassaneh et al. (2021)
10.	Responsibility and Accountability	Responsibility and Accountability refers to clearly defining who is responsible for what in the supply chain when it comes to sustainable practices and responsible behaviour	Parmigiani et al. (2011), Mares (2010)

use of renewable resources, and lowering the carbon footprint. Businesses may contribute to the preservation of natural resources and save the earth for future generations by implementing environmentally friendly practices (Yang and Sheu 2011). This all-encompassing approach cannot be driven without collaborative collaborations. Through collaboration with diverse stakeholders such as suppliers, customers, NGOs, and governments, corporations can advance optimal methodologies and catalyse group initiatives towards morally sound and environmentally responsible supply chain management. These collaborations may guarantee that all stakeholders are committed to humanising the supply chain, assist create industry standards, and exchanging information and ideas (Moharana et al. 2010).

6.5 Discussion and Conclusion

Supply chain management entails a lot of diverse things such as procuring raw materials, manufacturing a product efficiently and at a low cost, updating with the latest technologies available, delivering the product to the customer and much more. As a lot of sustainable products are in demand which means manufacturing a product from raw materials which do not harm the environment or procuring those kinds of raw materials which can be reused or refurbished into a new product. The sustainable raw materials are majorly procured by the companies through collaborations and involving stakeholders of the companies. This is being done just to reduce the environment impact by minimising the waste that has been generated over the years and not exploiting the nature that resulted in climate change. Also, to keep the supply chain systems updated with those technologies that can generate efficiency and reduce the carbon emissions that the machines generate. These technological updations can also keep customer satisfied and help the supply chain systems to handle any kind of external as well as internal events in an efficient manner. There are a lot of inventions that are happening when it comes to technologies for the supply chain systems so the companies should adopt them as it will help them only and boost the business due to this the customers will get the right product at the right time and at the right place. These technologies might also help employees working in these supply chain systems by increasing their efficiency as well. Also, it is required by the companies of the supply chain systems to address any requirement or concern that any employee might have to ensure a smooth flow (Fig. 6.3).

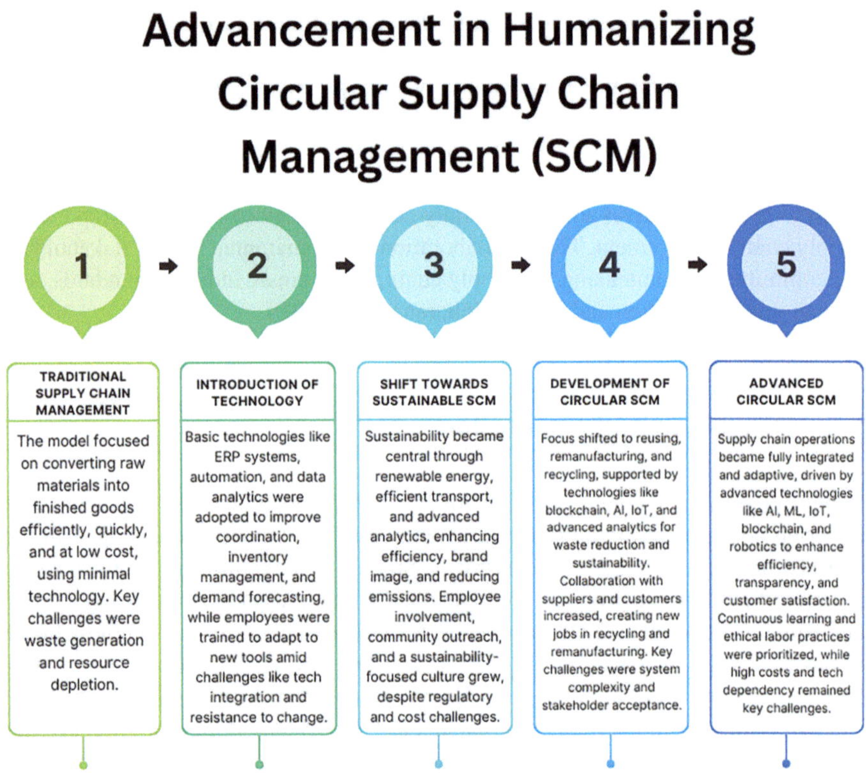

Fig. 6.3 Advancement in humanizing circular supply chain management

6.6 Conclusion

The process of humanising supply chain management comes with several problems and difficulties, such as how to incorporate ethical practices, improve working conditions in the supply chain systems, and guarantee sustainable practices. Effectiveness and employee well-being must coexist harmoniously, which frequently necessitates essential changes to current processes and organisational culture. Addressing labour rights, promoting transparency, addressing employees' needs and controlling supply chain operations' effects on the environment are important challenges. Businesses must manage this complexity or the challenges while remaining competitive, which frequently calls for creative thinking and a dedication to long-term, comprehensive policies that put the welfare of people and the environment ahead of profit.

6.6.1 *Implications*

The implications of this study on humanization of Circular Supply Chain Management (CSCM) are discussed as follows: Building on these implications, practitioners and scholars can have a deeper insight into how human-centred concepts can improve the effectiveness and sustainability of circular supply chain.

Managerial Implications Managers must address human factors that affect CSCM by synchronizing them to address both strategic and ethical requirements aimed at improving focal organizations' operations while strengthening the support of multiple stakeholders in the long-term.

Societal Implications Humanising aspect of CSCM aligns to social justice, workers' welfare and protection of the environment thus promoting sustainable business and promoting responsible business practices.

Theoretical Implications It opens new avenues for building theory that connects solutions that are human centric with circular economy practice and encourages cross disciplinary research on sustainable supply chain configurations.

References

Agarwal S, Kant R, Shankar R (2020) Modeling the enablers of humanitarian supply chain management: a hybrid group decision-making approach. Benchmarking 28(1):166–204. https://doi.org/10.1108/bij-03-2020-0093

Ahamed M, Rahim, Ahamed A (2022) Humanizing the fourth industry revolution in sustainable supply chain management. https://www.researchgate.net/publication/368465767_Humanizing_the_Fourth_Industry_Revolution_in_Sustainable_Supply_Chain_Management

Akbari M (2023) Revolutionizing supply chain and circular economy with edge computing: systematic review, research themes and future directions. Manag Decis 62:2875–2899. https://doi.org/10.1108/md-03-2023-0412

Apte UM, Viswanathan S (2002) Strategic and technological innovations in supply chain management. Int J Manuf Technol Manag 4(3/4):264. https://doi.org/10.1504/ijmtm.2002.001449

Attaran M (2020) Digital technology enablers and their implications for supply chain management. Supply Chain Forum 21(3):158–172. https://doi.org/10.1080/16258312.2020.1751568

Baah C, Acquah ISK, Ofori D (2021) Exploring the influence of supply chain collaboration on supply chain visibility, stakeholder trust, environmental and financial performances: a partial least square approach. Benchmarking 29(1):172–193. https://doi.org/10.1108/bij-10-2020-0519

Bag S, Dhamija P (2022) Research progress on working conditions in supply chains: a comprehensive literature review and future research propositions. TQM J 35(8):2282–2303. https://doi.org/10.1108/tqm-09-2022-0290

Bag S, Wood LC, Telukdarie A, Venkatesh VG (2021) Application of Industry 4.0 tools to empower circular economy and achieving sustainability in supply chain operations. Prod Plan Control 34(10):918–940. https://doi.org/10.1080/09537287.2021.1980902

Birkel HS, Hartmann E (2020) Internet of Things—the future of managing supply chain risks. https://www.emerald.com/insight/content/doi/10.1108/SCM-09-2019-0356/full/html

Blanchard E, Bown C, Johnson R (2016) Global supply chains and trade policy. https://doi.org/10.3386/w21883

Boiko A, Shendryk V, Boiko O (2019) Information systems for supply chain management: uncertainties, risks and cyber security. Procedia Comput Sci 149:65–70. https://doi.org/10.1016/j.procs.2019.01.108

Brun A, Karaosman H, Barresi T (2020) Supply chain collaboration for transparency. Sustainability 12(11):4429. https://doi.org/10.3390/su12114429

Casciani D, D'Itria E (2024) Fostering directions for digital technology adoption in sustainable and circular fashion: toward the circular fashion-tech lab. Systems 12. https://doi.org/10.3390/systems12060190

Cerullo G, Guizzi G, Massei C, Sgaglione L (2016) Efficient supply chain management: traceability and transparency. https://ieeexplore.ieee.org/abstract/document/7907551

Chen YJ, Sheu JB (2009) Environmental-regulation pricing strategies for green supply chain management. Transp Res E Logist Transp Rev 45(5):667–677. https://doi.org/10.1016/j.tre.2009.04.010

Chen JY, Slotnick SA (2015) Supply chain disclosure and ethical sourcing. Int J Prod Econ 161:17–30. https://doi.org/10.1016/j.ijpe.2014.11.001

Closs DJ, Bolumole YA, Rodammer F (2014) Supply chain management opportunities for regional economic development. Transp J 53(4):453–498. https://doi.org/10.5325/transportationj.53.4.0453

Co HC, Barro F (2009) Stakeholder theory and dynamics in supply chain collaboration. Int J Oper Prod Manag 29(6):591–611. https://doi.org/10.1108/01443570910957573

Craighead CW, Ketchen DJ, Darby JL (2020) Pandemics and supply chain management research: toward a theoretical toolbox*. Decis Sci 51(4):838–866. https://doi.org/10.1111/deci.12468

Davies J, Sharifi H, Lyons A, Forster R, Elsayed OK (2024) Non-fungible tokens: the missing ingredient for sustainable supply chains in the metaverse age? Transp Res E Logist Transp Rev 182:103412. https://doi.org/10.1016/j.tre.2024.103412

De Angelis R, Howard M, Miemczyk J (2018) Supply chain management and the circular economy: towards the circular supply chain. Prod Plan Control 29(6):425–437. https://doi.org/10.1080/09537287.2018.1449244

De Camargo Fiorini P, Jabbour CJC, De Sousa Jabbour ABL, Ramsden G (2021) The human side of humanitarian supply chains: a research agenda and systematization framework. Ann Oper Res 319(1):911–936. https://doi.org/10.1007/s10479-021-03970-z

Dong H, Keates S, Clarkson PJ (2004) Inclusive design in industry: barriers, drivers and the business case. In: Lecture notes in computer science, pp 305–319. https://doi.org/10.1007/978-3-540-30111-0_26

Giudice MD, Chierici R, Mazzucchelli A, Fiano F (2020) Supply chain management in the era of circular economy: the moderating effect of big data. https://www.emerald.com/insight/content/doi/10.1108/IJLM-03-2020-0119/full/html

Gonzalez M (2014) Advancing human rights in supply chains through community engagement and investment. https://pmworldlibrary.net/wp-content/uploads/2014/03/pmwj20-mar2014-gonzalez-advancing-human-rights-FeaturedPaper.pdf

Gupta M, Gupta S (2019) Influence of national cultures on operations management and supply chain management practices—a research agenda. Prod Oper Manag 28(11):2681–2698. https://doi.org/10.1111/poms.13100

Heise S, Pierece JR, Schaur G, Schott PK (2015) Trade policy and the structure of supply chains. https://citeseerx.ist.psu.edu/document?repid=rep1&type=pdf&doi=8013c296eb45d82e377feab2d3a440db19ecfc8a

Ivanov D, Dolgui A, Sokolov B (2022) Cloud supply chain: integrating Industry 4.0 and digital platforms in the "Supply Chain-as-a-Service." https://www.sciencedirect.com/science/article/abs/pii/S1366554522000709

Jadallah N, Bhatti F (2020) Political instability and sustainable green supply chain management. http://m.growingscience.com/beta/msl/3581-political-instability-and-sustainable-green-supply-chain-management.html

Jenkins B, Henley G (2014) Collaborative management: community engagement as the decision-making process. https://search.informit.org/doi/abs/10.3316/informit.410600285061426

Kassaneh TC, Bolisani E, Cegarra-Navarro JG (2021) Knowledge management practices for sustainable supply chain management: a challenge for business education. Sustain For 13(5):2956. https://doi.org/10.3390/su13052956

Kharsun L, Kavun-Moshkovska O, Kotova M, Nechyporuk A (2022) Adaptation of risk management in the supply chains of e-commerce enterprises under the conditions of political instability. Eastern Eur J Enterpr Technol 5(13 (119)):6–20. https://doi.org/10.15587/1729-4061.2022.265649

Krishnan VV, Lee J, Mnyshenko O, Shin H (2019) Inclusive innovation: product innovation in technology supply chains. Manuf Serv Oper Manag 21(2):327–345. https://doi.org/10.1287/msom.2018.0746

Kumar RK (2016) Global trends in supply chain management. https://www.indianjournals.com/ijor.aspx?target=ijor:zijbemr&volume=6&issue=1&article=009

Kumar S, Luthra S, Haleem A (2014) Critical success factors of customer involvement in greening the supply chain: an empirical study. Int J Logist Syst Manag 19(3):283. https://doi.org/10.1504/ijlsm.2014.065498

Kumar A, Choudhary S, Garza-Reyes JA, Kumar V, Khan SR, Mishra N (2021) Analysis of critical success factors for implementing Industry 4.0 integrated circular supply chain—moving towards sustainable operations. Prod Plan Control 34(10):984–998. https://doi.org/10.1080/09537287.2021.1980905

Lamba A, Singh S, Balvinder S, Dutta N, Rela S (2017) Analyzing and fixing cyber security threats for supply chain management. Soc Sci Res Netw. https://doi.org/10.2139/ssrn.3492687

Maheshwari S, Gautam P, Jaggi CK (2020) Role of Big Data Analytics in supply chain management: current trends and future perspectives. Int J Prod Res 59:1875–1900. https://www.tandfonline.com/doi/abs/10.1080/00207543.2020.1793011

Mangla SK, Luthra S (2020) When challenges need an evaluation: for operational excellence and sustainability orientation in humanitarian supply and logistics management. https://www.tandfonline.com/doi/abs/10.1080/09537287.2020.1834129

Mares R (2010) The limits of supply chain responsibility: a critical analysis of corporate responsibility instruments. Nordic J Int Law 79(2):193–244. https://doi.org/10.1163/157181010x12668401898995

Marshall D, McCarthy L, McGrath P, Claudy M (2015) Going above and beyond: how sustainability culture and entrepreneurial orientation drive social sustainability supply chain practice adoption. Supply Chain Manag 20(4):434–454. https://doi.org/10.1108/scm-08-2014-0267

Mason SJ, Ribera PM, Farris JA, Kirk RG (2003) Integrating the warehousing and transportation functions of the supply chain. Transp Res E Logist Transp Rev 39(2):141–159. https://doi.org/10.1016/s1366-5545(02)00043-1

Mattsson L (2003) Reorganization of distribution in globalization of markets: the dynamic context of supply chain management. Supply Chain Manag 8(5):416–426. https://doi.org/10.1108/13598540310500259

Menon ST (2012) Human resource practices, supply chain performance, and wellbeing. Int J Manpow 33(7):769–785. https://doi.org/10.1108/01437721211268311

Miller T, De Matta R (2008) A global supply chain profit maximization and transfer pricing model. J Bus Logistics 29(1):175–199. https://doi.org/10.1002/j.2158-1592.2008.tb00074.x

Moharana H, Murty JS, Senapati SK, Khuntia K (2010) Coordination, collaboration and integration for supply chain management. https://core.ac.uk/download/pdf/480907759.pdf

Mollenkopf DA, Ozanne LK, Stolze HJ (2020) A transformative supply chain response to COVID-19. J Serv Manag 32(2):190–202. https://doi.org/10.1108/josm-05-2020-0143

Mollenkopf DA, Esper TL, Stolze HJ, Ozanne LK (2024) Transformative supply chain research: a new frontier for SCM scholars. J Bus Logist 45(3). https://doi.org/10.1111/jbl.12389

Moosavi J, Naeni LM, Fathollahi-Fard AM, Fiore U (2021) Blockchain in supply chain management: a review, bibliometric, and network analysis. https://link.springer.com/article/10.1007/s11356-021-13094-3

Ni D, Xiao Z, Lim MK (2019) A systematic review of the research trends of machine learning in supply chain management. Int J Mach Learn Cybern 11(7):1463–1482. https://doi.org/10.1007/s13042-019-01050-0

OriginTrail (2023) OriginTrail. origintrail.io/. Accessed 30 Aug 2023

Pagell M, Parkinson M, Veltri A (2024) Putting worker safety at the heart of supply chain management. In: Springer eBooks, pp 679–696. https://doi.org/10.1007/978-3-031-19884-7_98

Parmigiani A, Klassen RD, Russo MV (2011) Efficiency meets accountability: performance implications of supply chain configuration, control, and capabilities★. J Oper Manag 29(3):212–223. https://doi.org/10.1016/j.jom.2011.01.001

Pateman H, Hughes K, Cahoon S (2013) Humanizing humanitarian supply chains: a synthesis of key challenges. Asian J Shipping Logist 29(1):81–102. https://doi.org/10.1016/j.ajsl.2013.05.005

Petrudi SHH, Tavana M, Abdi M (2020) A comprehensive framework for analyzing challenges in humanitarian supply chain management: a case study of the Iranian Red Crescent Society. Int J Disaster Risk Reduct 42:101340. https://doi.org/10.1016/j.ijdrr.2019.101340

Queiroz MM, Telles R, Bonilla SH (2019) Blockchain and supply chain management integration: a systematic review of the literature. https://www.emerald.com/insight/content/doi/10.1108/SCM-03-2018-0143/full/html

Rasi RZ, Ismail H, Shahbaz MS, Sundram VPK (2023) Interdisciplinary challenges in the circular supply chains: a systematic literature review. https://www.ncbi.nlm.nih.gov/pmc/articles/PMC10113859/

Rejeb A, Simske S, Rejeb K, Treiblmaier H, Zailani S (2020) Internet of things research in supply chain management and logistics: a bibliometric analysis. Internet of Things 12:100318. https://doi.org/10.1016/j.iot.2020.100318

Rinaldi M, Murino T, Gebennini E, Morea D, Bottani E (2022) A literature review on quantitative models for supply chain risk management: can they be applied to pandemic disruptions? Comput Ind Eng 170:108329. https://doi.org/10.1016/j.cie.2022.108329

Roberts S (2003) J Bus Ethics 44(2):159–170. https://doi.org/10.1023/a:1023395631811

Rothengatter T, Petronella VDD (2022) How Industry 4.0 technologies can support a human-centric supply chain: an exploratory multiple case study with manufacturers and service providers. DIVA. https://www.diva-portal.org/smash/record.jsf?pid=diva2%3A1664797&dswid=-9826

Sahay B (2003) Supply chain collaboration: the key to value creation. Work Study 52(2):76–83. https://doi.org/10.1108/00438020310462872

Sainathuni B, Parikh PJ, Zhang X, Kong N (2014) The warehouse-inventory-transportation problem for supply chains. Eur J Oper Res 237(2):690–700. https://doi.org/10.1016/j.ejor.2014.02.007

Salinas-Navarro DE, Mejia-Argueta C, Montesinos L, Rodriguez-Calvo EZ (2022) Experiential learning for sustainability in supply chain management education. Sustainability 14(20):13133. https://doi.org/10.3390/su142013133

Saroha M, Garg D, Luthra S (2021) Identification and analysis of circular supply chain management practices for sustainability: a fuzzy-DEMATEL approach. Int J Prod Perform Manag 71(3):722–747. https://doi.org/10.1108/ijppm-11-2020-0613

Shang C, Saeidi P, Goh CF (2022) Evaluation of circular supply chains barriers in the era of Industry 4.0 transition using an extended decision-making approach. J Enterp Inf Manag 35(4/5):1100–1128. https://doi.org/10.1108/jeim-09-2021-0396

Sharma R, Shishodia A, Gunasekaran A, Min H, Munim ZH (2022) The role of artificial intelligence in supply chain management: mapping the territory. Int J Prod Res 60(24):7527–7550. https://doi.org/10.1080/00207543.2022.2029611

Shayganmehr M, Gupta S, Laguir I, Stekelorum R, Kumar A (2021) Assessing the role of Industry 4.0 for enhancing swift trust and coordination in humanitarian supply chain. Ann Oper Res 335(3):1053–1085. https://doi.org/10.1007/s10479-021-04430-4

Singh N (2003) Emerging technologies to support supply chain management. Commun ACM 46(9):243–247. https://doi.org/10.1145/903893.903943

Soundararajan V, Wilhelm MM, Crane A (2021) Humanizing research on working conditions in supply chains. https://pure.rug.nl/ws/portalfiles/portal/173114959/jscm.12260.pdf

Tirkolaee EB, Sadeghi S, Mooseloo FM, Vandchali HR, Aeini S (2021) Application of machine learning in supply chain management: a comprehensive overview of the main areas. Math Probl Eng 2021:1–14. https://doi.org/10.1155/2021/1476043

Vechain (2023) Web3 for better: Whitepaper 3.0. www.Vechain.org/assets/whitepaper/whitepaper-3-0.pdf. Accessed 30 Aug 2023

Yang C-L, Sheu C (2011) The effects of environmental regulations on green supply chains. https://academicjournals.org/journal/AJBM/article-full-text-pdf/A864BA415707

Younis H, Sundarakani B, Alsharairi M (2022) Applications of artificial intelligence and machine learning within supply chains: systematic review and future research directions. https://www.emerald.com/insight/content/doi/10.1108/JM2-12-2020-0322/full/html

Zelbst PJ, Yang L, Green KW, Sower VE (2023) Linkages between technologies and supply chain performance: benefits, power and risk reduction. Supply Chain Manag 29(1):207–218. https://doi.org/10.1108/scm-03-2023-0131

Chapter 7
Sustainable Value Chain Analysis Using ISM and MICMAC Methods: A Comprehensive Approach

Radha Rani, Sushma Sharma, and Poonam Gahlot

Abstract The background has provided valued understandings into competitive advantage ever since Michael Porter introduced the concept of value chain in 1985. This paper gives an original view by integrating Interpretative Structural Modeling (ISM) and MICMAC (Matrice d' Impacts Croises Multiplication Appliquee a un Classment methodologies to analyse workable practices with in the value chain. The purpose of the study is to create a model that categorises and evaluate the interconnectedness among sustainable key variables. The ISM technique is used to develop a classifies construction viewing how these variables move and depend on each other in the value chain. MICMAC study further orders these variables based on their influence and necessity, offering a calculated viewpoint on their role in nurturing sustainability.

The results showed that traditional value chain reflect efficiency and cost effectiveness, a sustainability focussed approach requires a thorough analysis of how various factors contribute to and support green practices. The study accomplishes with an argument of the implications of the ISM and MICMAC analysis for value chain management and suggests future research directions to further develop and refine integrated frameworks for sustainability.

Keywords Value chain analysis · ISM · MICMAC · Sustainability · Green practices · Procurement · Low carbon production · Green supply chains

R. Rani · S. Sharma (✉) · P. Gahlot
SRM University, Sonepat, Haryana, India
e-mail: sushma.s@srmuniversity.ac.in

K. Mathiyazhagan et al. (eds.), *Humanizing Circular Supply Chain Management*, Circular Economy and Sustainability,
https://doi.org/10.1007/978-3-031-86562-6_7

7.1 Introduction

The idea of the "value chain" familiarised by the Porter (1985), defines the series of activities involved in generating a product or service from preliminary activities through creation, supply and final disposal. Porter's model highlights that value is added at each stage of the chain and these accumulations can be leveraged to attain competitive advantage (Porter 1985). As Hellin and Meijer (2006) noted that the creation thorough various phases create incremental change and hence develop framework (Brown 1997). This framework supports as a tactical to gain competitive edge.

Over the decades, this concept has a noteworthy figure of literature in management, highlighting its broad applicability and significance (Abecassis-Moedas 2006).

Since its beginning, the avenues of value chain analysis have extended small firms to big industries and global value chains. Sturgeon et al. (2012) define how global value chain incorporate activities dispersed across numerous countries, generating complex global linkages. They define commercial advancement as a step towards valuable product and services through resourceful worldwide linkages (Ernst 2004). This thought emphasizes the character of global outcomes with in the value chain, with recent studies determining whether monetary upgrading aligns with social advancement (Lee et al. 2011).

Recent studies conducted by Kaur et al. (2021) determine the impact of value chain framework in achieving sustainability. Researcher highlighted that sustainability not only include sustainability but also social and commercial dimensions. It was found that sustainability aspects enhance value chain performance. Likewise, Glover et al. (2022) observe the procedure of blockchain technology to improve transparency and effectiveness. It observes how blockchain can be traceable, reduce fraud, and adoptive sustainable practices with in the value chain, offering a current tool that complements traditional value chain models. On the other hand, Pagell and Wu (2009) defined that integrating sustainability into supply chain management is central for achieving competitive advantage. Researchers highlights that value chains must stabilise financial, ecological and societal goals. Moreover, Dubey et al. (2019) explore how green supply chain management practices can advance company performance. Additionally, Matos and Silvestre (2023) determine the combination of circular economic values into the value chain, conferring how these values can be fixed to achieve sustainability. The research illustrates that implementing circular economic activities within the value chain can be the key factor. Despite all these, there remains a significant gap in applying analytical methods like ISM and MICMAC. Existing tools often lack comprehensive approaches whereas existing one is thoroughly evaluating the interdependence among variables and their influence on value chain performance.

The purpose is to develop a robust framework that characterise the association between important sustainability factors including that characterises the relationship among factors such as low-carbon production, sustainable material sourcing, green supply chains, and logistics, green marketing strategies, recycling and reconstitution of products, and people resources, technology, and infrastructure. By using MICMAC to categorise these factors into driving, linkage, and dependent variables and ISM to build a digraph, the study aims to throw light on the complication of these sustainability factors and offer guidance for enhancing green practices throughout value chains. About prior research on the topic, Sect. 7.3 goes on to define the literature review analysis carried out to regulate and measure the variables influencing value chain analysis. The findings of two important analyses are presented in Sect. 7.4: the fuzzy Matriced Impacts of Croise's Multiplication Appliquée a UN Classement (MICMAC) and the Interpretative Structural Modelling (ISM) studies. Understanding the linkage between the components that contribute to value chain analysis was made possible thanks in large part to these analyses. The results of MICMAC and ISM are discussed in Sect. 7.5, with an emphasis on the key variables manipulating sustainable practices along the value chain. The outcomes are placed in perspective with in the larger context of value chain management and sustainability in this debate. Based on the study findings, Sect. 7.6 offers organisations useful advice. It makes approvals for tactics and steps businesses may take to improve their sustainability practices and get better results for the environment and their operations. In Sects. 7.7 and 7.8 the key findings are summarised; deficiencies are discussed and suggestions for future research directions are made. Sections 7.7 and 7.8 the paper's key decisions are summarised, its deficiencies are discussed, and suggestions for future research directions are made. This section highlights the research contributions to the field of sustainable value chain management. The purpose of this paper structure is to provide a detailed analysis of sustainability factors along with the value chain, together with theoretical understandings and applied suggestions for enhancing sustainability practices into organisations.

7.2 Research Methodology

Interpretive structural modeling (ISM), a methodology presented by Warfield in 1976, assists as an influential tool for creating and analysing essential relationships among variables within a system. This method enables the progress of models that represent the interconnectedness between diverse variables, thereby changing complex, unclear phenomena into understandable and explainable models of the system (Ahuja et al. 2009). In this framework, ISM is applied to discover and classify nine key variables linked with sustainable value chains. The aim of the study is to determine these variables, regulate their relationships, and categorize them as powerful factors, linkage variables, and dependent variables. The diagraph created through ISM provides a graphic image of these relationships, which is indispensable for considerate the complications and impact of each variable in the framework of sustainable practices. The variables for this study were acknowledged through brainstorming techniques, a detailed review of the literature, and professional talks. These variables were then characterised to prevent idleness and create a systematic framework. For this study, nine variables were identified as significant factors influencing sustainability with in the value chain.

7.3 Identification of Variables

A preliminary review of studies is used to identify variables in order to find out the current trends. This incorporates a survey of studies and identified variables that impact the effectiveness of value chains. There are many occurencies which has given the beautiful insights of the current research. For examples, Barton et al. (2022) stated that the impact of technology and green marketing in improving the performance of supply chain. In a similar way, researcher highlighted the success of sustainable practices. Moreover, Johnson and Smith (2023a, b) also exemplified the role of low carbon production for achieving sustainable goals. Researcher stressed the role of these practices in reducing the environmental impact. In the same way, Lee et al. (2024) also discovered the value of infrastructure and human resources is facilitating supply chain activities. Moreover, Chen and Wang (2021) analysed the linkage between sustainability variables for analysing their influence on value chain performance. After having thorough discussion with experts, following are the variables as drivers of value chain.

7.3.1 Procurement

Procurement means arrangement of resource which is the first step of value chain. A study initiated by Martin and Jones (2023) discovered the role of procurement in enhancing sustainable practices. Researcher identified that procurement initiates the sustainability of value chain.

7.3.2 Human Resources

Garcia and Lopez (2023) demonstrated that how training and development can progress employee engagement and performance in achieving green initiatives. They highlight the status of human resource practices and role in elevating the sustainable practices in organisation. Similarly, Patel et al. (2022) conferred the impact of leadership and culture in the implementation of supply chain activities.

7.3.3 Technology

Revolutions and high-tech advancements play a important role in defining sustainable practices. A study initiated by Johnson and Smith (2023a, b) discussed how the technologies such as artificial intelligence, blockchain and IoT, can expand transparency, efficiency and sustainability across supply chain operations. These technologies enable monitoring and optimization of resources, reducing waste and emissions. Additionally, scientific innovations support the expansion of ecofriendly goods and services. Overall, the acceptance of technology is essential for driving sustainable practices and achieving long term environmental goals.

7.3.4 Infrastructure

The physical and organisational structures essentials are sustainable operations play a key role in enhancing the overall sustainability of value chains (Lee et al. 2024) emphasize the infrastructure, including energy efficient buildings. Renewable energy systems, and advanced waste management facilities buildings, renewable energy systems, and advanced waste management facilities, is indispensable for minimizing the environmental impact of creation and supply processes.

7.3.5 Sourcing Sustainable Material

A study conducted by Williams et al. (2022) emphasize that the biodegradable materials and raw inputs, can significantly reduce the track of manufacturing processes. They observe that using sustainable materials helps greenhouse gas emissions, promotes biodiversity, and reduce dependency on non-renewable resources. Additionally, the study points out the transparent and ethical sourcing practices increase corporate reputation and meet growing consumer expectations.

7.3.6 Low Carbon Production

It is indispensable that effective policies should be made for reducing carbon emissions in manufacturing process that will in turn help in achieving sustainability goals. Brown and Green (2021) emphasized the importance of implementing low carbon production methods. These methods may include energy-efficient technologies and renewable energy sources and these can substantially decrease the bad environmental impacts of manufacturing. It was also revealed that companies that implement these strategies are benefitted due to lower operational costs and better legal compliance. Also, Smith et al. (2023) stressed the role of carbon capture and storage technologies in decreasing the carbon emissions from economic activities. Their findings suggested that combining these advanced technologies into production processes not only decreases greenhouse gas emissions but also improves the level of sustainability. Moreover, Johnson and White (2022) discussed the role of optimization and waste reduction techniques in achieving low carbon production.

7.3.7 Green Supply Chains and Logistics

To achieve sustainability, it is must to management supply chain activities in a eco-friendly manner. A study initiated by Rodriguez and Williams (2022) stated that applications of green logistics practices i.e., optimizing transportation routes and using electric vehicles can significantly reduce carbon emissions and operational costs. Study demonstrate that applying environmentally friendly logistics not only reduces ill environmental influence but also recovers supply chain efficiency. Chen and Liu (2023) discussed the benefits of implementing green supply chain management practices.

7.3.8 Green Marketing Techniques

If an organisation practices green marketing techniques, then it can result into better consumer trust and achieving higher level of competitive advantage. Bansal and Desai (2021) suggested that green marketing, which includes eco-labelling, sustainable packaging, and highlighting the environmental benefits of products, can enhance the brand reputation and customer loyalty. Also, Kumar and Singh (2022) explored the impact of digital marketing promotions that focus on green initiatives. It revealed that social media and other online platforms are significant tools for increasing awareness about sustainable products and practices. Moreover, emphasized the role of causal marketing, where companies collaborate with environmental institutions to promote environment sustainability. Overall, green marketing tools are important for engaging consumers and developing a market for sustainable products and services.

7.3.9 Recycle and Reconstitute Products

The practices for managing waste and increasing the product life cycles are crucial for extending sustainability in value chains. Brown and Green (2022) suggested that ‘Recycling and Reconstituting Products’ such as converting waste materials into

Table 7.1 represents the variables used in the study to arrive at the model

Variables	Abbreviation
Procurement	P
Human resources	HR
Technology	T
Infrastructure	I
Sourcing sustainable material	SSM
Low carbon production	LCP
Green supply chains and logistics	GSCL
Green marketing techniques	GMT
Recycle and reconstitute products	RRP

usable products can significantly reduce landfill waste and save natural resources. Their research suggests that implementing focussed recycling programs can lead to significant environmental benefits and cost savings. Additionally, Lee and Kim (2023) suggested that if circular economy principles are adopted then the economic and environmental advantages can be obtained where products are designed for the purpose of durability, reuse, and recycling (Table 7.1).

7.4 Analysis

7.4.1 Structured Self-Interaction Matrix

A framework is used to analyse the relationships and results derived from the model. This framework uses a symbol to define the nature of relations between each pair of variables (i, j):

V: Variable i can influence variable j.
A: Variable j influences variable i.
X: Variables i and j mutually influence each other.
O: Variables i and j do not influence each other.

On the basis of this framework, researcher can regulate the driving forces behind each variable and comprehend the changing aspects of their connections within the model. This method offers understandings into how each variable affects others and helps to simplify the composite interdependences that outline the complete system. The following table demonstrates these interdependences using the specified symbols (Table 7.2).

This table represents the interdependencies and relationships among the variables using the symbols described, offering a structured view of how each variable influences or is influenced by others in the study context.

Table 7.2 Variable relationship structural self-interaction matrix (SSIM)

Variable	P	HR	T	I	SSM	LCP	GSCL	GMT	RRP
P		V	V	V	V	V	V	A	V
HR			V	V	V	V	V	V	V
T				V	V	V	V	V	V
I					V	V	V	V	V
SSM						V	V	V	V
LCP							V	V	V
GSCL								V	V
GMT									V
RRP									

Table 7.3 Reachability matrix

Variables	1	2	3	4	5	6	7	8	9	Driving Power
P	1	1	1	1	1	1	1	0	1	8
HR	0	1	1	1	1	1	1	1	1	8
T	0	0	1	1	1	1	1	1	1	7
I	0	0	0	1	1	1	1	1	1	6
SSM	0	0	0	0	1	1	1	1	1	5
LCP	0	0	0	0	0	1	1	1	1	4
GSCL	0	0	0	0	0	0	1	1	1	3
GMT	1	0	0	0	0	0	0	1	1	3
RRP	0	0	0	0	0	0	0	0	1	1
Dependence power	2	2	3	4	5	6	7	7	9	

7.4.2 Reachability Matrix

See Table 7.3.

7.4.3 Final Reachability Matrix

The final matrix offers a complete view of the associations between the variables in the study. It identifies how each variable influences and is influenced by the other variables (Table 7.4).

7.4.3.1 Key Observations

Human Resources (HR) and Procurement (P) are important in the system, manipulating influenced by all other variables, stress their contribution in the sustainable value chain. Technology act as a major driver with high interdependence, highlighting its key role in encouraging sustainability. Variables such as Infrastructure (I), Sourcing Sustainable Materials (SSM), Low Carbon Production (LCP), Green

Table 7.4 Final reachability matrix

Variables	1	2	3	4	5	6	7	8	9	Driving power
P	1	1	1	1	1	1	1	1*	1	9
HR	1*	1	1	1	1	1	1	1	1	9
T	1*	1*	1	1	1	1	1	1	1	9
I	1*	1*	1*	1	1	1	1	1	1	9
SSM	1*	1*	1*	1*	1	1	1	1	1	9
LCP	1*	1*	1*	1*	1*	1	1	1	1	9
GSCL	1*	1*	1*	1*	1*	1*	1	1	1	9
GMT	1	1*	1*	1*	1*	1*	1*	1	1	9
RRP	0	0	0	0	0	0	0	0	1	1
Dependence power	8	8	8	8	8	8	8	8	9	

* The Final Reachability Matrix is significant in the determining relationships among variables, forming the basis for hierarchical modeling. It helps classify elements based on their driving and dependence power, guiding the structure of the ISM model. This matrix also serves as the foundation for level partitioning and further MICMAC analysis

Table 7.5 Level partitioning

Elements	Reachability set	Antecedent set	Intersection set	Level
1	1, 2, 3, 4, 5, 6, 7, 8	1, 2, 3, 4, 5, 6, 7, 8	1, 2, 3, 4, 5, 6, 7, 8	2
2	1, 2, 3, 4, 5, 6, 7, 8	1, 2, 3, 4, 5, 6, 7, 8	1, 2, 3, 4, 5, 6, 7, 8	2
3	1, 2, 3, 4, 5, 6, 7, 8	1, 2, 3, 4, 5, 6, 7, 8	1, 2, 3, 4, 5, 6, 7, 8	2
4	1, 2, 3, 4, 5, 6, 7, 8	1, 2, 3, 4, 5, 6, 7, 8	1, 2, 3, 4, 5, 6, 7, 8	2
5	1, 2, 3, 4, 5, 6, 7, 8	1, 2, 3, 4, 5, 6, 7, 8	1, 2, 3, 4, 5, 6, 7, 8	2
6	1, 2, 3, 4, 5, 6, 7, 8	1, 2, 3, 4, 5, 6, 7, 8	1, 2, 3, 4, 5, 6, 7, 8	2
7	1, 2, 3, 4, 5, 6, 7, 8	1, 2, 3, 4, 5, 6, 7, 8	1, 2, 3, 4, 5, 6, 7, 8	2
8	1, 2, 3, 4, 5, 6, 7, 8	1, 2, 3, 4, 5, 6, 7, 8	1, 2, 3, 4, 5, 6, 7, 8	2
9	9	1, 2, 3, 4, 5, 6, 7, 8, 9	9	1

Supply Chains and Logistics (GSCL), and Green Marketing Techniques (GMT) exhibit high interdependence, demonstrating their important contributions to sustainability goals. Recycle and Reconstitute Products figure out with high urgency but minimal driving power, signifying it as an outcome rather than a direct influencer.

The background is critical for strategic planning and decision-making, as it recognizes the most significant variables and those needing thoughtful management due to their high dependence, providing a basis for creating a classified model of value chain.

7.4.3.2 Level Partitioning

See Table 7.5.

The partitioning table above is resulting from the reachability matrix and categorizes the variables based on their levels of influence and dependence. Each variable is linked with a reachability set, antecedent set, intersection set, and level. Variables 1–8 have identical reachability and antecedent sets, indicating that they influence and are influenced by all variables. This process helps in understanding the classified structure of the variables, enabling stronger insights into which elements are drivers, linkages, or dependent factors within the system. This classification is crucial for tactical planning and decision-making in value chain analysis.

7.4.4 Diagraph

The diagraph depicts a graphic representation of the relationship among the elements with the value chain. Each node characterises a variable, while arrows reflect the impact of one variable on other. Variables like Procurement (P), Human Resources (HR), and Technology (T) are central, with numerous outgoing and incoming arrows, illustrating their significant driving power and interdependence. Variables such as Infrastructure (I), Sourcing Sustainable Materials (SSM), Low Carbon Production (LCP), Green Supply Chains and Logistics (GSCL), and Green Marketing Techniques (GMT) form a tightly interconnected cluster, reflecting their critical roles and mutual influences in achieving sustainability. Recycle and Reconstitute Products (RRP) has several incoming arrows but fewer outgoing ones, highlighting its position as an outcome influenced by other variables rather than a direct driver. This digraph effectively maps out the complex dependencies and driving forces within the value chain, aiding in the identification of key leverage points for strategic interventions and policy formulations (Fig. 7.1).

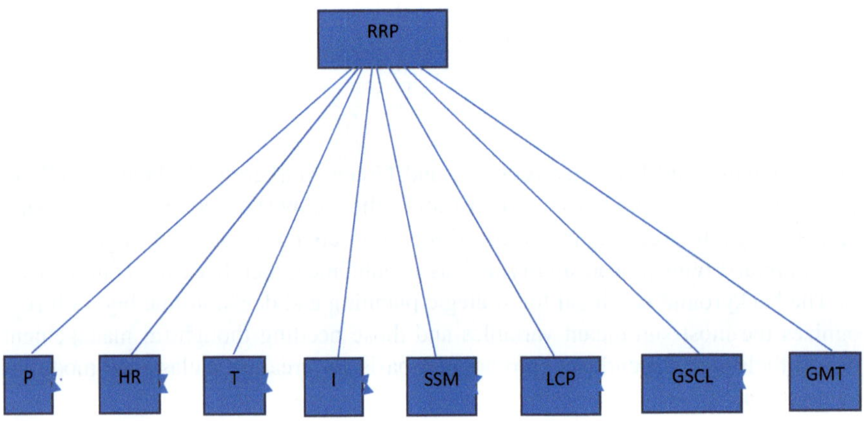

Fig. 7.1 A digraph representation of value chain variables

7.4.5 MICMAC Analysis

With the help of MICMAC analysis, it was concluded that Procurement, Human Resources, and Technology are major drivers, influencing the overall system significantly. Infrastructure, Sourcing Sustainable Materials, Low Carbon Production, Green Supply Chains and Logistics, and Green Marketing Techniques show high interdependence, indicating they are serious components for achieving sustainability. Recycle and Reconstitute Products, being highly dependent but not driving, suggests that recycling efforts are largely a result of the other sustainability initiatives (Fig. 7.2).

This analysis is vital for identifying where to focus strategic efforts. By understanding which variables drive the system and which are consequences, organizations can prioritize actions that will have the most significant impact on enhancing sustainability within the value chain.

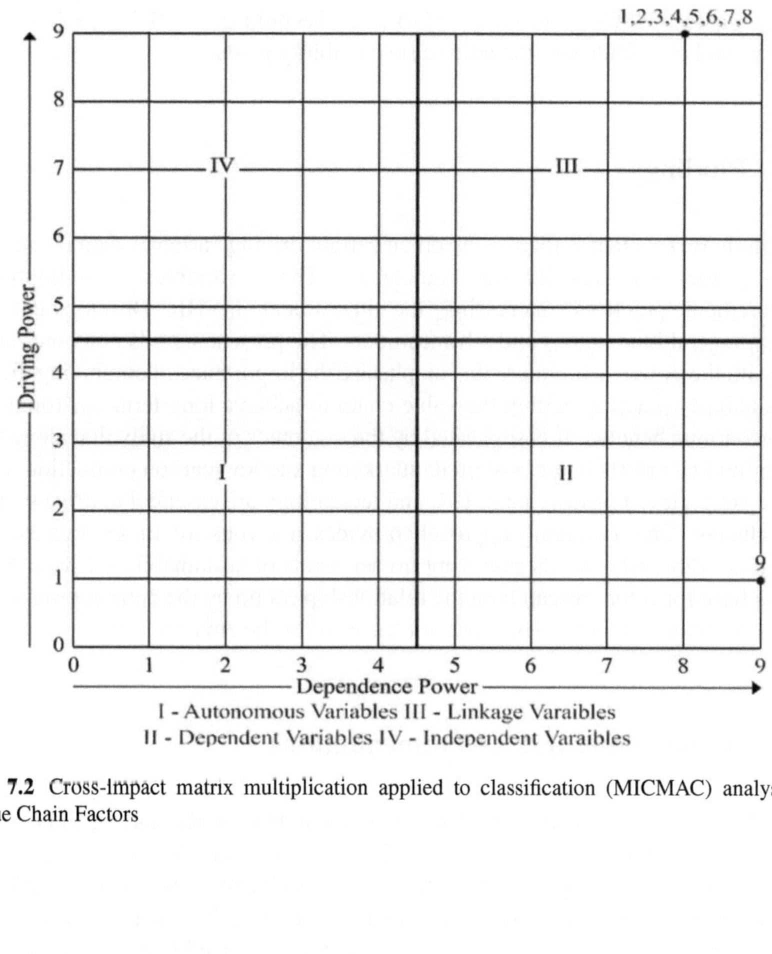

Fig. 7.2 Cross-impact matrix multiplication applied to classification (MICMAC) analysis of Value Chain Factors

7.5 Discussions

The present study provides important outcomes regarding the dynamics of sustainability in the value chain of an organisation. Procurement (P), Human Resources (HR), and Technology (T) are the variables that have been identified as key drivers because of their high driving power and impact on other elements in the value chain system. Procurement is the basic requirement for the implementation of the sustainable practices as human resources play an important role in cultivating a sustainable and technologically advanced culture in the organisation. Technology has a substantial influence and it facilitates the development and implementation of innovative environment friendly practices. On the other hand, variables like Recycle and Reconstitute Products (RRP) are more dependent variables the explains that recycling efforts are result of efficient management practices and these do not pose any direct influence. At last, it is also highlighted that the Infrastructure (I), Sourcing Sustainable Materials (SSM), Low Carbon Production (LCP), Green Supply Chains and Logistics (GSCL) and Green Marketing Techniques (GMT) simultaneously support and contribute significantly to sustainability goals.

7.6 Findings

The study reveals that if focus is given on certain driving variables then it can positively affect the sustainability outcomes. These variables are—Improving Procurement practices, increasing the investment in HR Development and Technological innovations and advancements. The present study's outcomes are in line with the previous literature that emphasise the importance of combining diverse sustainability practices within the value chain to achieve long-term environmental and economic benefits. It is suggested by the outcomes of the study that along with the individual efforts in areas of green marketing and low-carbon production, innovative strategies in procurement, HR, and technology are essential to drive systematic change. This emerging approach provides a layout for those organisations which are directed towards achieving higher levels of sustainability. It would provide a base for future research on the relationship set up by the current research for such economic outcomes which are not harmful for the environment.

7.7 Implications and Recommendations

The MICMAC analysis suggests that procurement, Human Resources, and technology are significant variables in the sustainable value chain. These variables influence other variables significantly and also get influenced by other variables. A significant focus is required on these variables for achieving greater sustainability. The analysis shows that Infrastructure, Sourcing Sustainable Materials, Low Carbon

Production, Green Supply Chains & Logistics and Green Marketing Techniques are highly interdependent which emphasizes the requirement for an integrated approach to achieve a level of comprehensive sustainability. One variable 'Recycle and Reconstituting products' stand out as an indirect driver which indicates that recycling efforts are dependent upon the successful implementation of other practices leading to sustainability. These findings align with currently available literature that stresses upon a systemic approach to sustainability where technological advancements, human resource development, and strategic procurement play crucial roles.

It is recommended to the organizations to enhance procurement strategies keeping the sustainability criteria in focus in order to fulfil the environmental concerns. Investing in human resource development through training and leadership programs can strengthen sustainability initiatives. The use of technology is essential for progressive, sustainable practices, including the need to explore new manufacturing technologies and waste reduction. Supporting sustainable practices requires investment in infrastructure and environmentally friendly products and improve resource efficiency. Spreading green marketing practices to emphasize the environmental benefits will engage consumers in sustainability efforts. Finally, recycling efforts and restructuring efforts ensure that sustainability practices are successfully completed and combined with other practices. Addressing these recommendations will allow organizations to leverage their sustainability resources, manage dependencies, and create a robust strategy for achieving their goals and it will be sustainable.

7.8 Conclusion

The primary focus of this study is the interrelationship between major variables of the study i.e. Procurement, Human Resources, Technology and Infrastructure which employs ISM and MICMAC methods to analyse a sustainable value chain framework. The outcomes of the study highlight Procurement, Human Resources, Technology, and other variables like Low-Carbon production, Green Supply Chains, and Green Marketing strategies as significant sustainability drivers. The review suggests that reusing and reconstitution are results of effective management practices. It gives valuable suggestions for sustainability efforts across sectors and recommends improving procurement, investing in human resources, and utilizing technology strategically to drive sustainability.

References

Abecassis-Moedas C (2006) Integrating design and retail in the clothing value chain: an empirical study of the organisation of design. Int J Oper Prod Manag 26(4):412–428

Ahuja V, Yang J, Shankar R (2009) Study of ICT adoption for building project management in the Indian construction industry. Autom Constr 18(4):415–423

Bansal R, Desai A (2021) Green marketing strategies and their impact on brand reputation. J Sustain Mark 15(2):123–134
Barton M, Johnson T, Smith R (2022) Technology and green marketing in sustainable supply chains. Int J Environ Sci 27(4):567–589
Brown SL (1997) Competitive advantage: the key to global leadership. Int J Bus 2(3):56–67
Brown D, Green H (2021) Low carbon production methods and their effects on manufacturing sustainability. J Clean Prod 35(1):201–215
Brown D, Green H (2022) Recycling and reconstituting products for sustainable waste management. Waste Manag Res 40(3):245–258
Chen J, Liu S (2023) Benefits of green supply chain management practices. J Sustain Supply Chain Manag 22(1):33–50
Chen L, Wang J (2021) Interrelationships among sustainability variables in value chains. J Bus Res 125:56–72
Dubey R, Gunasekaran A, Papadopoulos T, Childe SJ (2019) Green supply chain management: theoretical framework and further research directions. BIJ 26(1):124–146
Ernst D (2004) Global production networks in East Asia's electronics industry and upgrading perspectives in Korea and Taiwan. In: Yusuf S, Altaf MA, Nabeshima K (eds) Global production networking and technological change in East Asia. Oxford University Press, pp 89–157
Garcia M, Lopez R (2023) Human resource practices for sustainability: training and employee engagement. Hum Resour Manag Rev 31(2):210–225
Glover JL, Champion D, Daniels KJ, Dainty AJD (2022) Using blockchain technology to enhance the application of digital marketing. J Bus Res 144:234–245
Hellin J, Meijer M (2006) Guidelines for value chain analysis. FAO Agricultural and Food Marketing Association Technical Report
Johnson T, Smith R (2023a) Low-carbon production and sustainable material sourcing in value chains. J Sustain Manuf 30(2):145–160
Johnson M, Smith L (2023b) Technological innovations and sustainable practices. J Sustain Technol 15(2):102–117
Johnson T, White K (2022) Process optimization for low carbon production. J Ind Ecol 26(4):1010–1025
Kaur P, Singh S, Mittal G (2021) Value chain analysis for sustainable supply chain management. J Clean Prod 278:123–136
Kumar N, Singh V (2022) Digital marketing campaigns and green initiatives. J Interact Mark 28(3):205–218
Lee J, Kim H (2023) Circular economy principles in product design. Sustain Dev J 40(1):91–108
Lee J, Gereffi G, Barrientos S (2011) Global value chains, upgrading and poverty reduction. Int J Technol Learn Innov Dev 4(1–3):123–152
Lee M, Chang P, Kim H (2024) Role of infrastructure in sustainable supply chains. J Infrastruct Dev 35(1):12–28
Martin A, Jones P (2023) Strategic procurement and sustainability in supply chains. Supply Chain Manag Rev 29(2):78–95
Matos S, Silvestre BS (2023) Integrating circular economy principles into the value chain. J Ind Ecol 27(1):34–51
Pagell M, Wu Z (2009) Building a more complete theory of sustainable supply chain management using case studies of 10 exemplars. J Supply Chain Manag 45(2):37–56
Patel S, Green J, White P (2022) Leadership and organizational culture in green supply chain initiatives. J Sustain Bus Pract 19(3):176–189
Porter ME (1985) Competitive advantage: creating and sustaining superior performance. Free Press
Rodriguez J, Williams P (2022) Green logistics practices and their impact on supply chain efficiency. Logist Transp Rev 18(2):89–104

Smith SM, Geden O, Nemet GF, Gidden MJ, Lamb WF, Powis C, Bellamy R, Callaghan MW, Cowie A, Cox E, Fuss S, Gasser T, Grassi G, Greene J, Lück S, Mohan A, Müller-Hansen F, Peters GP, Pratama Y, Repke T, Riahi K, Schenuit F, Steinhauser J, Strefler J, Valenzuela JM, & Minx JC. (2023). The State of Carbon Dioxide Removal – 1st Edition. The State of CDR. https://www.stateofcdr.org

Sturgeon TJ, Linden G, Zhang J (2012) Global value chains and economic upgrading: a framework for analysis. Rev Int Polit Econ 19(1):99–130

Warfield JN (1976) Societal systems: planning, policy, and complexity. Wiley-Interscience

Williams K, Johnson R, Smith T (2022) Sourcing sustainable materials for manufacturing. Mater Today Sustain 14:102–118

Chapter 8
Working Capital Management and Its Impact on Emerging Sustainability Economic Growth in Context of Pharmaceutical Industry in India

Priyanka Tomar and Supriya Srivastava

Abstract There is the most commonly used framework which is the ESG (Environmental, Social and Governance) one to address the issue of corporate sustainability. For a few years now, the consumers, the legislators, and the groups have grown more conscious about the sustainability of the pharmaceutical sector. The exacerbation of attention to this topic can be explained by concerns regarding the implementation of sustainability concepts into the creation of novel distribution methods, environmentally friendlier products, the absence of waste, conservation of water, eco-friendlier manufacturing processes, and eco-friendly packaging. Scholars from the engineering, business, chemistry, environmental sciences and management fields have also shown interest in the sustainability of the pharmaceutical industry. Pharmaceutical firms are facing an increased level of management complexity. To assure environmental, economic, and social sustainability while controlling expenses, economic reasoning has been incorporated into sustainability concerns management. Within the framework of India's pharmaceutical sector, the goal of this study is to evaluate how WCM (working capital management) affects sustainability and economic growth. The increasing complexity in managing the sector, driven by high research and development costs, stringent regulatory compliance, and evolving market dynamics, necessitates effective WCM. By exploring the relationship between WCM and sustainability, the study aims to provide insights into how firms can align financial objectives with ESG goals, fostering long-term value creation and resilience. Through a rigorous empirical examination of the existing literature, the purpose of this study is to further academic research as well as useful decision-making in the pharmaceutical sector, ultimately fostering sustainable and inclusive economic development in India.

P. Tomar · S. Srivastava (✉)
Faculty of Management Sciences, Shoolini University of Biotechnology and Management Sciences, Solan, Himachal Pradesh, India

K. Mathiyazhagan et al. (eds.), *Humanizing Circular Supply Chain Management*, Circular Economy and Sustainability,
https://doi.org/10.1007/978-3-031-86562-6_8

Keywords Working capital management · Sustainability · Economic growth · Pharmaceutical industry · Circular economy

8.1 Introduction

Presently valued at over USD 50 billion, the "pharmaceutical sector in India" is the third biggest globally based on the volume. Pharmaceuticals are exported to over 200 countries and territories of the world, including the most selective markets like the USA, the United Kingdom, the countries of the European Union, Canada, etc. (NPP 2023; IBEF 2023). The WCM is one of the core components of any organization that affects its financial stability and performance (Deloof 2003; Festa et al. 2021). It involves the handling of short-term assets and responsibilities to support the everyday functioning of an organization in the most effective way possible as well as make the most of the money. Regarding the pharmaceutical sector, where "research and development (R&D) costs are high, regulatory compliance is stringent, and market dynamics are constantly evolving", adept management of working capital assumes heightened significance (Deloof 2003; Rahman et al. 2019).

The pharmaceutical sector operates within a dynamic landscape characterized by rapid "technological advancements, changing regulatory frameworks, and evolving consumer preferences". These dynamics mean that there is a need to carry out "R&D, manufacturing capabilities as well as distribution networks" (Akhtar 2013). Therefore, pharmaceutical firms experience "sizeable capital demands, something that makes efficient working capital management necessary in order to fund operations, growth strategies, and competitiveness" (James and Kumar 2021). The "pharmaceutical industry" in India is one of the key pillars of the country's economy with huge potential in the advancement of health care delivery (Datta et al. 2024). It is expected that the market size of the sector will reach "USD 65 billion" in 2024; however, its importance extends beyond its economic significance to being a key area that can solve many social and ecological issues (FICCI 2024). When considering the key success determinants of the sector's development, the management of working capital is seen as one of the influential aspects that define the future of the industry and its roles in the economy.

ESG has recently become more important to the business society and thus pharmaceutical firms are compelled to embrace sustainable business strategies (Sarkis et al. 2011). Examination of the WCM and sustainability can open up the possibilities of how firms can link financial targets with ESG concerns and create sustainable value in the long run. Pharmaceutical firms nowadays have to compete in a global environment that has become highly competitive keeping in mind the costs and the expected profitability (Nenni et al. 2014). The management working capital can provide competitive advantages because of the factors such as "improved operational efficiency, low financing cost and identification of possible investment opportunities that support the implementation of sustainable solutions." An important function in the context of the regulation of the sector is performed by policymakers determining the conditions for the functioning of the pharmaceutical industry (Jaberidoost et al. 2013). Appreciating the complexity of how working capital interplays with capital management sustainability and economic growth can help in formulation of policies that can 'support innovation, encourage initiatives in sustainability, and encourage inclusiveness'.

The interplay among "capital management, sustainability, and economic growth" within the Indian pharmaceutical business has, however, not received much attention. Several aspects highlighted the necessity of doing a thorough investigation of this connection. Dwivedi (2023) asserts that the pharmaceutical business is vital to India's GDP and exports, making its steady growth essential to the nation's economic stability. To assist "policymakers, industry stakeholders, and investors," it is critical to understand how working capital management methods impact ESGs outcomes. These considerations lead this study to discuss the impact of WCM on developing sustainable economic growth in the Indian pharmaceutical sector, with the goal of adding to the body of existing work. Hence, this research contributes to the existing literature on WC management techniques and economic performance and sustainability, and it offers relevant policy implications and practical implications to policymakers and industry practitioners.

8.2 Literature Review

In the context of India's pharmaceutical sector, this review of the literature aims to compile previous research on WCM and its implications for the recently developing sustainable economic growth. The specifics of WCM, the overall concept of "sustainability and economic growth", earlier research on the Indian pharmaceutical business, and previous literature addressing the connection between WCM, and economic growth are all included in this review.

8.2.1 Working Capital Management

Due to this, company, academics, and scholars have accorded a lot of importance to this aspect of financial management which is the working capital management (WCM) in the quest to discover the correlation that exists between the two; the success of the financial managers as well as company's performance (Farhan et al. 2021). Pratap Singh and Kumar (2014) noted that there isn't much advancement in the WCM literature currently in publication. In addition, working capital management research has been devoted to examining the determinants of WCM, the causes of WCM, and firms' long-term financing (Mansoori and Muhammad 2012; Nazir and Afza 2009; Narender et al. 2008; Chiou et al. 2006). However, there are not many works that focus on the field of working capital that explore the profitability and liquidity of the business (Gill and Obradovich 2012; Abuzayed 2012; Deloof 2003).

The literature on emerging economies is scarce, despite the fact that most of the research that found effectiveness of WCM were performed in developed nations, such as the "United States and Australia." According to the "World Bank," India falls under the category of the low-middle income country in Asia (Somvanshi 2016). Numerous studies by Kazimoto (2016), Ajibolade and Sankay (2013), and Wire (2015) demonstrated a favourable relationship between WCM factors and company performance. WCM is used to describe the "management of the company's current assets and current obligations". A company may make sure it has enough "cash on hand" to cover its short-term obligations and other operating costs by managing its working capital (Deloof 2003). Thus, working capital includes "payables, receivables, cash, inventory, or short-term debts, among other things". To ensure that operating cash flow remains sufficient and there are no disruptions in the organization's operations, it is important to manage these components effectively (Richards and Laughlin 1980; Sogomi et al. 2024).

Several research works have investigated the connection between WCM and worldwide economic growth (Berg et al. 2024). For example, Shin and Soenen (1998) found that the profitability and growth rate of different industries are significantly and favourably impacted by the working capital efficiency. WCM has a major influence on the profitability and expansion of businesses, according to several earlier studies done in the Indian context (Ghosh and Maji 2004). Improved organizational performance and economic development have been linked to efficient receivables, inventories, and payables management.

8.2.2 Concept of Sustainability in Economic Growth

A certain amount of economic production within an economy is referred to as economic sustainability. It incorporates practices that will sustain a country's economic base in the future without negatively affecting resources (Goodland 1995; Hariram et al. 2023). Some of them are sustainable levels of "debts, sustainable rates of inflation, budgetary balance, and optimum resource use". These indicators help to guarantee that sustained economic growth is possible in the future without a negative impact on generations to come (Solow 1991; Alshaib et al. 2023). The pharmaceutical industry is involved in "discovering, manufacturing, and selling drugs, vaccines, and treatments for various diseases", including the usual and the exceptional ones. This is due to the gradual aging of the world's population and enhanced health technologies that are compelling the pharmaceutical business to become more mindful of development sustainability in addition to providing a diverse range of pharmaceutical products. As a result, due to the effects of the pharmaceutical industry's operations on sustainability, attention has been focused more on this sector (Milanesi et al. 2020).

The literature has identified many approaches for communicating and evaluating sustainability. Governments use the ESG framework in their sustainability laws, and it is also frequently used in the commercial environment to describe a company's performance on the three sustainability pillars: E, S, and G (Bade et al. 2023). An ideal or a perfect supply chain as described by Debnath et al. (2023) is a "data-driven, accurate, responsive and agile supply chain" through a concept of an end-to-end information and material flow.

8.2.3 Indian Pharmaceutical Industry

India's chemical and pharmaceutical industries have extremely high levels of competitiveness, which are characterized by fierce "price competition and price controls" imposed by the government (Vijayalakshmi and Srividya 2015). Wright (2009) claims that one of the largest manufacturers of generic drugs worldwide is the "Indian pharmaceutical industry", which has a substantial impact on the global drug supply. "Regulations, pricing structures, and the ever-increasing demands for new innovations" are a few of the dangers. Possibilities include the rising cost of healthcare, the prevalence of ailments, and the steadily rising need for generic medications (KPMG 2011).

Scholarly investigations have validated that proficient handling of working capital enhances the efficacy of pharmaceutical enterprises and promotes long-term expansion. For instance, a study by Sharma and Kumar (2010), particularly in the context of "Indian pharmaceutical enterprises", demonstrated a substantial relationship between WCM methods and organizations' financial success. The evaluations of working capital in the "Indian pharmaceutical industry" that are currently available are reviewed. Bhunia and Das (2015), Jain (2014), Mehra (2013), Viswanathan et al. (2016), Joshi and Ghosh 2012, and Vijayalakshmi and Srividya (2015) as in the case of each of these studies, the working capital was measured with different proxy variables and its implications for the Indian firms' performance.

8.2.4 Theoretical Perspectives

WCM is crucial for maintaining a firm's liquidity and business flow by managing current assets and liabilities. It impacts a firm's ability to fund short-term needs and invest in future growth. The working capital ratio, particularly in chemical industries like pharmaceuticals, affects cost and profitability. Liquidity management involves balancing liquidity with profitability. Models like the "Baumol and Miller-Orr" models explain cash management. Liquidity is especially important in the pharmaceutical industry due to seasonality and constant product development investment (Baumol 1952; Miller and Orr 1966; Deloof 2003; Filbeck and Krueger 2005; Saini and Bansal 2020).

The pharmaceutical industry is increasingly recognizing the importance of sustainability in business, particularly in the "Triple Bottom Line" (TBL). Implementing WCM principles that incorporate sustainability principles can enhance organizational sustainability and society's well-being. "Stakeholder theory" emphasizes the importance of considering all stakeholders' claims, not just shareholders. Effective WCM can improve stakeholder "value, trust, and corporate reputation", particularly in the pharmaceutical industry. The RBV posits that heterogeneous resources and capabilities provide a competitive advantage, and effective management of WCM can contribute to sustainable economic growth (Freeman 1984; Barney 1991; Mosakowski 1993; Elkington 1999; Akbar et al. 2021; Habib et al. 2024).

The "dynamic capabilities theory" emphasizes a firm's strategic adaptation to market changes. In the pharmaceutical industry, firms can create dynamic capabilities by managing regulatory changes and sustainability factors. WCM, particularly in the emerging Indian economy, significantly impacts macroeconomic factors, such as GDP and employment, by effectively controlling working capital. This creates a conducive environment for development (Teece et al. 1997; Zahra and George 2002; Ghosh and Maji 2004). Previous research indicates that high WCM is linked to ROA and ROE, indicating a correlation between financial performance and sustainable growth in the pharmaceutical industry. The integration of these theoretical perspectives enhances understanding of WCM's impact on financial performance and sustainable economic growth in India, providing suggestions for industry actors (Kafeel et al. 2020).

8.2.5 Sustainability and Circular Economy in the Pharmaceutical Industry

The pharmaceutical industry is embracing "sustainability and a circular economy" to reduce environmental impact. Unlike linear economies, this aspect focuses on "resource use, waste removal, and recycling", promoting environmental conservation and sustainable development. This approach is crucial for the growth of the economy (Webster 2017). The pharmaceutical industry in India aims for sustainability through "resource conservation, waste reduction, and environmental themes" in production and distribution processes. This aligns with sustainable healthcare systems and promotes economic efficiency, social and environmental impact. Sustainable practices include renewable energy use and limiting environmental effects, such as pharmaceuticals' negative effects on ecosystems (Singh et al. 2016). The circular economy in the pharmaceutical supply chain focuses on "product recovery, recycling, and sustainable procurement, reducing resource usage and waste generation". In India, aiming to become the world's pharmacy hub, incorporating circular economy principles is crucial for sustainable development and equity in the industry (Khan and Ali 2022).

8.2.6 *Relationship Between Working Capital Management and the Circular Supply Chain*

WCM is a crucial strategic factor in the pharmaceutical industry, particularly in "inventory control, receivables, and liquidity". It allows a company to sustain short-term goals while making investments that will benefit the company in the future. Integrating WCM with circular supply chain principles offers strategic benefits that support sustainability and improved financial performance. Circular supply chain management principles can improve resource utilization and reduce waste, contributing to sustainability initiatives. By using recycled or refurbished materials instead of virgin material, pharmaceutical companies can free up capital tied up in raw material inventory, allowing for innovative research or expansion (Javaid, et.al., 2024). WCM also helps control returns and reverse logistics effectively, fostering sustainability initiatives. Proper handling of returns, such as refurbishing pharmaceutical products or selling refurbished items, brings back cash, enhancing operational productivity (Bakås et al. 2021). Incorporating WCM and circular supply chain in the pharmaceutical industry creates both economic viability and environmental responsibility. As the Indian pharmaceutical industry expands to meet global sustainability standards, such strategies will be of paramount importance to ensure the industry remains competitive and sustainable (Faggioni et al. 2023).

8.3 Objectives

- To study the connection between the pharmaceutical business in India and sustainable economic growth through working capital management strategies through existing literature.

8.4 Capital Management in the Indian Pharmaceutical Industry

India's domestic drug markets underperformed in the "2016–17 fiscal year" due to government efforts to reduce the cost of medicine and make it more accessible to the vast majority of its citizens, despite the fact that the country's "pharmaceutical industry" is third in the world and recognized as the "fastest-growing" drug market globally. Sales in the industry were low throughout the previous two quarters until the end of September 2016, at which point there was only a 2% increase in sales. Nine percent in the quarter ending in December 2016. Historically, as compared to other industries, this one has had the poorest working capital performance (Farhan et al. 2021).

The pharmaceutical business in India positions itself as a global center of industry and research. Thus, there are a number of scientists and engineers in India today who can take the industry to a highly developed and competent level. Based on available data, the pharmaceutical business in India leads other science-based industries with significant potential for medication manufacturing and technology. In fact, it is well-established and acknowledged as an organized industry (Padiya 2015). Additionally, Fig. 8.1 shows that over 50% of Indian medicines will rise over the next 32 years, indicating strong long-term development. It is projected to have a current value of "US$4.5 billion and to expand by 8–9%" annually. It supplies over 70% of the nation's needs for "bulk medicines, oral, tablet, formulations, drug intermediates, and capsule forms, as well as chemicals". According to estimates, the sector can produce almost 10% of the world's total output (Geethalakshmi and Jothi 2016; Pandey et al. 2016).

Among the pharmaceutical industries worldwide, India is a significant competitor. "Vaccines, bulk drums, contract research OTC medications, and manufacturing, biosimilars and biologics, and generic drugs" are a few of the main areas of the Indian pharmaceutical sector. In terms of "DPT, BCG, and measles vaccinations," India leads the world in their distribution. One of the world's primary providers of reasonably priced vaccines is India. Therefore, 60% of the world's vaccine production was borne by India, which also provided 90% of the measles vaccine and 70% of the WHO's requirement for "DPT, BCG, and tetanus shots" (Malik 2023; IBEF 2023). With "20% of the global market", the nation leads the globe in both the production of immunizations and the supply of generic drugs. India is home to over "3000 pharmaceutical firms" with a vast network of over "10,500 production units" and a robust personnel pool. The largest provider of generic medications is still India. It offers 20% of the world's generic need and manufactures over 60,000 distinct generic pharmaceuticals under 60 therapeutic areas.

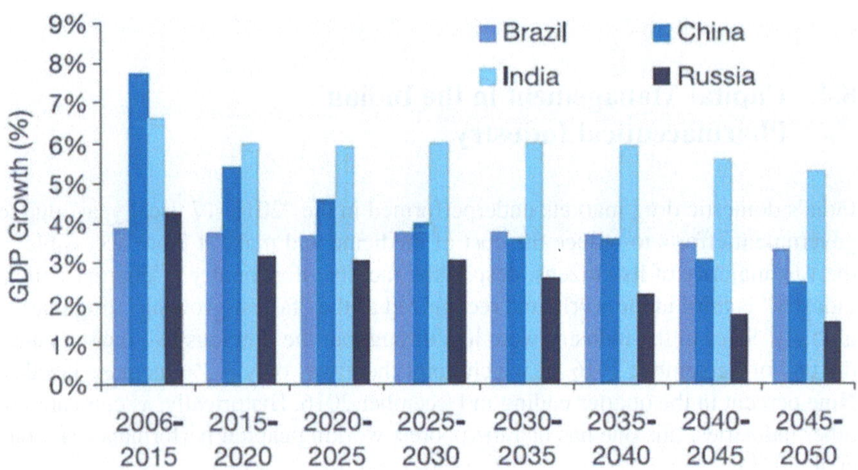

Fig. 8.1 Indian pharmaceutical industries forecasted growth (Mohiuddin et al. 2017)

One of the greatest medical success stories continues to be the accessibility of reasonably priced HIV medication from India. India is known for "pharmacy of the world" as its medications are in great demand worldwide due to their affordable prices and superior quality (Malik 2023; IBEF 2023). If the enterprise in the pharmaceutical sector is a green field project, FDI up to 100% is allowed using the automated method. The brown field pharmaceuticals business is accessible to foreign equity investment; 74% of the investment is approved automatically, with the remaining 25% requiring government permission. India is the global supplier of reasonably priced vaccines, and the country provides up to "20% of the global requirement of generic drugs". With "60% of all vaccines produced globally", India is the world's largest vaccine maker (Malik 2023; IBEF 2023).

8.5 Sustainability in Pharmaceutical Industry

Since they improve people's health, certain parts of sustainability might be considered as essential to pharmaceutical companies; nevertheless, other aspects—like "how medicines are distributed fairly, how pharmaceuticals affect the environment, and how unethical selling occurs are divisive topics" (López-Toro et al. 2021). These debates occur in MAHs ("marketing authorization holders") in a highly intricate system of customers, employers, investors, supervisors, physicians, and "statuary health insurance" (SHI). Since medical doctors are the professionals who prescribe drugs, pharmacists who sell the drugs and the SHIs who are the payers or the determinant of what will be paid for, MAHs have very little control over communicating directly with the consumers. In addition, the peculiarity of the regulatory environment is that the number of rules and regulations is sufficiently large, and they differ not only in the countries of the world but also in regions. Pharmaceutical companies are socially responsible, but there is need for development in other areas as well. Some examples are the "ecological sustainability of packaging, worker rights at manufacturing sites, and diversity on the board of directors" (Bade et al. 2023).

Therefore, actively seeking out more sustainable company practices has a positive economic impact across a range of industries. These gains may stem from improved action conformance to stakeholder preferences or reputational adjustments (Grove et al. 2018). As the criteria for tenders are being widened to include sustainability elements, pharmaceutical companies are finding that it is increasingly more vital to improve their performance on the specified sustainability dimensions (Bade et al. 2023). Upcoming legislative changes highlight the need to take action, and the research indicates that pharmaceutical businesses should be increasing their influence in all areas of sustainability. Given the significant role that MAHs play in sustainability, having this viewpoint and understanding what is needed to support change is very helpful. For example, 2019 research found that in 2015, the CO_2 emission of the "pharmaceutical business" was "55% greater" than that of the "automobile industry" (Forin and Scholz 2022; Okereke 2021; Belkhir and Elmeligi 2019).

Fewer actions can be taken today by the pharmaceutical industry's emission values per one million dollars of revenue are more or less equal to the average wastage rate of the automotive industry's "31.4 tons per one million dollars" of revenue due to digitalization and the utilization of the batch manufacturing process as opposed to a continuous process. Besides, it is one of the largest sectors of the world economy and the scope of its growth constantly expands (Okereke 2021), meaning that millions of people and their means of subsistence are impacted by its decisions and output. Hence, MAHs have a stake in raising awareness of sustainable maturity as it might enable them to progress on ESG issues and foster more confidence in their sector and institutions. It is especially crucial since, in spite of its beneficial effects on people's health and quality of life, the public rarely has faith in the pharmaceutical industry (Kessel 2014; Bade et al. 2023).

8.6 Strengthening the Relevance to Circular Supply Chain

Circular supply chain aims to "reduce waste, extend product life, and optimize resources" for sustainability. It significantly impacts WCM by managing inventory, avoiding stock accumulation, and increasing asset efficiency. For instance, just-in-time inventory systems can reduce excess stock in sectors like the pharmaceutical industry. Indian regulators can support sustainable WCM practices by developing policies that encourage companies to adopt these practices (Lacy and Rutqvist 2015). WCM strategies require effective management of reverse logistics, a crucial aspect of the circular supply chain. Regulators can encourage efficient reverse logistics in pharmaceutical companies by funding recycling plants or organizing the return and reuse of pharmaceutical packaging, as suggested by Govindan et al. (2015).

Circular supply chains aim to extend product and material lifespans, with WCM enhancing procurement and production activities to minimize resource use and waste. Government authorities can encourage pharma industries to adopt lifecycle management strategies combining WCM with the circular economy, including resource utilization standards and life cycle assessments (Kirchherr et al. 2017). Collaboration is crucial for establishing and operating circular supply chains. Regulators can encourage relationships between pharmaceutical "companies, suppliers, and recycling" organizations to create circular supply chain systems. This will enhance supply chain management practices and improve efficiency. Legal measures should be adopted to create an environment for circular business models within WCM frameworks, including policies and procedures for incorporating circular economy concepts. These frameworks can help pharmaceutical companies transition to sustainable supply chains and improve economic and environmental performance (Bocken et al. 2016; Geissdoerfer et al. 2017).

8.7 Discussion, Recommendations and Conclusions

8.7.1 Discussion

This study aims to establish the importance of WCM for organizations' financial health and operational efficiency, with a focus on pharmaceutical firms. It refers to the short-term operations of an organization, which entails managing of assets and liabilities that are short-term in nature. The industry faces significant capital requirements due to rapid technological advancements, changing regulatory frameworks, and evolving consumer preferences. Efficient WCM positively impacts profitability and growth across various industries, including the "Indian pharmaceutical industry". Economic sustainability is essential for long-term economic health without depleting resources. Some of the measures include—"sustainable level of debts, reasonable inflation rates, budgetary balance, and effective utilization of resources". Currently, India occupies the third place in the global pharmaceutical market by volume and actively participates in the country's economy. However, the challenges of the regulation and the pressure on the prices the industry is a promising industry because of the demand for the healthcare services, chronic diseases and the market for the generic drugs.

Optimal WCM fosters improved economic outcomes and contributes to the long-term development of the sector in terms of pharmaceuticals. With 70% of the nation's needs met by the Indian pharmaceutical industry—which is now thought to be the fastest-growing in the world—it provides "chemicals, bulk pharmaceuticals, drug intermediates, formulations for drugs, capsules, and oral tables." WCM best practice can improve the operation efficiency of the organizations, decrease the financing costs and facilitate the investment in sustainable practices. Mitigating environmental consequences, unequal distribution of medicines, and unethical promotions may lead to benefits in terms of tangible returns, enhanced stakeholders' relationship, and organizational image. The research focuses on the concept of WCM in the context of the Indian pharmaceutical industry and its positive impact on "operations, cost of financing, and sustainability". The research contributes to knowledge for "industry stakeholders, policy makers and academics", to inform the development of effective interventions in generating sustainable innovation within the context of sustainability and sustainable development for inclusive growth in the pharmaceutical sector.

The Indian pharmaceutical industry's working capital management (WCM) is crucial for the economy and its sustainability. To improve WCM practices, Indian regulators can require additional disclosures in financial statements, including information on stock, accounts receivables, and payables management. This transparency will help stakeholders evaluate the efficiency and sustainability of companies, boosting investor confidence. Additionally, the government can establish incentives to encourage companies to adopt sustainable practices in their WCM plans, such as tax credits, subsidies, or privileged rights to bid for government contracts (Farhan et al. 2021). This aligns with Shroff (2018) assertion that financial incentives can encourage green practices in operational strategies.

Regulators should promote training on WCM methods and sustainability to enhance financial and environmental performance in the pharmaceutical industry. Standard measures for WCM evaluation can bring coherence and assess sustainability performance. Collaboration between "regulators, pharmaceutical companies, financial institutions, and sustainability experts" can stimulate efficient WCM practices that create value for businesses and promote sustainable development. These partnerships can help address multifaceted problems and implement easily implemented solutions. Overall, promoting training and collaboration can lead to more sophisticated WCM practices in the pharmaceutical sector (Vijayalakshmi and Srividya 2015; Souza et al. 2021; Farhan et al. 2021).

8.7.2 Concrete Policy Recommendations

The suggestions are to require the provision of comprehensive WCM information within annual reports, offering financial incentives for sustainability, the creation of the standard WCM across the industries, establishment of the WCM training programs, and the promotion of cross-sectoral cooperation. These measures seek to address the existing gaps in the current WCM practices and ensure that these practices are in line with sustainable development goals that will foster growth of the economy and the environment as supported by the pharmaceutical industry. One of the ways that the sustainability can be incorporated in the financial management is through the provision of financial incentives like tax cuts or grants. It is recommended that training programs are available to the industry practitioners and that these training programs incorporate real life examples for implementing sustainability into WCM strategies. The above recommendations are intended to fill gaps within the current state of WCM practice, and advance sustainable practice within the pharmaceutical industry.

8.7.3 Conclusion

This article's goal is to investigate how WCM and sustainability relate to the Indian pharmaceutical sector. Because the pharmaceutical business is so vital to the national and international economies, effective WCM is crucial to the long-term viability of pharmaceutical companies' financial positions. This research identifies key challenges inherent in the "Indian pharmaceutical industry" which includes "high research and development costs, regulatory concerns, and market changes". Implementing good WCM practices that increase "liquidity and profitability", also fosters sustainable economic development. Within the pharmaceutical sector, the notion of sustainability encompasses not just "environmental, social, and governance" aspects, but much more. The coordination of ESG targets with the help of WCM and sustainable financial management results in long term value creation and industry's development, which is in line with the overall sustainability objectives. The production capabilities and market share of Indian pharma industry is strategically located because it supplies half of the global requirement of generic medicines and vaccines. Nevertheless, a peculiar literature gap can be distinguished, and it concerns the investigation of the WCM's influence on sustainability and economic development in the context of the "Indian pharmaceutical industry" in particular.

Further research works should focus on: the effects of WCM practices in the long run, WCM practices in different sectors and geographical regions, the role of new technologies in WCM, relationship between WCM and the existing regulations and sustainability. Future research should be directed toward examining longitudinal outcomes of the Waste Management (WCM) initiatives concerning "sustainability and economic development" of the "pharmaceutical industry". Multi-sector and cross-country comparisons can offer additional understanding of the benchmarks and novel solutions. Therefore, the examination of the application of technology and digitalization in improving the efficiency and sustainable impacts of WCM is also necessary. Hence, knowledge of the relationship between regulations, WCM and sustainability will assist policymakers and other industry players in developing better polices and strategies.

References

Abuzayed B (2012) Working capital management and firms' performance in emerging markets: the case of Jordan. Int J Manag Financ 8(2):155–179. https://doi.org/10.1108/17439131211216620

Ajibolade SO, Sankay OC (2013) Working capital management and financing decision: synergetic effect on corporate profitability. Papers.ssrn.com. https://papers.ssrn.com/sol3/papers.cfm?abstract_id=2367990

Akbar A, Akbar M, Nazir M, Poulova P, Ray S (2021) Does working capital management influence operating and market risk of firms? Risks 9(11):201. https://doi.org/10.3390/risks9110201

Akhtar G (2013) Indian pharmaceutical industry: an overview. IOSR J Humanit Soc Sci 13(3):51–66. https://doi.org/10.9790/0837-1335166

Alshaib BM, Ghazi M, Nuţă AC, Hamra M, Mishra P, Gautam R, Hussain S (2023) Fiscal sustainability and its implications for economic growth in Egypt: an empirical analysis. SAGE Open 13(4). https://doi.org/10.1177/21582440231215983

Bade C, Olsacher A, Boehme P, Truebel H, Bürger L, Fehring L (2023) Sustainability in the pharmaceutical industry—an assessment of sustainability maturity and effects of sustainability measure implementation on supply chain security. Corp Soc Responsib Environ Manag 31(1):224–242. https://doi.org/10.1002/csr.2564

Bakås O, Tveit SS, Thomassen MK (2021) Reverse logistics for improved circularity in mass customization supply chains. In: Lecture notes in mechanical engineering. Springer, pp 1031–1038. https://doi.org/10.1007/978-3-030-90700-6_118

Barney J (1991) Firm resources and sustained competitive advantage. J Manag 17(1):99–120. https://journals.sagepub.com/doi/10.1177/014920639101700108

Baumol WJ (1952) The transactions demand for cash: an inventory theoretic approach. Q J Econ 66(4):545. https://doi.org/10.2307/1882104

Belkhir L, Elmeligi A (2019) Carbon footprint of the global pharmaceutical industry and relative impact of its major players. J Clean Prod 214:185–194. https://doi.org/10.1016/j.jclepro.2018.11.204

Berg T, Gustafsson E, Wahlstrøm RR (2024) Cost management and working capital management: ebony and ivory in perfect harmony? J Manag Control 35:207–233. https://doi.org/10.1007/s00187-024-00368-3

Bhunia A, Das A (2015) Underlying relationship between working capital management and profitability of pharmaceutical companies in India. Am J Theor Appl Bus 1(1):27–36

Bocken NMP, de Pauw I, Bakker C, van der Grinten B (2016) Product design and business model strategies for a circular economy. J Ind Prod Eng 33(5):308–320. https://doi.org/10.1080/21681015.2016.1172124

Chiou JR, Cheng L, Wu HW (2006) The determinants of working capital management. J Am Acad Bus 10(1):149–155

Datta C, Kuruvilla SJ, Pande AS (2024) Unlocking pharma manufacturing agility: innovations for reliability, flexibility, and growth. The Economic Times. https://economictimes.indiatimes.com/industry/healthcare/biotech/pharmaceuticals/unlocking-pharma-manufacturing-agility-innovations-for-reliability-flexibility-and-growth/articleshow/107421498.cms?from=mdr

Debnath B, Saha J, Rahman M (2023) Assessing the critical success factors for implementing Industry 4.0 in the pharmaceutical industry: implications for supply chain sustainability in emerging economies. PLoS One 18(6):e0287149–e0287149. https://doi.org/10.1371/journal.pone.0287149

Deloof M (2003) Does working capital management affect profitability of Belgian firms? J Bus Financ Acc 30(3–4):573–588. https://doi.org/10.1111/1468-5957.00008

Dwivedi R (2023) Indian pharmaceutical market size, Report 2024–2032. www.imarcgroup.com. https://www.imarcgroup.com/india-pharmaceutical-market

Elkington J (1999) Cannibals with forks: the triple bottom line of 21st century business | Wiley. Wiley.com. https://www.wiley.com/en-in/Cannibals+with+Forks%3A+The+Triple+Bottom+Line+of+21st+Century+Business-p-9781841120843

Faggioni F, Valerio Rossi M, Sestino A (2023) Supply chain resilience in the pharmaceutical industry: a qualitative analysis from scholarly and managerial perspectives. Int J Bus Manag 18(1):129. https://doi.org/10.5539/ijbm.v18n1p129

Farhan NHS, Belhaj FA, Al-ahdal WM, Almaqtari FA (2021) An analysis of working capital management in India: an urgent need to refocus. Cogent Bus Manag 8(1):1924930. https://doi.org/10.1080/23311975.2021.1924930

Festa G, Kolte A, Carli MR, Rossi M (2021) Envisioning the challenges of the pharmaceutical sector in the Indian health-care industry: a scenario analysis. J Bus Ind Mark 37:1662–1674. https://doi.org/10.1108/jbim-07-2020-0365

FICCI (2024) FICCI: industry's voice for policy change. Ficci.in. https://ficci.in/api/sector_details/29

Filbeck G, Krueger TM (2005) An analysis of working capital management results across industries. Am J Bus 20(2):11–20. https://doi.org/10.1108/19355181200500007

Forin S, Scholz R (2022) Are medicines more greenhouse gas intensive than cars? Comment to Belkhir, L., Elmeligi, A., 2019: Carbon footprint of the global pharmaceutical industry and relative impact of its major players. J Clean Prod 331:129963. https://doi.org/10.1016/j.jclepro.2021.129963

Freeman RE (1984) Strategic management: a stakeholder approach. In: Google Books. Pitman. https://www.google.co.in/books/edition/Strategic_Management/4PUJAQAAMAAJ?hl=en

Geethalakshmi A, Jothi K (2016) A study on profitability position of pharmaceutical industry in India. Int J Adv Res Comput Sci Manag Stud 4(4):181–188

Geissdoerfer M, Savaget P, Bocken NMP, Hultink EJ (2017) The circular economy—a new sustainability paradigm? J Clean Prod 143(1):757–768. https://doi.org/10.1016/j.jclepro.2016.12.048

Ghosh S, Maji SG (2004) Working capital management efficiency: a study on the Indian cement industry. Manag Account:363–372

Gill AS, Obradovich JD (2012) Corporate governance, institutional ownership, and the decision to pay the amount of dividends: evidence from USA. Int Res J Financ Econ 97(1):60–71

Goodland R (1995) The concept of environmental sustainability. Annu Rev Ecol Syst 26(1):1–24. https://doi.org/10.1146/annurev.es.26.110195.000245

Govindan K, Soleimani H, Kannan D (2015) Reverse logistics and closed-loop supply chain: a comprehensive review to explore the future. Eur J Oper Res 240(3):603–626

Grove H, Clouse M, Schaffner LG (2018) Digitalization impacts on corporate governance. J Gov Regul 7(4):51–63. https://doi.org/10.22495/jgr_v7_i4_p6

Habib A, Anwar S, Hussain W, Fenyves V (2024) The role of sustainable investment practices in maintaining efficient working capital management. J Int Stud 17(2):206–219. https://doi.org/10.14254/2071-8330.2024/17-2/11

Hariram NP, Mekha KB, Suganthan V, Sudhakar K (2023) Sustainalism: an integrated socio-economic-environmental model to address sustainable development and sustainability. Sustainability 15(13):10682. https://doi.org/10.3390/su151310682

IBEF (2023) Pharma industry in India: pharma sector overview, market size, analysis...| IBEF. Www.ibef.org. https://www.ibef.org/industry/pharmaceutical-india

Jaberidoost M, Nikfar S, Abdollahiasl A, Dinarvand R (2013) Pharmaceutical supply chain risks: a systematic review. DARU J Pharm Sci 21(1). https://doi.org/10.1186/2008-2231-21-69

Jain M (2014) Critical analysis of management of working capital in pharmaceutical industry in Madhya Pradesh. Handle.net. http://hdl.handle.net/10603/236161

James TC, Kumar D (2021) Public policy and economic development case study of Indian pharmaceutical industry. https://www.ris.org.in/sites/default/files/Publication/Public%20Policy%20and%20Economic%20Development%20Case%20Study%20of%20Indian%20Pharmaceutical%20Industry-min.pdf

Javaid M, Haleem A, Khan IH, & Singh RP (2024). *Industry 4.0 and circular economy for bolstering healthcare sector: A comprehensive view on challenges, implementation, and futuristic aspects*. Biomedical Analysis, 1(2), 174–198

Joshi LK, Ghosh S (2012) Working capital management of Cipla limited: an empirical study. Int J Mark Financ Serv Manag Res 1(8):170–186

Kafeel, Ali J, Din MU, Waris A, Tahir M, Khan S (2020) Working capital management and firms' profitability: dynamic panel data analysis of manufactured firms. J Financ Risk Manag 9(4):720–726. https://doi.org/10.4236/jfrm.2020.94027

Kazimoto P (2016) Working capital management and financial performance of selected companies in Kampala. Int J Res IT 6:4. https://www.indusedu.org/pdfs/IJRIME/IJRIME_763_90809.pdf

Kessel M (2014) Restoring the pharmaceutical industry's reputation. Nat Biotechnol 32(10):983–990. https://doi.org/10.1038/nbt.3036

Khan F, Ali Y (2022) Implementation of the circular supply chain management in the pharmaceutical industry. Environ Dev Sustain 24:13705–13731. https://doi.org/10.1007/s10668-021-02007-6

Kirchherr J, Reike D, Hekkert M (2017) Conceptualizing the circular economy: an analysis of 114 definitions. Papers.ssrn.com. https://papers.ssrn.com/sol3/papers.cfm?abstract_id=3037579

KPMG (2011) Indian Pharma Outlook. https://www.in.kpmg.com/pdf/Indian%20Pharma%20Outlook.pdf

Lacy P, Rutqvist J (2015) Waste to wealth. Palgrave Macmillan. https://doi.org/10.1057/9781137530707

López-Toro AA, Sánchez-Teba EM, Benítez-Márquez MD, Rodríguez-Fernández M (2021) Influence of ESGC indicators on financial performance of listed pharmaceutical companies. Int J Environ Res Public Health 18(9):4556. https://doi.org/10.3390/ijerph18094556

Malik R (2023) Pharmaceuticals. www.investindia.gov.in. https://www.investindia.gov.in/sector/pharmaceuticals

Mansoori E, Muhammad D (2012) The effect of working capital management on firm's profitability: evidence from Singapore. Papersssrncom https://papers.ssrn.com/sol3/papers.cfm?abstract_id=2185840

Mehra P (2013) Effect of working capital management on the profitability of the Indian pharmaceutical sector. Int J Enhanc Res Manag Comput Appl 2(3):1–7

Milanesi M, Runfola A, Guercini S (2020) Pharmaceutical industry riding the wave of sustainability: review and opportunities for future research. J Clean Prod 261:121204. https://doi.org/10.1016/j.jclepro.2020.121204

Miller MH, Orr D (1966) A model of the demand for money by firms. Q J Econ 80(3):413. https://doi.org/10.2307/1880728

Mohiuddin M, Mazumder MNH, Chrysostome E, Su Z (2017) Relocating high-tech industries to emerging markets: case of pharmaceutical industry outsourcing to India. Transnatl Corp Rev 9(3):201–217. https://doi.org/10.1080/19186444.2017.1370808

Mosakowski E (1993) A resource-based perspective on the dynamic strategy-performance relationship: an empirical examination of the focus and differentiation strategies in. J Manag 19(4):819–839. https://doi.org/10.1016/0149-2063(93)90029-m

Narender V, Menon S, Shwetha V (2008) Sustainability of the Indian railways turnaround: a stage theory perspective. SSRN Electron J. https://doi.org/10.2139/ssrn.2144044

Nazir M, Afza T (2009) Impact of aggressive working capital management policy on firms' profitability. https://lahore.comsats.edu.pk/Papers/Abstracts/146-8588087906194227058.pdf

Nenni ME, Giustiniano L, Pirolo L (2014) Improvement of manufacturing operations through a lean management approach: a case study in the pharmaceutical industry. Int J Eng Bus Manag 6:24. https://doi.org/10.5772/59027

NPP (2023) Draft National Pharmaceuticals Policy 2023 for stakeholder consultation. https://Pharmaceuticals.gov.in/. https://pharmaceuticals.gov.in/sites/default/files/31.10.23%20Draft%20National%20Pharmaceuticals%20Policy.pdf

Okereke M (2021) How pharmaceutical industries can address the growing problem of climate change. J Climate Change Health 4:100049. https://doi.org/10.1016/j.joclim.2021.100049

Padiya HJ (2015) Working capital management of selected pharmaceutical units in Gujarat

Pandey NS, Sugumari G, Azhagaiah R (2016) The impact of working capital management on profitability of pharmaceutical industry in India. Pac Bus Rev Int 9(6):1–7

Pratap Singh H, Kumar S (2014) Working capital management: a literature review and research agenda. Qual Res Financ Mark 6(2):173–197. https://doi.org/10.1108/qrfm-04-2013-0010

Rahman S, Iqbal K, Nadeem A (2019) Effect of working capital management on firm performance: the role of ownership structure. Glob Soc Sci Rev IV(I):108–119. https://doi.org/10.31703/gssr.2019(iv-i).10

Richards VD, Laughlin EJ (1980) A cash conversion cycle approach to liquidity analysis. Financ Manag 9(1):32–38. https://doi.org/10.2307/3665310

Saini N, Bansal S (2020) Pharmaceutical companies & liquidity analysis: a review. Int J Econ Financ Issues 10(5):236–242. https://econjournals.com/index.php/ijefi/article/view/10465

Sarkis J, Zhu Q, Lai K (2011) An organizational theoretic review of green supply chain management literature. Int J Prod Econ 130(1):1–15. https://doi.org/10.1016/j.ijpe.2010.11.010

Sharma AK, Kumar S (2010) Effect of working capital management on firm profitability. Glob Bus Rev 12(1):159–173. https://doi.org/10.1177/097215091001200110

Shin HH, Soenen L (1998) Efficiency of working capital management and corporate profitability. Financ Pract Educ 8:37–45

Shroff S (2018) Working capital policy of Indian service industry. Pac Bus Rev Int 10:104–118

Singh RK, Kumar R, Kumar P (2016) Strategic issues in pharmaceutical supply chains: a review. Int J Pharm Healthc Mark 10(3):234–257. https://doi.org/10.1108/ijphm-10-2015-0050

Sogomi FC, Patrick MK, Kamau CG (2024) Exploring the relationship between working capital management, liquidity, and financial performance within the context of Kenyan SME'S. Afr J Commer Stud 4(2):113–120. https://doi.org/10.59413/ajocs/v4.i2.4

Solow R (1991) Sustainability: an economist's perspective. The eighteenth J. Seward Johnson lecture. Woods Hole Oceanographic Institution, Woods Hole, MA

Somvanshi KK (2016) World Bank to change classification of countries; India will now be called 'lower-middle income'. Economic Times

Souza HO, Costa RS, Quadra GR, Fernandez MAS (2021) Pharmaceutical pollution and sustainable development goals: going the right way? Sustain Chem Pharm 21:100428. https://doi.org/10.1016/j.scp.2021.100428

Teece DJ, Pisano G, Shuen A (1997) Dynamic capabilities and strategic management. Strateg Manag J 18(7):509–533. https://doi.org/10.1002/(SICI)1097-0266(199708)18:7<509::AID-SMJ882>3.0.CO;2-Z

Vijayalakshmi V, Srividya M (2015) A study on working capital management of pharmaceutical industry in India. J Manag Sci 5(3):248–269

Viswanathan M, Palanisamy A, Mahesh R (2016) A comparitive study on working capital management of selected pharmaceutical companies in India. IRA-Int J Manag Soc Sci 3(3). https://doi.org/10.21013/jmss.v3.n3.p11

Webster K (2017) The circular economy in detail. www.ellenmacarthurfoundation.org. https://www.ellenmacarthurfoundation.org/the-circular-economy-in-detail-deep-dive#:~:text=A%20circular%20economy%20is%20a

Wire WC (2015) Influence of working capital management practices on financial performance of small and medium manufacturing Enterprises in Nairobi County, Kenya. Ir.jkuat.ac.ke. http://ir.jkuat.ac.ke/handle/123456789/1866

Wright B (2009) Sudip Chaudhuri. The WTO and India's pharmaceuticals industry: patent protection, TRIPS, and developing countries. New Delhi: Oxford university press, 2005. Pp. xvi+358. $55.00 (cloth). Econ Dev Cult Chang 57(3):604–607. https://doi.org/10.1086/596603

Zahra SA, George G (2002) Absorptive capacity: a review, reconceptualization, and extension. Acad Manag Rev 27(2):185–203

Chapter 9
Employees' Skills Required for Efficient Circular Supply Chain Management

Arun Kaushal and Pallavi Dogra

Abstract Circular supply systems, as opposed to conventional linear supply chains, are designed to enhance resource usage and minimize waste. This is achieved by incorporating ideas such as recycling, remanufacturing, and refurbishment into the existing procedures. Sustainable processes related to the supply chain require a broader set of competencies that can be enhanced among employees to achieve better employee productivity. The present study attempts to identify the necessary skills required for employees to operate effectively without creating waste with the help of recycling. In the present work, a semi-structured interview of experts from diversified industries like IT, cement, and automobiles has been taken to explore the required skills. To operationalize the logical and systematic analysis of the recorded transcripts, QSR NVivo v.10 software was used to find the concrete findings of the study. After conducting the theme analysis, the present study proposed a list of skills that facilitate the execution of waste-free mechanisms. The identified skills are "sustainability" Literacy, systems thinking, innovative problem-solving, stakeholder engagement, ethical and social responsibility, technical and analytical skills, and Regulatory Compliance Knowledge. Outcomes of the research help the companies to identify the skills required for CSCM execution and they can effectively target their training programs. Educational institutions and professional training providers can incorporate these skills into their curricula, ensuring that current and future supply chain professionals are equipped to meet the demands of a circular economy.

Keywords Sustainable supply chain management · Systems thinking · Innovative problem-solving · Skills

A. Kaushal (✉)
Mittal School of Business, Lovely Professional University, Phagwara, Punjab, India

P. Dogra
Dr. B.R. Ambedkar University, Delhi, New Delhi, India

K. Mathiyazhagan et al. (eds.), *Humanizing Circular Supply Chain Management*, Circular Economy and Sustainability,
https://doi.org/10.1007/978-3-031-86562-6_9

9.1 Introduction

In recent times, circular economy (CE) strategies have become the prerequisites for attaining sustainable development goals proposed by 'The Club of Rome in 1972 (Halkos and Aslanidis 2024). The United Nations will accept these goals, and all these countries will strive to improve their operational efficiencies in their respective national economies (Dwivedi et al. 2023). However, sustainable circular practices have been impacted by the recent disruption to the world caused by the COVID-19 outbreak results to slow down the pace of implementation (Kaur et al. 2024; Piila and Sarja 2024; Piyathanavong et al. 2024). Additionally, Global warming, carbon emissions, and greenhouse effects force the government and regulatory bodies of various countries to make mandatory provisions for the manufacturing and services industries to adopt circular supply chain management practices (Rissman et al. 2020; Sawhney 2021). This requires integrating practices such as reducing, reusing, and recycling materials to minimize waste and improve sustainability. For example, the focus on Scope 3 emissions, encompassing all indirect emissions in a company's supply chain, drives firms to adopt more circular approaches to meet new regulatory requirements and achieve their ESG(environmental, social, and governance) goals (Quashie et al. 2024; Dwivedi et al. 2023).

Although in the international as well as domestic markets, organizations like Apple, Dow, Tata Consultancy Services (TCS), Infosys, ACC Limited, and Mahindra Group have taken a competitive advantage to become early adopters of circular supply chain management, there is still a significant gap in terms of its acceptability in every segment of the entire industries (Nyffenegger et al. 2024; Kannan et al. 2022).

This gap can be overcome by incorporating the necessary skills required for employees to operate for circular supply chain management. Recent studies emphasized the fact that there is a need for a diverse set of competencies that go beyond traditional supply chain management skills to include sustainability literacy, systems thinking, and innovative problem-solving abilities (Franco 2017; Tseng et al. 2018). Furthermore, employees must possess a deep understanding of environmental and sustainability issues, as well as they should have the ability to apply this knowledge to their roles within the supply chain. However, industry experts and practitioners of supply chain systems posited that to formulate aligned and informed decisions with the Circular economy, the human capital of the organizations strives to have sufficient basic Knowledge and struggles to make them fit for routine operations (Zamfir et al. 2017). Furthermore, for the sound execution of a sustainable supply chain system, it is advisable to have a holistic approach that includes various stages of the "product life cycle" along with internal association within the entire supply chain system. Subsequently, these inputs help the current authors to develop a "system thinking approach" that facilitates improvement of the design thinking of a successful circular supply chain system (Murray et al. 2017). Previous literature reveals that the conceptualization of competencies and skill sets required for the successful implementation of circular supply chain processes is at a nascent stage

(Luthra et al. 2020). After conducting qualitative research in this study, the authors of the present study have identified a bunch of proficiencies that can help practitioners to accelerate the circular supply chain processes in the existing system of the organizations. The outputs of the study significantly facilitate the stakeholders of the logistics and supply chain companies to enhance the skills of their employees in such a way that they collectively can contribute to developing robust product design innovatively for the overall success of the company's business models. Further, these identified competencies help the logistic experts to tailor their training programs where the maximum emphasis should be to uplift the skills of their trainees. Additionally, educational institutions involved in "supply chain education" also benefited from this study in that they can add additional concepts, principles, and techniques related to the circular supply chains in their curricula resulting in a better understanding of the circular economy at an educational level for emerging graduates. Finally, from the results of the current study human resource policy developers will also be able to draft sound skill set mapping programs with incentives in an efficient manner for their workforce. These initiatives help organizations retain competent workers for the successful implementation of Circular supply chain management.

9.2 Literature Review

An innovative approach to supply chain management that emphasizes the reusability of the existing resources for overall sustainability is termed circular supply chain management (Sudusinghe et al. 2024). Circular supply chain systems depend upon waste minimization and effective resource usage, which is opposite to linear conventional supply chains (Lahane et al. 2020). Successful execution of circular supply processes needs redesign, remanufacturing, renovation, and recycling in the existing process (Farooque et al. 2019). Apart from this, companies can make significant changes in their system for resource allocation, improve their market reach, and decrease their adverse effects by extending product life cycles (De Angelis et al. 2018). Thereafter in such scenarios, global companies have sensitized the value of the circular supply chain and integrate their entire business operations with sustainability. Gartner (2022) reported in his study that across the globe 74% of the supply chain giants have realized the gravity of the circular economy and expect significant profits after executing circular principles in their business mechanism in the last 3 years. Previous literature on supply chains posits that an effective supply chain strategy also needs circular processes for the successful implementation of network design, warehousing policies, and sustainable supply chains (Lengyel et al. 2021). In the contemporary situation, post-pandemic has brought a new approach for organizations in which they have to adopt pollution-free environmental resources that can help them to work for sustainable development goals suggested by the United Nations (Lengyel et al. 2021). This alteration in the business approach will be possible with the help of a circular supply chain system. Extant literature suggests that circular principles surely improve natural resources like air, water, and soil resulting in the upgradation of the entire ecosystem of the planet (Geissdoerfer et al. 2017).

Past studies stated that the execution of a circular supply chain system needs a wider perspective in terms of employees' competencies for overall employees' productivity in a well-organized organizational setup (Farooque et al. 2019; Lengyel et al. 2021). Nowadays companies try to prioritize eco-friendly environmental culture in their working system and this further depends upon the skill sets of the people in the organizations. In this regard, companies are highlighting the relevance of new skills to their employees for mutual benefit (Geissdoerfer et al. 2017). Furthermore, the literature reveals that the transformational shift from linear to circular supply chain networks is confronted with various hindrances that require significant adjustments in diverse areas of supply chain administration (Borms et al. 2023; Kaur et al. 2024; Tseng et al. 2018). Various methods to implement this new transition include Overwhelming cultural resistance, creating new skill sets, working on collaborative partnerships, adopting emerging technologies, redesigning product designs, and aligning the existing operational system with a regulatory framework (Franco 2017; Luthra et al. 2020). These alterations surely provide more strengths to the supply chain practitioners and with these initiatives, they can move ahead toward sustainability goals.

9.3 Research Methodology

In the current study, an attempt has been made to investigate the "skill set" required to implement circular supply chain management with the help of a qualitative approach. The primary data for this study was collected with the help of semi-structured interviews. The reason for selecting a semi-structured interview is due to its ability to fetch in-depth insights, opinions, and thoughts of the participants in a better way and provide root issues that have to be addressed for developing skills among the workforce in organizations for effective implementation of circular supply chain processes (Kaur et al. 2024; Luthra et al. 2020). The purposive sampling method was used to select the participants in this study to ensure that the sample should consist of different levels of employees working in the hierarchy supply chain structure and having diversified experience in different industries. Their exact details are given in Table 9.1. Further, the method adopted to approach the participants is with the help of a professional social media platform (LinkedIn) followed by communicating with them by sending an official email to invite them for discussion on the Google Meet platform. In the e-mail, it is requested that they give their opinions on the desired topic without any worry as this discussion will be used for research purposes only. After receiving their consent, semi-structured interviews were conducted online.

Table 9.1 Details of participants for semi-structured interview

Participant name	Company	Total experience (years)	Designation
P1	ACC limited	10	SupplyChain manager
P2	Tata consultancy services	12	Circular supply chain manager
P3	Mahindra group	20	Head of sustainability
P4	ACC limited	14	Operations officer
P5	ACC limited	14	Supply chain manager
P6	Mahindra group	16	Senior consultant
P7	Tata consultancy services	11	Supply chain manager
P8	Tata consultancy services	17	Supply chain officer

9.4 Data Analysis and Discussion

After getting a detailed description through a semi-structured interview, transcripts of recorded interviews were used for further investigation. As a component of theme analysis (Braun and Clarke 2006), the collected qualitative data were operationalized through deductive and inductive coding approaches. Theme analysis was carried out with the QSR NVivo v.10 software to identify the major skills and tools for transitioning the existing supply chain management to a circular supply chain suggested by participants for the implementation of circular supply chain management systems in the organizations and a detailed description of the same is available in Table 9.2.

The present study attempts to explore the necessary skills required for organizations to implement effective processes used for the supply of the material from point of origin to point of consumption along with minimum wastage by using a qualitative research approach. "Circular economy" principles have become a need in today's business world, including applying sustainable practices to perform various operations (Lengyel et al. 2021; Taddei et al. 2024). In such a situation, the current study provides insights related to the application aspects of a circular economy, and it has been conducted to investigate the human core skills requirements that convert the circular supply chain management principles into actual execution. For this study, a semi-structured interview of experts from diversified industries like IT, cement, and automobiles has been taken to explore the required skills. To operationalize the logical and systematic analysis of the recorded transcripts, QSR NVivo v.10 software was used to find the concrete findings of the study. After conducting the theme analysis, the present study proposed a list of skills and some barriers that facilitate as well as create hurdles respectively for the successful execution of effective "circular supply chain management" in organizations. The identified skills are sustainability Literacy, systems thinking, innovative problem-solving, stakeholder engagement, ethical and social responsibility, technical and analytical skills, and regulatory compliance knowledge. Similarly, hurdles shared by participants are instructional and structural barriers A detailed explanation of all these skills and

Table 9.2 Results of the qualitative study

Major skills and tools for transitioning to CSCM	Sample extracts from targeted participants	Literature support
• Innovative problem-solving (skills) • Developing new skills and competencies (tool for transitioning to CSCM)	*"We had to rethink our entire packaging process to use biodegradable materials, which required out-of-the-box thinking and collaboration with new suppliers" (P4, P7)*	De Jesus et al. (2019), Kazancoglu et al. (2023), Taddei et al. (2024)
• Sustainability literacy (skills) • Redesigning products and processes (tool for transitioning to CSCM)	*"To implement effective circular strategies, we need a deep understanding of sustainability concepts and how they apply to our operations"(P6, P1)*	De los Rios and Charnley (2017), Korhonen et al. (2018)
• Stakeholder engagement (skills) • Building collaborative networks (tool for transitioning to CSCM)	*"Regular meetings with our stakeholders help us stay aligned on sustainability goals and address any concerns promptly" (P2, P5)*	Herremans et al. (2016), Liu et al. (2018)
• System thinking (skills) • Adopting a circular business model (tool for transitioning to CSCM)	*"We must see the big picture and how each part of the supply chain affects the whole to implement circular strategies effectively as a system" (P3, P8)*	Alem Fonseca et al. (2024)
• Ethical and social responsibility (skills) • Adopting a circular business model (tool for transitioning to CSCM)	*"We ensure fair labor practices and community benefits are part of our sustainability initiatives" (P3, P5)*	Doh and Tashman (2014)
• Technical and analytical skills (skills) • Implementing advanced technologies (tool for transitioning to CSCM)	*"We rely on advanced analytics to monitor our resource use and optimize our processes continually"(P2, P7)*	Lahane et al. (2020), Gawankar et al. (2020)
• Regulatory & Compliance Knowledge (skills) • Aligning with regulatory framework (tool for transitioning to CSCM)	*"Keeping up with regulatory changes is crucial to avoid fines and maintain our sustainability commitments" (P1, P6)*	Hastig and Sodhi (2020), Rizos et al. (2016)

hurdles has been given below. Further, all these skills and the required designated person in the organizational hierarchy have been shown in Fig. 9.1.

9.4.1 Innovative Problem-Solving

This study's results proposed a shift to circular processes often involving complex challenges that demand creative and innovative solutions. Many employees need skills to think outside the box and develop novel approaches to circular supply chain

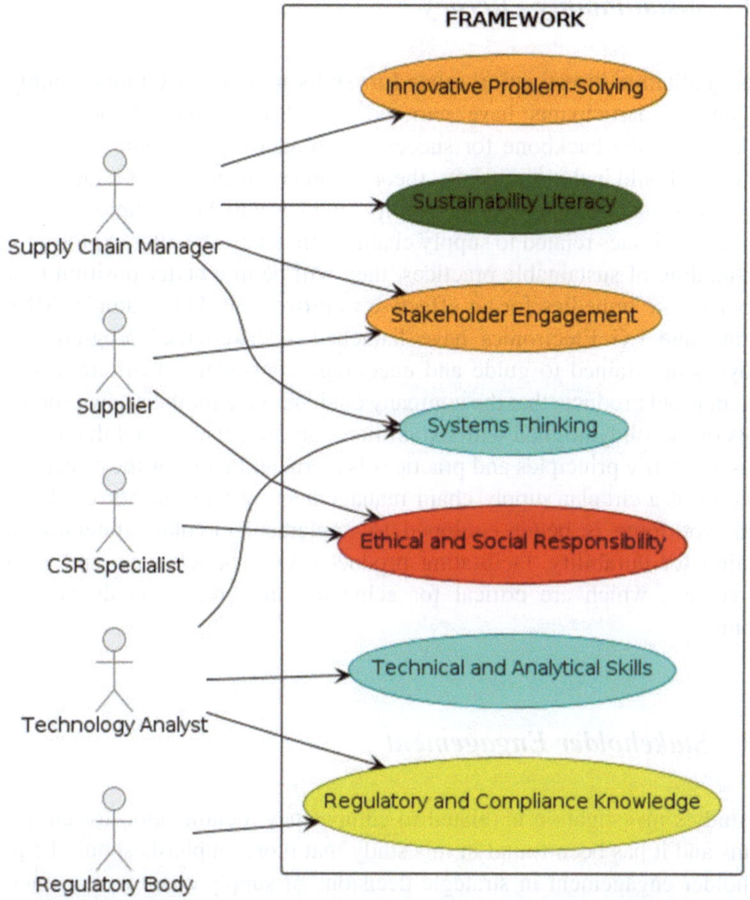

Fig. 9.1 Proposed "skills set" for circular supply chain management implementation

challenges. The results are consistent with previous literature that stated that for any organization to survive in a competitive market, there is a need for skills to develop and implement innovative solutions (Bocken et al. 2023). Subsequently, the organization should initiate new initiatives to effectively train employees in creative product design and sustainable process selection. Assessing skills required for circular supply chain implementation will help companies enhance the core competencies of their employees and engage them in a more sustainable business culture. This results in the development of innovative problem-solving skills; most multinational companies have started investing in their research and development departments and trying to draft their operation strategy by aligning cost minimization, operational efficiency, and resource preservation for sustainable growth.

9.4.2 Sustainability Literacy

In this qualitative investigation related to skills needed for circular supply chain management, participants have narrated that "sustainability-related conceptual knowledge" is the backbone for successful circular supply chain execution. This knowledge should include concepts, theories, and basic principles related to circular supply chain management. Sustainability literacy will help employees to understand the real issues related to supply chain inefficiency and after exploring the vast understanding of sustainable practices, they will be in a better position to suggest the appropriate remedies for an effective improvement. For example, MNCs like Samsung and LG Electronics have launched a "take-back" program in which employees are trained to guide and encourage consumers of different regions to return their old products that the company could recycle for their new products. The outputs of the study aligned with extant literature that also argued that understanding sustainability principles and practices is a crucial factor for the effective implementation of a circular supply chain management system (Zamfir et al. 2017). A skilled workforce is better equipped to implement circular strategies, such as designing for durability, facilitating product take-back schemes, and optimizing resource use, which are critical for achieving the broader goals of a circular economy.

9.4.3 Stakeholder Engagement

This study's investigation is related to competency identification for circular ecosystems and it has been found in this study that more emphasis should be given to stakeholder engagement in strategic decisions of supply chain management. It is suggested by participants that organizations should empower employees at higher levels to develop their skills for stakeholder engagement so that they can collaborate with multiple suppliers, customers, technology service providers, and regulatory bodies for the overall success of the supply chain. For example, Adidas, as one of the major brands in the footwear segment, has launched the "Futurecraft Loop" sneaker along with collaboration with its technology service providers and claimed its commitment to reducing plastic waste. Further, they narrated that the ability to develop and retain a better professional relationship with stakeholders is often lacking among supply chain professionals, and this results in creating bottlenecks for the sound implementation of circular supply chain initiatives. This identified dimension of the current study is consistent with previous literature that posited proficiency in retaining good supplier relationships and managing various relationships with professional bodies plays a significant role in the overall success of circular supply chain execution (Batista and Francisco 2018). These associations surely help the employees develop strategic partnerships with business partners and improve their negotiation skills for procurement and purchasing in a health supply chain system. Identifying the necessary skills helps foster a culture of collaboration and effective communication, which is essential for implementing circular initiatives.

9.4.4 System Thinking

Participants in the study stated that implementing CSCM requires a holistic perspective that considers the entire lifecycle of products and the interdependencies within the supply chain. Nevertheless, supply chain professionals need more systems thinking skills, limiting their ability to design and manage circular systems effectively. Previous literature also concluded that there should be an ability among supply chain professionals to view the supply chain as an interconnected system, considering the entire lifecycle of products and the interdependencies between different components (Murray et al. 2017). For example, the CSCM system of Renault company has enhanced supply chain resilience by creating closed-loop systems that reduce dependence on raw material supplies and mitigate the risks associated with supply chain disruptions. Similarly, another player in the automobile industry (BMW)has adopted the "design for disassembly" approach and attempted to create the design of cars in such a way that it makes it easier for the production team to recover its parts sub-assembled components, metals, aluminum, and plastics, confirming these machinery parts can be reintroduced again for future manufacturing process. Apart from this, manufacturing companies like Volvo have adopted "circular product service models" and "leasing and subscription models" which have benefited the company by providing the opportunity to have complete control over the entire life cycle of their vehicles. This mechanism helps Volvo to refurbish and recycle the entire vehicle after a specific lease period.

9.4.5 Ethical and Social Responsibility

The study's results suggested that supply chain stakeholders should make decisions that balance economic performance with social and environmental responsibility. Furthermore, the literature supported that understanding labor standards and human rights issues within supply chains ensures fair and ethical treatment of workers (Locke et al. 2013). Ethical employees can reduce costs through improved resource efficiency, create new revenue streams from recycled materials and remanufactured products, and enhance their brand reputation by demonstrating a commitment to sustainability among each other.

9.4.6 Technical and Analytical Skills

The present study has concluded that familiarity with technologies that enhance supply chain sustainability, such as renewable energy and resource-efficient manufacturing processes, is critical to a sound circular supply chain management system and results are consistent with the previous literature (Ahi and Searcy 2015). Nowadays, companies are interested in adopting and integrating new technologies that enable circular practices, such as tracking and tracing systems, which require significant effort and investment. Additionally, advancing recycling technologies to efficiently process a broader range of materials is essential but often challenging. Hence, the skills employees require to work for these technologies require core competencies to cope with these dynamic technologies. Apart from that, proficiency in data analytics is essential for monitoring and optimizing circular supply chains. Despite its importance, there is a significant gap in data analytics capabilities among supply chain professionals, limiting the ability to leverage data to improve circular practices.

9.4.7 Regulatory and Compliance Knowledge

As per the discussion with participants in the current study, it has been found that understanding environmental laws and regulations that impact supply chain activities and the ability to ensure that supply chain operations comply with relevant environmental and social standards are the primary benchmarks for the execution of effective circular supply chain system in an organization. Consistency of this with past literature can be seen from the fact that Navigating the complex landscape of regulations related to waste management, recycling, and product standards can be difficult suggested by Walker and Jones (2012) and the lack of supportive policies and incentives for businesses to adopt circular practices can hamper the transition of traditional supply chain system into "circular supply chain system" (Fig. 9.2).

Hurdles/Barriers for Circular Supply Chain Management During the discussion with participants related to competencies required to implement circular supply chain systems in organizations, they also highlighted a few operational hurdles or barriers that create destruction for implementation of circular supply chain management in organizations, and a description of the same is given below.

Institutional Barriers/Structural Barriers Initially most of the participants narrated that a lack of appropriate government support, cultural practices, absence of compulsory recycling laws, and inappropriate circular infrastructure create institutional barriers to the successful execution of circular supply chain systems in companies. For example: MNCs like Nestlé, Coca-Cola, and Unilever have taken a few initiatives to transform the traditional supply chain system into a circular system but these obstacles have created a lot of issues for them. They further emphasize the fact

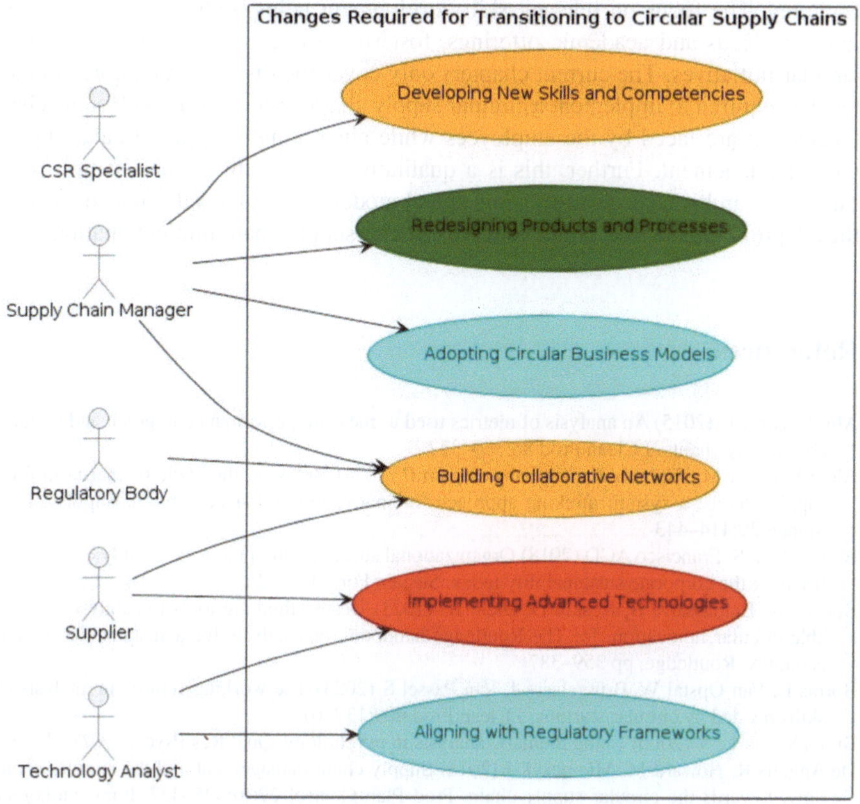

Fig. 9.2 Proposed changes required to transiting traditional SCM to circular SCM

that most of the Indian waste management system is fragmented and entirely managed by unorganized players. Apart from that, limited recycling facilities create a lot of challenges to applying circular economy principles in the country.

9.5 Managerial Implications

The results of the current study proposed a set of skills that the organizations can use to invest in building the necessary ecosystem for CSCM to gain a competitive advantage. They can reduce costs through improved resource efficiency, create new revenue streams from recycled materials and remanufactured products, and enhance their brand reputation by demonstrating a commitment to sustainability. Furthermore, identifying the skills required for CSCM informs the development of targeted education and training programs. Educational institutions and professional training providers can incorporate these skills into their curricula, ensuring that current and future supply chain professionals are equipped to meet the demands of a circular economy. The focus on relevant skill development helps bridge the gap between industry needs and academic offerings, fostering a talent pipeline ready to drive circular initiatives. The current chapters only cover the discussion on core competencies required to implement a circular supply chain system along with some challenges that are faced by the employees while attempting to apply circular supply chain management. Further, this is a qualitative study with a small sample size. Future researchers can develop a conceptual model that explains the antecedents for the adoption intention of employees for circular supply chain implementation.

References

Ahi P, Searcy C (2015) An analysis of metrics used to measure performance in green and sustainable supply chains. J Clean Prod 86:360–377

Alem Fonseca M, Tsolakis N, Kittipanya-Ngam P (2024) Breaking the cycle of inertia in food supply chains: a system thinking approach for innovation and sustainability. Supply Chain Manag 29:414–443

Batista AADS, Francisco ACD (2018) Organizational sustainability practices: a study of the firms listed by the corporate sustainability index. Sustain For 10(1):226

Bocken N, Baldassarre B, Keskin D, Diehl JC (2023) Design thinking tools to catalyse sustainable circular innovation. In: The Routledge handbook of catalysts for a sustainable circular economy. Routledge, pp 359–387

Borms L, Van Opstal W, Brusselaers J, Van Passel S (2023) The working future: an analysis of skills needed by circular startups. J Clean Prod 409:137261

Braun V, Clarke V (2006) Using thematic analysis in psychology. Qual Res Psychol 3(2):77–101

De Angelis R, Howard M, Miemczyk J (2018) Supply chain management and the circular economy: towards the circular supply chain. Prod Plan Control 29(6):425–437. https://doi.org/10.1080/09537287.2018.1449244

De Jesus A, Antunes P, Santos R, Mendonça S (2019) Eco-innovation pathways to a circular economy: envisioning priorities through a Delphi approach. J Clean Prod 228:1494–1513

De los rios IC, Charnley FJ (2017) Skills and capabilities for a sustainable and circular economy: the changing role of design. J Clean Prod 160:109–122

Doh JP, Tashman P (2014) Half a world away: the integration and assimilation of corporate social responsibility, sustainability, and sustainable development in business school curricula. Corp Soc Responsib Environ Manag 21(3):131–142

Dwivedi A, Chowdhury P, Paul SK, Agrawal D (2023) Sustaining circular economy practices in supply chains during a global disruption. Int J Logist Manag 34(3):644–673

Farooque M, Zhang A, Thürer M, Qu T, Huisingh D (2019) Circular supply chain management: a definition and structured literature review. J Clean Prod 228:882–900

Franco MA (2017) Circular economy at the micro level: a dynamic view of incumbents' struggles and challenges in the textile industry. J Clean Prod 168:833–845

Gartner (2022) Information. Retrieved from https://www.gartner.com/en/documents/4000917

Gawankar SA, Gunasekaran A, Kamble S (2020) A study on investments in the big data-driven supply chain, performance measures and organisational performance in Indian retail 4.0 context. Int J Prod Res 58(5):1574–1593

Geissdoerfer M, Savaget P, Bocken NM, Hultink EJ (2017) The circular economy—a new sustainability paradigm? J Clean Prod 143:757–768

Halkos GE, Aslanidis PSC (2024) How waste crisis altered the common understanding: from fordism to circular economy and sustainable development. Circ Econ Sustain 4(2):1513–1537

Hastig GM, Sodhi MS (2020) Blockchain for supply chain traceability: business requirements and critical success factors. Prod Oper Manag 29(4):935–954

Herremans IM, Nazari JA, Mahmoudian F (2016) Stakeholder relationships, engagement, and sustainability reporting. J Bus Ethics 138(3):417–435. https://doi.org/10.1007/s10551-015-2634-0

Kannan D, Solanki R, Kaul A, Jha PC (2022) Barrier analysis for carbon regulatory environmental policies implementation in manufacturing supply chains to achieve zero carbon. J Clean Prod 358:131910

Kaur M, Palazzo M, Foroudi P (2024) Circular supply chain management in post-pandemic context. A qualitative study to explore how knowledge, environmental initiatives and economic viability affect sustainability. Qual Mark Res Int J. https://www.emerald.com/insight/content/doi/10.1108/QMR-10-2023-0140/full/html

Kazancoglu Y, Ozkan-Ozen YD, Sagnak M, Kazancoglu I, Dora M (2023) Framework for a sustainable supply chain to overcome risks in transition to a circular economy through Industry 4.0. Prod Plan Control 34(10):902–917. https://doi.org/10.1080/09537287.2021.1980910

Korhonen J, Nuur C, Feldmann A, Birkie SE (2018) Circular economy as an essentially contested concept. J Clean Prod 175:544–552

Lahane S, Kant R, Shankar R (2020) Circular supply chain management: a state-of-art review and future opportunities. J Clean Prod 258:120859

Lengyel P, Bai A, Gabnai Z, Mustafa OMA, Balogh P, Péter E, Tóth-Kaszás N, Németh K (2021) Development of the concept of circular supply chain management—a systematic review. Processes 9(10):1740

Liu L, Zhang M, Hendry LC, Bu M, Wang S (2018) Supplier development practices for sustainability: a multi-stakeholder perspective. Bus Strateg Environ 27(1):100–116. https://doi.org/10.1002/bse.1987

Locke T, Alcorn N, O'Neill J (2013) Ethical issues in collaborative action research. Educ Action Res 21(1):107–123

Luthra S, Kumar A, Zavadskas EK, Mangla SK, Garza-Reyes JA (2020) Industry 4.0 as an enabler of sustainability diffusion in supply chain: an analysis of influential strength of drivers in an emerging economy. Int J Prod Res 58(5):1505–1521

Murray A, Skene K, Haynes K (2017) The circular economy: an interdisciplinary exploration of the concept and application in a global context. J Bus Ethics 140:369–380

Nyffenegger R, Zehendner A, Quarshie AM, Leuschner R (2024) Change agents' cognitive maps of circular supply chain transition—an investigation of barriers, actions, and outcomes. J Purch Supply Manag 30:100906

Piila N, Sarja M (2024) Extraordinary supply chain disruptions and the circular economy transition in the construction industry—an opportunity within crisis? Sustain Prod Consum 47:71–86

Piyathanavong V, Huynh V-N, Karnjana J, Olapiriyakul S (2024) Role of project engagement on sustainable supply chain development through Industry 4.0 technologies and circular economy during the COVID-19 pandemic: a multiple case study of Thai metals industry. Oper Manag Res 17(1):13–37. https://doi.org/10.1007/s12063-022-00283-7

Quashie R, Fugar FD, Antwi-Afari P, Ng ST (2024) Evaluating the key competency skills of construction professionals for the attainment of circular construction in developing economies. Clean Prod Lett 6:100060

Rissman J, Bataille C, Masanet E, Aden N, Morrow WR III, Zhou N, Elliott N, Dell R, Heeren N, Huckestein B (2020) Technologies and policies to decarbonize global industry: review and assessment of mitigation drivers through 2070. Appl Energy 266:114848

Rizos V, Behrens A, Van der Gaast W, Hofman E, Ioannou A, Kafyeke T et al (2016) Implementation of circular economy business models by small and medium-sized enterprises (SMEs): barriers and enablers. Sustain For 8(11):1212

Sawhney A (2021) Striving towards a circular economy: climate policy and renewable energy in India. Clean Techn Environ Policy 23(2):491–499. https://doi.org/10.1007/s10098-020-01935-7

Sudusinghe JI, de Lima FA, Seuring S, Genovese A (2024) Conceptualizing circular supply chains: a theory building approach. In: The Palgrave handbook of supply chain management. Springer International Publishing, Cham, pp 201–220
Taddei E, Sassanelli C, Rosa P, Terzi S (2024) Circular supply chains theoretical gaps and practical barriers: a model to support approaching firms in the era of Industry 4.0. Comput Ind Eng 190:110049
Tseng M-L, Tan RR, Chiu AS, Chien C-F, Kuo TC (2018) Circular economy meets Industry 4.0: can big data drive industrial symbiosis? Resour Conserv Recycl 131:146–147
Walker H, Jones N (2012) Sustainable supply chain management across the UK private sector. Supply Chain Manag 17(1):15–28
Zamfir A-M, Mocanu C, Grigorescu A (2017) Circular economy and decision models among European SMEs. Sustainability 9(9):1507

Chapter 10
Human Entrepreneurship in Circular Supply Chain Management: A Synopsis of Theory, Practice and Future Pathways

Poonam Gulati and Sakshi Arora

Abstract In modern supply chain management systems and practices, human entrepreneurship is the fulcrum with respect to fostering and activating innovation, attaining efficiency, and therefore leading toward the path of sustainability. This book chapter highlights the interplay of human entrepreneurship and supply chain management, exploring myriad themes, case studies, theoretical frameworks, and findings from a review of the literature. It exemplifies how entrepreneurial thinking caps and implementation impact the dynamics of the supply chain, from acquisition and possession to distribution and stockpiling. It also delves into the impact of entrepreneurial leadership skills on resilience, acrobatic agility, and alteration/adjustment to the volatile market scenario. This successful integration of human-centric entrepreneurial strategies bolsters and optimizes the supply chain performance and competitiveness. The review concludes with the main legal provisions that support sustainable human entrepreneurship, policy implications, and future directions in this ever-evolving human entrepreneurship shaping and striving with the foot on the pedal toward creative and sustainable supply chain strategies.

Keywords Entrepreneurship · Supply chain management · Technological innovation · Policy initiatives

10.1 Introduction

Human entrepreneurship and the adjective resilience to an organization are two very crucial aspects towards research and development in the realm of commerce, economics, and various studies involving management. As the entrepreneurship ambit

P. Gulati (✉)
Amity Law School, Amity University, Mohali, India

S. Arora
Amity School of Social Sciences, Amity University, Mohali, India

K. Mathiyazhagan et al. (eds.), *Humanizing Circular Supply Chain Management*, Circular Economy and Sustainability,
https://doi.org/10.1007/978-3-031-86562-6_10

hovers around creativity and advancement of novice and antecedent endeavours (Kraus et al. 2021), the strength depicted by the organization zooms in on the ability of different institutions to adapt and refurbish in these ever-changing events (Hillmann and Guenther 2021).

Recent studies have suggested also putting forth their respective similar viewpoints. An old study by Covin and Slevin (1989) also exemplified that an entrepreneurial vision has the possibility of improving the ability of the institution to respond to unforeseen circumstances by tapping on its risk-plunging capability and innovation aligned. A more recent study by Miller (2011) has revealed that by providing flexibility in the organisation, the respective organisation flourishes even more and sails smoothly through the rough waters.

Therefore, developing an entrepreneurial orientation is a pivotal effort for companies to exploit the dynamics that permeate their environments (Hillmann and Guenther 2021). As in this dynamic world, we have been seeing that technology is changing rapidly, competition exponentially increasing, all the firms have to circumvent the problems and take risks and innovate wherein supply chain management techniques play a central role.

10.2 Review of Literature

An important organizational framework constitutes a philosophy involving management techniques followed by a pursuit, thereafter a detailed process is worth mentioning here (Mentzer et al. 2001). Firstly, the management philosophy aspect of Supply Chain Management is networked a standalone entity, clubbing of all parties including consumers (Ultimate receivers) here. Secondly, the techniques include all the information that is imperative to be disseminated and experiences, and learning's shared. Thirdly, the last perspective as a pursuit that is articulated in a manner where all the information is combined and vetted, and later executed.

In the process of reviewing the legal provisions relating to CSR, the study by Mohapatra and Mohanty (2023) titled "An analysis of corporate social responsibility policies and practices in India" highlights the Company Law provisions in India, post-independence. The Act of 2013 and amendments of 2019 focus on creating accountability for complying with the CSR regulations and the results of non-compliance, corporate misconduct, etc. are discussed. The study acumens to contribute to the broader address on the progression of CSR practices and regulatory frameworks in determining corporate behaviour towards sustainability and inclusive societal growth.

10.3 Research Gaps

1. There is a need to further explore how the entrepreneurial firms can effectively utilize the ever-emerging technologies such as blockchain and the like to enhance transparency, efficiency, and elasticity in the supply chain management.
2. A Cross-sectional investigation on how different entrepreneurial strategies differ across various markets within India amidst varied policies, practices and regulatory environments have not been analyzed in detail.
3. Delving deeper to assess the impression of human entrepreneurial initiatives on supply chain effectiveness and analyzing further the road leading to sustainability remains an area that has not been touched upon.

10.4 Research Objectives

1. To explore how blockchain and other emerging technologies can be integrated by entrepreneurial firms to enhance supply chain transparency and efficiency.
2. To explore the cross-sectional features inducing the ratification and triumph of entrepreneurial strategies in different global supply chain environments amidst varied policies, practices, and regulatory environments.
3. To understand the various metrics for assessing and accreditation of the impact of human entrepreneurial actions on supply chain management operations.

10.5 Key Themes in Human Entrepreneurship and Supply Chain Management

1. **Decentralisation:** The human entrepreneurs lead to more efficiency with the division of labour, and workflow streamlining leading to specialization and cost-effectiveness.
2. **Digitalization and Innovation**: Entrepreneurial ability seeks to research and innovate in the circular supply chain to bring about new refined products. (Jones et al. 2019).
3. **Efficiency and Optimization**: Entrepreneurial actions often lead to process efficiencies through streamlined workflows, better utilization of resources, and enhanced cost-effectiveness (Gielnik et al. 2021).
4. **Sustainable Green Supply Chains**: Entrepreneurs play a pioneering role in promoting green supply chains leading to sustainable development, inclusive of reducing carbon footprints, adapting to environmentally pro techniques and tools, and the like (Robinson and Carter 2020).

Table 10.1 Tabulation of the key themes involved in Human Entrepreneurship and its implications

Theme	Description
Innovation and creativity	Entrepreneurial innovation in SCM: New product development, process improvement
Efficiency and optimization	Streamlined workflows, resource optimization, cost-effectiveness
Sustainability and green SCM	Environmental practices, carbon footprint reduction, waste minimization
Impact on SCM resilience	Entrepreneurial leadership fostering agility, adaptability to market changes, and disruptions
Future directions	Integration of digital technologies, global SCM dynamics, and regulatory implications

Author's tabulation

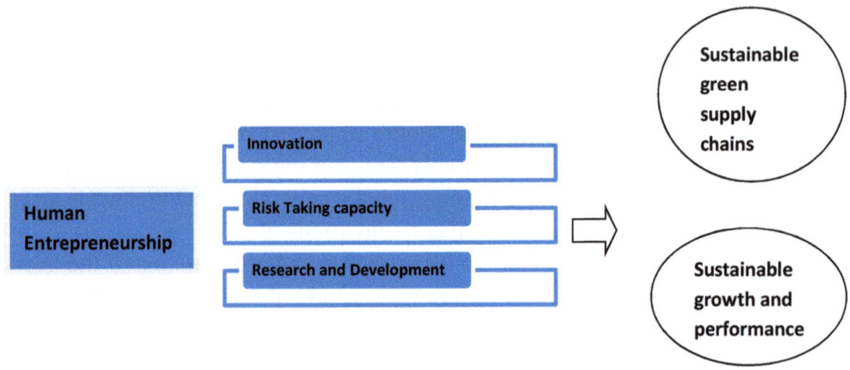

Flowchart 10.1 Conceptualisation of the themes. (Source: Authors illustration based on Hayek (1964) and Freiling (2008))

Table 10.1 and Flowchart 10.1 provide a detailed explanation of human entrepreneurship accolades, impact, and myriad supply chain strategies proven resilient. The table summarizes key themes and implications for "Supply Chain Management" research, practice, and implications.

10.6 Case Studies for Empirical and Applicability Insights

Case studies from various prevalent Industries here have showcased a successful amalgamation of human entrepreneurial practices and strategies. It has demonstrated how business acumen, innovation, and aforementioned themes have been put to acrobatic capabilities supporting sustainable Supply Chain Management:

- **Optimization of Logistics**: Optimization of routes taken while using transportation, switching to biofuel, e-fuel. Various case studies have been looked into like in case of DHL, which has provided a global logistic advice and has followed the route towards sustainability which also led to the enhancement of their reputation in the economy.

- **Innovative Procurement Strategies**: Confederation of Indian Industry has provided various reports wherein industries have reduced their costs with better procurement solutions and strategies using Innovation. Tata Consultancy services, Larsen and Toubro (used in construction projects) are few of the prime exemplar for the same. Furthermore, Public private partnership also exemplifies this impact in many cases.
- **Sustainable Practices in FMCG**: Various companies selling FMCG are following myriad practises, Hindustan Unilever Limited has delved into sustainable packing of the product, ITC into sustainable sourcing of the raw materials and the like.

10.7 Pivotal Techniques and Strategies That Entrepreneurs Delve In

In recent entrepreneurial development various techniques and strategies have been adopted which bolster the sustainable supply chain management. The following tabular presentation describes the same techniques along with it's unique description.

Table 10.2 provides an in-depth exploration of various entrepreneurial techniques aligned with circular economy principles that drive creativity leading to innovation and sustainability within supply chain management. Table 10.2 depicts various strategies and their implications for enhancing SCM resilience, efficacy, and efficiency.

Table 10.2 Techniques and strategies with their description

Technique	Description
Closed-loop supply chains	Designing products for reuse, remanufacturing, or recycling to minimize waste
Product life extension	Extending product lifespan through repair, refurbishment, or resale
Resource sharing platforms	Facilitating collaborative consumption and resource optimization
Waste-to-value innovations	Converting waste materials into valuable resources or products
Case studies	Examples of successful entrepreneurial initiatives integrating circular economy in SCM
Future directions	Policy frameworks, technological advancements, and research gaps in circular SCM practices

Author's tabulation

10.8 Schemes and Initiatives to Promote Sustainable Human Entrepreneurship in India

Numerous Schemes bolstering startups and entrepreneurship have paved the way in India and a considerable number line up with the ideologies of the circular economy. For instance; a few schemes and initiatives are mentioned herein:

1. **"Start-up India":**
   - **Objective:** Startup India was launched in 2016 aiming to construct an ecosystem favorable for startups by including incentives such as simplified compliance, tax exemptions, support for funding via the Fund of Funds, etc.
   - Startups may rely upon the scheme thus focusing on innovative solutions to enhance sustainability and efficiency resulting in contribution to the circular economy goals.
2. **"Atal Innovation Mission (AIM)":**
   - **Objective:** With the help of Atal Incubation Centers, Atal Tinkering Labs, and Atal Community Innovation Centers, this scheme ensures innovation and entrepreneurship across schools, universities, and industries.
   - By promoting innovation, AIM helps in achieving solutions related to encouraging environmental concerns while promoting circular practices in industries, including management of waste and renewing energy resources.
3. **"National Mission on Interdisciplinary Cyber-Physical Systems (NM-ICPS)":**
   - **Objective:** This initiative helps indulgence in technological Research and development such as, IoT, AI, and robotics hence creating an ecosystem and talent pool for tech-driven startups.
   - This further is associated with the creation of resource management through smart solutions and sustainable production processes resulting in practices supporting circular economy.
4. **"Swachh Bharat Mission (SBM)":**
   - **Objective:** SBM was introduced as a mass mission throughout India as a cleanliness and sanitation drive. This has led to the use of innovative technologies sanitisation and waste management.
   - By converting waste into useful reusable products and manure etc., Startups help in developing models that promote circular economy systems and implement sustainable solutions.
5. **"Smart Cities Mission":**
   - **Objective:** This mission focuses on developing smart cities hence indulging in sustainable and inclusive growth with the help of innovation and tech-driven solutions.

- Using circular economy principles in infrastructure development and urban planning, the contribution of Startups in the promotion of resource efficiency and reduction of environmental impact may prove commendable.

6. **"Pradhan Mantri MUDRA Yojana (PMMY)":**

- **Objective:** Easy availability of credit for micro-enterprises and startups is possible through PMMY thereby promoting financial inclusion and entrepreneurship.
- MUDRA loans allow Startups to facilitate sustainable manufacturing or eco-friendly products by scaling up operations that endorse circular practices.

The above-mentioned initiatives have been implemented by the government to keep the progress of entrepreneurship in India with a view to the larger objective of achieving sustainable solutions and a circular economy across innumerable sectors. Following the incentives of these schemes the business industry especially the start-ups prefer to align with the ideologies of the circular economy thus resulting in economic and environmental sustainability (Table 10.3).

The role of the above-mentioned policies is significant in promoting a cordial ecosystem for entrepreneurial progress in circular supply chain management. They support various governmental initiatives and in turn, reduce regulatory burdens in

Table 10.3 The following tabular presentation of government schemes to boost entrepreneurship

Scheme	Objective	Link to circular economy
Startup India	Foster a conducive ecosystem for startups through tax incentives and funding	Encourages startups to develop sustainable solutions, promoting resource efficiency and waste reduction
Atal Innovation Mission (AIM)	Promote innovation and entrepreneurship through Atal Tinkering Labs and more	Supports startups developing technologies for balanced practices, such as waste management and renewable energy
"National Mission on Interdisciplinary Cyber-Physical Systems (NM-ICPS)"	Develop R&D in emerging technologies to boost tech-driven startups	Enables startups to integrate IoT, AI, etc., into circular economy solutions like smart resource management
Smart Cities Mission	Develop sustainable and inclusive urban environments through technology	Encourages startups to contribute circular economy solutions in urban planning and infrastructure development
Swachh Bharat Mission	Achieve universal sanitation and promote innovations in waste management	Drives startups to create circular economy models by converting waste into valuable resources
Pradhan Mantri MUDRA Yojana (PMMY)	Provide access to finance for micro-enterprises and startups	Supports startups in scaling up operations that promote circular practices, such as eco-friendly products

Author's tabulation

Special economic zones (SEZs) and drive tech-based innovations. Naming a few are; the 'Make in India,' the 'Digital India,' or the 'Startup India' initiatives.

10.9 Pivotal Legal Provisions Support Sustainable Human Entrepreneurship in India: The Companies Act, 2013 & the Amended Act, 2019

The most important provision under section 135 discussed in the 2013 Companies Act provides for the liability of companies registered under the Act bringing them under Corporate Social Responsibility (CSR). A class of companies mandatorily allot 2% of the average, of the previous 3 years, net profit in performing activities encompassing CSR. This contribution goes a long way in improving overall global sustainability initiatives.

In 2019, the amended Act introduced four new provisions that filled up lacunas related to unspent CSR allocation funds and proposed hefty fines ranging between 50,000 and 5 Lakh Indian Rupees with or without imprisonment up to the tune of 3 Years, in Sections 5(ii), 6 and 7 respectively. Furthermore, Section 8 empowers the Central Government to notify directions for CSR compliance as per Section 135 of the Companies registered under the Act.

Moreover, Schedule VII of the Act details an activity list of eligible CSR activities within such ambit. These activities focus on the eradication of hunger poverty, child mortality, gender inequality, etc., and encourage improved maternal health, employment rates, environmental sustainability, etc. within the scope of CSR initiatives (MCA 2013).

10.9.1 National Green Tribunal of India

In 2010, the National Green Tribunal (NGT) that was established under the National Green Tribunal Act of 2010 ensured effective and quick solutions to legal matters related to the environment, including forests, and other similar resources. It also handles the enforcement of environmental rights and provides compensatory relief to individuals and/or property. As a dedicated organisation with the proficiency to manage complex disputes related to environmental litigation, the NGT operates based on principles of natural justice, rather than being bound by the procedures outlined in the Code of Civil Procedure, 1908.

The NGT's focused jurisdiction on environmental issues aims to deliver prompt justice in environmental cases, thereby alleviating the caseload of higher courts. The NGT is required to strive for the final resolution of petitions within 6 months of filing. The NGT operates from five locations, adopting a circuit procedure to enhance accessibility. The principal location is in New Delhi, with additional benches in Bhopal, Pune, Kolkata, and Chennai. (Ministry of Environment, Forest and Climate Change, Government of India).

Several legislations in India focus on environmental protection laws. To mention a few

1. That the "Air (Prevention and Control of Pollution) Act, 1981," purposes to avert, regulate, and decrease air pollution. It establishes Boards with the authority and responsibilities to achieve these goals and address related matters. Following the decisions made in June 1972, at the "United Nations Conference on the Human Environment" in Stockholm, India took steps to preserve natural resources, including air quality and pollution control. This Act is necessary to implement those decisions, focusing on maintaining air quality and controlling air pollution.
2. The "Water (Prevention and Control of Pollution) Act," enacted in 1974, intends to stop and decrease water pollution and maintain or restore the purity of water in India. Additionally, the "Water (Prevention and Control of Pollution) Cess Act," of 1977, imposes a cess on water consumption by certain industrial activities to fund the "Central and State Boards" accountable for damage contributing to water pollution.
3. Rapid industrialization in India has led to resource depletion and increased pollution, generating large quantities of hazardous waste that pose severe environmental and health risks. The "Ministry of Environment, Forest, and Climate Change" introduced the "Hazardous Waste (Management and Handling) Rules" in 1989, updated in 2008, 2009, 2010, and 2016, to manage this waste effectively. Key causes of resulting in wastes of a hazardous nature include the use of hazardous substances, outdated technologies, and end-of-pipe treatment of emissions. India ratified the "Basel Convention in 1992" to control the cross-border movement of hazardous waste, promoting sound environmental management. ["Hazardous and Other Wastes (Management and Transboundary Movement) Rules, 2016"]

Easing the understanding of legal provisions for environmental and statutes for the universal population can significantly enhance public engagement. Additionally, implementing sound mechanisms for effectiveness and transparency both in the courts and the community can greatly improve the functioning of the administrative, legislative, and judicial systems. Law is often viewed as a "Utopian system of action;" making it amicable and community-friendly facilitates smoother transitions for the public. Remedies like "Public Interest Litigations (PILs)" are designed to enable the public to obtain information from the government system regarding important matters. The law is increasingly focused on providing easier access to appeals and enhancing the integration of public interest in judgments and cases (Mohapatra and Mohanty 2023).

10.10 Institutional Support for Entrepreneurs in India

Entrepreneurship in India is influenced by various factors, including individual, environmental, socioeconomic, and institutional setups. The institutional support system plays a vital role in guiding entrepreneurship development and influencing entrepreneurial behavior. The Indian government has established several institutions to support entrepreneurship through technical guidance, financial support, and training. Given that 93% of the workforce is associated with the unorganized sector, maintaining support for this sector is essential for its survival.

10.11 Key Institutions and Initiatives

- "National Commission for Enterprises in the Unorganised Sector (NCEUS)": Established on 20 September 2004 to warrant the worker's welfare, especially in the unorganized sector. It recommends providing technical, marketing, and credit support to enterprises in this sector (Kannan et al. 2008).
- "National Bank for Agriculture and Rural Development (NABARD)": Originated in 1982 to uplift "Rural India." NABARD plays a significant role in attracting rural youth and supporting rural entrepreneurship development programs (REDPs) (NABARD 2015)
- "Industrial Development Bank of India (IDBI)" and its subsidiary "Small Industries Development Bank of India" (SIDBI): Make available financial and non-financial assistance to the small-scale sector.
- "Small Industries Development Organization (SIDO)": Established in 1954 to promote small sectors. SIDO manages various institutes and centers, together with:
  - "Small Industries Service Institutes (SISI)"
  - "Product-cum-Process Development Centres (PCPDC)"
  - "Regional Training Centres (RTC)"
  - "National Institute for Small Industries Extension Training (NISIET), Hyderabad"
  - "National Institute for Entrepreneurship and Small Business Development (NIESBUD), New Delhi"
  - "National Small Industries Corporation (NSIC)": Offers machinery, financial assistance, raw material arrangement, marketing assistance, etc.
  - "Technical Services Centres (TSC)": Support the small-scale sector through institutions like the "Entrepreneurship Development Institute of India (EDI)."
  - "National Institute for Micro, Small and Medium Enterprises (NI-MSME), Hyderabad"
  - "National Institute for Entrepreneurship and Small Business Development (NIESBUD), NOIDA (NIESBUD)"
  - "Indian Institute of Entrepreneurship (IIE), Guwahati"

- To strengthen research and development at the village level, the "Mahatma Gandhi Institute for Rural Industrialization (MGIRI)" associated with the "Indian Institute of Technology (IIT), Delhi." These institutions conduct training programs, research, and consultancy to promote entrepreneurship.

10.12 Respective Alignment with the Circular Economy

The Indian entrepreneurial landscape has integrated circular economy principles with the help of the above-mentioned initiatives. For instance, Startup India fosters sustainable practices by offering exemptions to taxpayers and providing funding support thus incentivizing eco-friendly novelties (Government of India 2016). In the same way, the Atal Innovation Mission nurtures a culture that innovates and drives startups to focus on developing efficient resource utilization and waste management by developing new technologies (NITI Aayog n.d.).

Furthermore, technologies like IoT and AI may be integrated into circular economy solutions by Startups with the backing of the "National Mission on Interdisciplinary Cyber-Physical Systems (NM-ICPS)," thus enhancing the sustainability and efficiency of resources. ("Ministry of Electronics, and Information Technology, Government of India n.d."). Similarly, the Smart Cities Mission focuses on promoting sustainable urban development as it encourages startups by innovating in renewable energy and smart infrastructure, etc. ("Ministry of Housing and Urban Affairs, Government of India n.d.-a").

The "Swachh Bharat Mission" has successfully promoted circular economy models by converting waste into valuable energy sources and products hence catalyzing innovations in sanitation and waste management and technologies ("Ministry of Housing and Urban Affairs, Government of India n.d.-b"). Additionally, the "Pradhan Mantri MUDRA Yojana (PMMY)" helps in scaling up operations by providing necessary financial assistance to startups, focusing on eco-friendly products and services ("Micro Units Development and Refinance Agency Ltd. (MUDRA) n.d.").

The above-mentioned government schemes have gone a long way in nurturing entrepreneurship in India while endorsing sustainable development through circular economy practices facilitating economic growth and addressing environmental challenges. Innovative startup technologies resulting in resource efficiency and waste reduction pave a long way in the expansion of circular economy principles.

10.13 Impact of Entrepreneurial Leadership on SCM

The trend of innovation within establishments empowers entrepreneurial leaders and teams to respond rapidly to market changes and customer needs hence promoting belief in technology (Brown 2018). This improves the adaptability of SCM and preparedness for unexpected disturbances hence contributing to inclusive business resilience.

10.14 Future Directions

The way had led to the achievement of significant milestones in the contribution of human entrepreneurship in SCM, however, research gaps remain:

- **Need to Integrate Digital Technologies**: Future research must strive for solutions to recent technologies like blockchain and IoT that can be utilised by entrepreneurial firms in creating transparency and efficiency in supply chains.
- **Global Dynamics for Supply Chain Management**: Investigation of the entrepreneurial strategies that differ across global markets and regulatory environments shall avail insights into operative practices for international SCM (Smith et al. 2023).

10.15 Conclusion

Modern circular supply chain management is significantly affected by human entrepreneurship and is a source of efficiency, innovation, and sustainability. Organisations usually strengthen their competitive position consequence of adopting entrepreneurial principles and simultaneously develop resilience in CSCM hence controlling fierce market conditions. Though success has been achieved in bridging gaps overall it is pertinent to record that there is an imminent need for a survey into the integration of digital technology and dynamics of the global circular supply chain management, the development of associated frameworks required for advancement, understanding and application in this emergent field.

10.16 Policy Implications

Promotion of Digital Literacy and Technology Adoption: Industry and Government establishments must motivate entrepreneurial businesses to initiate the use of digital technologies like blockchain and IoT, etc. This may be possible by promoting via incentives, training programs, and infrastructure assistance and additionally boosting the circular supply chain in an efficient, resilient, and transparent manner.

International Collaboration and Knowledge Sharing: Policy and practice must encourage international collaboration within entrepreneurial houses for the adoption of necessary regulatory environments and promote innovation with the target of improving circular supply chain performance the world over.

Long-term Sustainability Goals: For the evaluation of the long-term effect of entrepreneurial actions on supply chain sustainability, policies need to prioritize the expansion of standardized frameworks for metrics and measurements, focusing on sustainable practices and supporting environmental stewardship.

Support for Research and Development: Research initiatives may be government-funded to report gaps identified in studies and analysis of entrepreneurial strategies throughout various regions and industries thus advancing knowledge in circular supply chain management and informing evidence-based policies.

Summarily, it may be recorded that the process of nurturing human entrepreneurship in circular supply chain management improves organizational competitiveness and efficiency and adds to sustainable economic growth. A crucial role is upon the shoulders of policymakers who create an environment that fosters entrepreneurial innovation, and collaboration, and promotes global sustainable practices for circular supply chains. Continued research and policy support are essential to leverage entrepreneurial potential for resilient and adaptive supply chain strategies in the future.

References

Barney JB (1991) Firm resources and sustained competitive advantage. J Manag 17(1):99–120. https://doi.org/10.1177/014920639101700108

Brown WS (2018) Socially responsible entrepreneurship as innovative human resource practice. J Appl Behav Sci 54(2):129–152. https://doi.org/10.1177/0021886317752146

Cohen WM, Levinthal DA (1990) Absorptive capacity: a new perspective on learning and innovation. Adm Sci Q 35(1):128–152. https://doi.org/10.2307/2393553

Covin JG, Slevin DP (1989) Strategic management of small firms in hostile and benign environments. Strateg Manag J 10:75–87. https://doi.org/10.1002/smj.4250100107

Eisenhardt KM, Martin JA (2000) Dynamic capabilities: what are they? Strateg Manag J 21(10–11):1105–1121. https://doi.org/10.1002/1097-0266(200010/11)21:10/11<1105::AID-SMJ133>3.0.CO;2-E

Freiling J (2008) Resource-based view and entrepreneurial orientation: an integrated model. Manag Rev 19(1):1–22

Gielnik MM, Frese M, Graf JM, Kampschulte A (2021) The psychology of entrepreneurship: action and process. Annu Rev Organ Psych Organ Behav 8:275–303. https://doi.org/10.1146/annurev-orgpsych-120920-055646

Government of India (2016) Startup India action plan. https://www.startupindia.gov.in/content/sih/en/startup-scheme.html

Hart SL (1995) A natural-resource-based view of the firm. Acad Manag Rev 20(4):986–1014. https://doi.org/10.5465/amr.1995.9512280021

Hayek FA (1964) The theory of complex phenomena. In: Bunge M (ed) The critical approach to science and philosophy. Free Press, New York, NY, pp 332–349

Helfat CE, Peteraf MA (2003) The dynamic resource-based view: capability lifecycles. Strateg Manag J 24(10):997–1010. https://doi.org/10.1002/smj.332

Hillmann J, Guenther E (2021) Organizational resilience: a valuable construct for management research? Int J Manag Rev 23(1):7–44. https://doi.org/10.1111/ijmr.12239

Jones P, Ratten V, Klapper R, Fayolle A (2019) Entrepreneurial identity and context: current trends and an agenda for future research. Int J Entrep Innov 20(1):3–7. https://doi.org/10.1177/1465750319825745

Kannan G, Pokharel S, Kumar PS (2008) A hybrid approach using ISM and fuzzy TOPSIS for the selection of reverse logistics provider. Resour Conserv Recycl 54(1):28–36. https://doi.org/10.1016/j.resconrec.2009.06.004

Kraus S, Palmer C, Kailer N, Kallinger FL, Spitzer J (2021) Digital entrepreneurship: a research agenda on new business models for the twenty-first century. Int J Entrep Behav Res 27(1):43–63. https://doi.org/10.1108/IJEBR-06-2020-0380

MCA (2013) Ministry of Corporate Affairs, SCHEDULE VII, Section 135, Companies Act 2013 https://upload.indiacode.nic.in/schedulefile?aid=AC_CEN_22_29_00008_201318_1517807327856&rid=79

Mentzer JT, DeWitt W, Keebler JS, Min S, Nix NW, Smith CD, Zacharia ZG (2001) Defining supply chain management. J Bus Logist 22(2):1–25

Micro Units Development & Refinance Agency Ltd. (MUDRA) (n.d.) Pradhan Mantri MUDRA Yojana (PMMY). https://www.mudra.org.in/

Ministry of Electronics & Information Technology, Government of India (n.d.) National mission on interdisciplinary cyber-physical systems (NM-ICPS). https://meity.gov.in/nm-icps

Ministry of Housing and Urban Affairs, Government of India (n.d.-a) Smart Cities Mission. https://smartcities.gov.in/

Ministry of Housing and Urban Affairs, Government of India (n.d.-b) Swachh Bharat Mission. https://swachhbharaturban.gov.in/

Mohapatra B, Mohanty T (2023) An analysis of corporate social responsibility policies and practices in India. SN Bus Econ 3:194. https://doi.org/10.1007/s43546-023-00571-9

NABARD (2015) National Bank for agriculture and rural development, 2015. https://www.nabard.org/events-past-article.aspx?id=25&cid=550&EID=4

NITI Aayog (n.d.) Atal Innovation Mission (AIM). https://www.aim.gov.in/

Smith T, Johnson M, Brown A (2023) Human entrepreneurship in supply chain management: a review. J Bus Logist 42(5):210–225

Teece DJ, Pisano G, Shuen A (1997) Dynamic capabilities and strategic management. Strateg Manag J 18(7):509–533. https://doi.org/10.1002/(SICI)1097-0266(199708)18:7<509::AID-SMJ882>3.0.CO;2-Z

Chapter 11
Assessment of Digital Practices for Enhancing the Sustainability of Humanized Supply Chains Using f-DEMATEL Approach

Mohit Sharma, Mohit Tyagi, and R. S. Walia

Abstract In the current landscape, a circular economy (CE) has garnered substantial focus to address sustainability challenges. Sustainable supply chain management (SSCM) is a vital conceptual foundation for businesses, it attempts to combine economic, social, and environmental aspects. Similarly, humanized supply chain management (HSCM) focuses on aligning an organization's activities with ecological, economic, and social considerations. The rapid advancement of digital technologies has transformed SC operations through the integration of various systems. These systems interconnect physical objects for real-time monitoring and decision support. While machine-to-machine communication is crucial, human involvement remains essential for monitoring and configuring digital devices, as well as interacting with digital systems. This enhances industrial intelligence and sustainability within HSC by adhering to CE principles (reuse, recycling, and remanufacturing). In the present study, various digital practices related to the humanized supply chain (HSC) have been identified and analyzed by use of the mathematical algorithm f-DEMATEL technique. The results of the study may be obliging for managers and researchers to assess digital practices in enhancing the sustainability of HSC.

Keywords Circular economy · Digital practices · Humanized supply chain management · f-DEMATEL · Sustainable supply chain management

M. Sharma · M. Tyagi (✉) · R. S. Walia
Department of Production and Industrial Engineering, Punjab Engineering College (Deemed to be University), Chandigarh, India

K. Mathiyazhagan et al. (eds.), *Humanizing Circular Supply Chain Management*, Circular Economy and Sustainability,
https://doi.org/10.1007/978-3-031-86562-6_11

11.1 Introduction and Background

The idea of a CE has been very popular in the last many years as a solution to the pressing need for sustainable resource management and environmental stewardship. The production of a single product often requires numerous natural resources, depleting the treasury of environmental assets and this depletion occurs because resource extraction happens at a rate much faster than the natural regeneration capacity of these resources (Tyagi et al. 2015b). In contrast to the conventional linear "take-make-dispose" approach, a CE aims to reduce waste and prolong the longevity of resources.

This approach reduces resource consumption, minimizes environmental impact, and promotes economic resilience. Over the last decade, the concept of sustainability has been more significant in SCM. Two essential concepts provide a framework for organizations to achieve the Sustainable Development Goals established by the UN. Corporate Sustainability (CS) may be enhanced by people management (PM) to improve organizational performance and foster sustainability (Bernstein 2017). The second notion is the Triple Bottom Line, which is the basis of the equilibrium among three fundamental pillars—economic, social, and environmental (Elkington 1987). Genuine sustainability in supply chain management integrates environmental, economic, and social elements, attaining a balance that concurrently addresses all three areas. Social sustainability specifically emphasizes the human dimension of sustainability (Koca and Mathiyazhagan 2020). A company's supply chain and HR policies and procedures need to be in sync for better business outcomes (Agarwal et al. 2022). A humanitarian supply chain (HSC) must demonstrate efficacy owing to the distinct risks and challenges it encounters (Bag et al. 2023). A humanitarian supply chain (HSC) must demonstrate efficacy owing to the distinct risks and challenges it encounters The HSC ensures staff safety by providing necessary safety equipment and medical supplies. It ensures continuous operations by promptly supplying equipment and resources, even in distant and challenging situations. HSC ensures the industry's long-term survival and enhances its overall sustainability and prosperity via comprehensive risk management and business continuity assurance. Since 2011, the development of Industry 4.0 technology has led humanitarian organizations to increasingly use digital tools for conducting surveys and assessments in crisis-affected regions, facilitating rapid data collecting and analysis (Gupta et al. 2022). An HSC must demonstrate efficacy owing to the distinct risks and challenges it encounters. Geographic Information System (GIS) technology helps humanitarian organizations map and analyze spaces, improving catastrophe planning and response.

The Industrial Revolution profoundly altered production and consumption methods, substantially improving human living standards (Agarwal et al. 2023). Digital technologies are seen as a potential solution due to their increasing prevalence and capacity to enhance supply chain processes (Panghal et al. 2024). There is significant potential to mitigate vulnerabilities via the use of sophisticated digital technologies such as Big Data Analytics (BDA) and the Internet of Things (IoT), in conjunction with the establishment of collaborative platforms and digital networks (Sharma et al. 2024), hence enhancing SC sustainability. By utilizing digital technologies, humanitarian organizations can improve sustainability and resilience in their SCs (Dubey et al. 2018). Resource allocation optimization, coordination mechanism improvement, crisis response management, and thorough monitoring of intervention effects may achieve this enhancement (Dubey et al. 2021). As a result, these initiatives lead to SCs that demonstrate improved efficiency, greater transparency, and a superior ability to respond effectively during emergencies. Employing digital technology to oversee the movement of materials, components, and products is likely a crucial practice for a circular economy. Data from this monitoring is produced at different stages of the industrial life cycle to enhance decision-making and resource management (Nobre and Tavares 2017; Antikainen et al. 2018; Bressanelli et al. 2018). Previous studies by Lai et al. (2018), and Yu et al. (2021) emphasize the significance of BDA skills. They confirm that BDA serves as an innovative tool aiding policymakers in accurately planning and executing development activities, thereby enhancing SC performance. The sophisticated advancement of IoT technology has revolutionized modern supply chain management, enabling real-time control and oversight. Real-time operational data, including warehouse and transportation environmental factors, is made possible via standardized IoT connection protocols, to be visualized on IoT dashboards (Tsang et al. 2022). Authors such as Shah et al. (2017) have demonstrated that human capital aspects influence employee attitudes, actions, beliefs, and intentions, motivating their involvement in innovative processes within the SC. Debates among policymakers, researchers, and academics will continue regarding the significance of humanized factors in predicting SC performance (Jaouadi 2022). Three primary characteristics may be used to group humanized aspects to predict SC innovativeness and sustainable performance: Employee empowerment, engagement, and development (Gupta et al. 2020). Management can enhance employee development through training programs, ensuring equal opportunities for learning and fostering new ideas to boost the effectiveness of sustainable SC (Fera et al. 2020; Halvarsson Lundkvist and Gustavsson 2018).

This study, considering the aforementioned aspects, focuses on evaluating digital practices to increase the sustainability of humanized SCs. The study's objectives are to list and evaluate important digital practices. To achieve this objective, 9 key practices were recognized through a review of literature and expert consultations. These digital practices, outlined in Table 11.1, have been evaluated within a fuzzy environmental framework using the f-DEMATEL approach. The overview of the present work is as follows—Part 2 explains a step-by-step procedure of f-DEMETAL approach. Numerical representation is given in Part 3. Part 4 delineates the outcomes and discussion, while Part 5 explores the conclusions and future scope, respectively.

Table 11.1 A compilation of key digital practices

Sr. no.	Digital practices	Descriptions'	References
1	Production powered by robotics and material recovery **(DP1)**	Robotics can help with material recovery while improving production accuracy and precision Furthermore, artificial intelligence-integrated robots can interact with the real world and perform unprogrammed activities	Awan et al. (2021), Javaid et al. (2022)
2	Supply chain visibility, traceability, and transparency **(DP2)**	Blockchain and digital twins' distinct capacities to track a product's lifecycle provide an advantage to enterprises for process monitoring and guaranteeing adherence to CE objectives	Kayikci et al. (2022)
3	Social sustainability and legitimacy **(DP3)**	Employee safety may be ensured in SC by using digital technologies such as digital twins and blockchain to identify & track dangerous items or non-compliant circular materials	Kayikci et al. (2022)
4	Information technology proficiency **(DP4)**	From the perspective of the circular economy, information technology's capacity to handle massive data sets and process them instantly offers substantial corporate value	Kayikci et al. (2022)
5	System for making decisions based on big data **(DP5)**	By detecting problems in systems and operations in real-time, BDA helps manufacturers plan and solve problems. It optimizes difficult judgments and substantially impacts corporate decisions, particularly when faced with unanticipated circumstances and possible dangers	Asees Awan and Ali (2019), Kumar et al. (2018)
6	Digitally updated procedures and quality control **(DP6)**	Digital transformation is the process of creating or changing consumer experiences, industrial processes, and cultural norms using digital technology. By putting in place a digitalized quality check, assessments can be completed more quickly and continuously, offering precise and comprehensive data that helps evade SC interruptions	Jazdi (2014), Rajput and Singh (2020), Shayganmehr et al. (2021)
7	Intelligent equipment for increased adaptability and large scale customization **(DP7)**	Real-time response to changing conditions is possible for manufacturing systems that are outfitted with smart devices that are interconnected and cooperative. Mass customization is made easier and the system's versatility is increased by this feature	Nascimento et al. (2019), Shayganmehr et al. (2021), Fragapane et al. (2022)

(continued)

Table 11.1 (continued)

Sr. no.	Digital practices	Descriptions'	References
8	Interoperability and system integration **(DP8)**	Interactions between industries are facilitated by system integration, which permits the sharing of data and services. The capacity of industrial networks to communicate with one another is referred to as interoperability	Bag et al. (2018), Gupta et al. (2020), Rosa et al. (2020)
9	Using augmented reality, provide advice and support remotely **(DP9)**	Enhanced communication between experts and field personnel, facilitated by augmented reality technology, reduces downtime and accelerates problem detection. Operations are rendered more effective as a consequence	Bag et al. (2018), Pham et al. (2019), Shayganmehr et al. (2021)

11.2 Research Methodology

In real-world complicated decision-making situations, data often display imprecision and ambiguity. To navigate these complications, decision-makers use language phrases such as “very low,” “low,” “medium,” “high,” and “very high” to articulate their assessments. Subsequently, these language concepts are converted into fuzzy integers for further evaluation. Fuzzy sets, originating from fuzzy logic, allocate a membership degree to each integer inside the interval [0, 1]. A subset of a discourse universe X consisting of ordered pairs with a membership function $\mu_{\tilde{A}}(X)$ is referred to as a fuzzy set $\tilde{A}$. This function maps items from X to the interval [0, 1]. A membership value, indicating the extent to which an element x is associated with $\tilde{A}$, is represented by the value of $\mu_{\tilde{A}}(X)$ attributed to it in $\tilde{A}$. The membership function $\mu_{\tilde{A}}(X)$ ranges from 0 to 1, where $\mu_{\tilde{A}}(X) = 1$ indicates complete membership in $\tilde{A}$, and $\mu_{\tilde{A}}(x) = 0$ signifies non-membership.

$$A = \left\{\left(x, \mu_{\tilde{A}}(x)\right)\right\}, \quad x \in X \tag{11.1}$$

Here $X = \{x\}$ denotes a set including elements x, whereas $\mu_{\tilde{A}}(x)$ functions as a membership function. Membership functions for fuzzy numbers, including bell-shaped, trapezoidal, triangular, and singleton forms, serve to classify these numbers. Due to their computational simplicity, Triangular Fuzzy Numbers (TFNs) are often used in practical applications. A TFN can be represented as $N_a = \left(l_a, m_a, n_a\right)$. A TFN's membership function $\mu_{\tilde{A}}(x)$ can be written as follows:

$$\mu_{\tilde{A}}(x) = \left\{ \begin{array}{l} 0, \quad x \le l_a \dfrac{x - l_a}{m_a - l_a}, \\ l_a \le x \le m_a \dfrac{n_a - x}{n_a - m_a}, \quad m_a \le x \le n_a 0, \quad x > n_a \end{array} \right\} \tag{11.2}$$

where l_a, m_a and n_a are the real numbers with relationship $l_a > m_a > n_a$.

11.2.1 f-DEMATEL Approach

An advanced framework for decision-making makes the importance of the f-DEMATEL method clear. This method can be defined step-by-step as follows.

11.2.1.1 Average Direct Relation Matrix Formation

Assessing the direct impact of the ith element on the jth element is crucial for creating of a direct relation matrix. Fuzzy-based linguistic terms like "no influence (No)", "very low influence (VL)", "low influence (L)", "high influence (H)", & "very high influence (VH)" were included in a questionnaire that was designed to achieve this. After that, experts were asked to use this questionnaire to provide their opinions on the given criteria. Triangular Fuzzy Numbers (TFNs) were created by converting the linguistic responses and human evaluations that were gathered.

Fuzzy aggregation techniques including a defuzzification process are required as human judgments are inherently a degree of uncertainty. In the current study, the Converting Fuzzy Data into Crisp Scores (CFCS) defuzzification technique was applied to address this uncertainty. By using the CFCS technique, fuzzy ranges for maximum and minimum values were determined, resulting in a crisp score represented as a weighted average value.

Consider, $\tilde{N}_{ij}^{p} = \left(l_{ij}^{p}, m_{ij}^{p}, n_{ij}^{p}\right)$, where $p = 1,2,\ldots,r$, the mean influence of ith criterion over j^{th} criterion.

$$\tilde{N}^{n} = \begin{bmatrix} 0\,\tilde{N}_{12}^{p}\,\tilde{N}_{13}^{p}\ldots\tilde{N}_{1h}^{p}\,\tilde{N}_{21}^{p}\,0\,\tilde{N}_{23}^{p}\ldots\tilde{N}_{2h}^{p} \\ \ldots\ldots\ldots\tilde{N}_{h1}^{p}\,\tilde{N}_{h2}^{p}\,\tilde{N}_{h3}^{p}\ldots 0 \end{bmatrix}, \quad h = 1,2,\ldots,q \tag{11.3}$$

where r is the number of responders and h is the number of criteria.

The below mentioned algorithm can be used to explain the CFCS approach:

1. **Normalization**

$$Xl_{ij}^{p} = \frac{\left(l_{ij}^{p} - n_{ij}^{p}\right)}{\Delta_{\min}^{\max}} \tag{11.4}$$

$$Xm_{ij}^{p} = \left(\frac{m_{ij}^{p} - n_{ij}^{p}}{\Delta_{\min}^{\max}}\right) \tag{11.5}$$

$$n_{ij}^{p} = \frac{\left(n_{ij}^{p} - n_{ij}^{p}\right)}{\Delta_{\min}^{\max}} \tag{11.6}$$

where $\Delta_{\min}^{\max} = \max l_{ij}^{p} - \min n_{ij}^{p}$

2. **Compute the low (ls) and high (rs) normalized values**

$$x_{nkij}^{p} = \frac{x_{mij}^{p}}{1 + x_{mij}^{p} - x_{nij}^{p}} \tag{11.7}$$

$$x_{lkij}^{p} = \frac{x_{lij}^{p}}{1 + x_{lij}^{p} - x_{mij}^{p}} \tag{11.8}$$

3. **Compute the total normalized crisp values**

$$x_{ij}^{p} = \frac{\left[xnk_{ij}^{p}\left(1 - xnk_{ij}^{p}\right) + xlk_{ij}^{p} * xlk_{ij}^{p}\right]}{\left[1 - xnk_{ij}^{p} + xlk_{ij}^{p}\right]} \tag{11.9}$$

4. **Compute crisp values**

$$D_{ij}^{p} = \min n_{ij}^{p} + x_{ij}^{p*} \Delta_{\min} \tag{11.10}$$

5. **Integrate crisp values**

$$D_{ij} = \frac{1}{r}\left(D_{ij}^{1} + D_{ij}^{2} + - - - - - + D_{ij}^{r}\right) \tag{11.11}$$

Consequently, $D = [D_{ij}]$ is the average direct relation matrix generated.

11.2.1.2 Normalization of Direct Relation Matrix

Equations (11.12) and (11.13), when used in conjunction with the established average direct-relation matrix, make it simple to construct a normalized direct-relation matrix Z.

$$Z = \lambda^{*} D \tag{11.12}$$

$$\lambda = \min \left[\frac{1}{1 \le i \le h max \sum_{j=1}^{h} \left|D_{ij}\right|}, \frac{1}{1 \le i \le h max \sum_{i=1}^{h} \left|D_{ij}\right|} \right], \quad i, j = 1, 2, \ldots, h \tag{11.13}$$

11.2.1.3 Determination of Total Relation Matrix

Equation (11.14) can be used to build a total relation matrix easily.

The following process may be used to obtain the entire relation matrix $A_{h \times h}$:

$$A = \sum_{p=1}^{\infty} D_i = A + A^2 + A^3 + \text{-------} + A^p$$
$$= A\left(I + A + A^2 + A^3 + \text{---------} + A^{p-1}\right) \tag{11.14}$$
$$A = A\left(I - A\right)^{-1}$$

where I denotes the identity matrix.

The complete matrix's rows (T) and columns (U) are totaled up, and the resulting values of T and U are combined to get the value (T + U), also known as "Prominence." It provides the relevant criterion's relative relevance. Similarly, for every criterion, (T − U) values designated as 'Relation' are determined.

$$A = \left[a_{ij}\right]_{h \times h}, \quad i, j = 1, 2, \ldots, h \tag{11.15}$$

$$T = \left[\sum_{j=1}^{h} a_{ij}\right]_{h \times 1} = \left[a_i\right]_{h \times 1} \tag{11.16}$$

$$U = \left[\sum_{i=1}^{h} a_{ij}\right]_{1 \times h} = \left[a_j\right]_{1 \times h} \tag{11.17}$$

11.2.1.4 Casual Diagram Formation

The process of mapping all values of (T + U, T − U) results in the causal diagram. It provides insightful information for decision-making and demonstrates the relationships between the enablers under consideration. Criteria with positive (T − U) values belong in the cause group; in contrast, criteria with negative (T − U) values belong in the effect group.

11.3 Numerical Representation

After reviewing the literature and consulting with experts, significant digital practices in enhancing have been identified as: Production powered by robotics and material recovery (DP1), Supply chain visibility, traceability, and transparency (DP2), Social sustainability and legitimacy (DP3), Information technology proficiency (DP4), System for making decisions based on big data (DP5), Digitally updated procedures and quality control (DP6), Intelligent equipment for increased adaptability and large scale customization (DP7), Interoperability and system integration (DP8), and Using augmented reality, provide advice and support remotely (DP9) are listed using literature evidence in Table 11.1.

To start the analysis, four groups of experts belonging to the manufacturing sectors, humanitarian organizations, and academia were made for discussion on the considered digital practices. The experts have a relevant background in sustainable SCM, digital transformation, and humanized SC. Each group has five to seven experts from which data has been collected through questionnaires and brainstorming sessions, collected data in the form of influence ratings is shown in Table 11.2. Their respective opinions were then collected in linguistic form using Eq. (11.3). Expert judgment is crucial in MCDM processes, but biases can influence outcomes. To address potential biases, expert feedback is collected anonymously.

Because of spatial constraints and the intention to minimize repetitive table entries, the linguistic data collected from experts are not depicted here. Additionally, for analytical purposes, the linguistic terms acquired are transformed into corresponding TFNs, resulting in fuzzy inputs. These fuzzy inputs are then combined using the CFCS defuzzification method [Eqs. (11.4)–(11.10)], yielding the defuzzified direct-relation matrix for expert 1. This process is iterated for every linguistic input received from specialists, resulting in a defuzzified direct-relation matrix designed for every specialist individually. The average direct-relation matrix displayed below (Table 11.3) is then derived by combining the defuzzified crisp direct-relation matrices from each expert with the help of Eq. (11.11).

Normalizing an average direct relation matrix is the next phase in the f-DEMATEL technique, which is used to standardize the variation in the collected data. Equations (11.12) and (11.14) have been utilized to get a normalized direct relation matrix (Table 11.4).

A whole relation matrix was then formed in the next phase using Eq. (11.15), which is displayed below (Table 11.5). In addition, the values of "prominence" (T + U) & "relation" (T − U) have been calculated to examine the characteristics of the enablers under consideration. After that, the total relation matrix's rows (T) and columns (U) sums were calculated using Eqs. (11.16) and (11.17) to get the CSC enablers' (T + U) or "prominence" and (T − U) or "relation" values. The complete summary of these values is shown in Table 11.6. The computed values of the "prominence" (T + U) and "relation" (T − U) are used to generate a cause-and-effect diagram, displayed below (Fig. 11.1).

Table 11.2 Fuzzy-based linguistic scale for evaluation

Sr. no	Linguistic term	Influence score	TFNs
1	No influence (NO)	0	(0,0.1,0.3)
2	Very low influence (VL)	1	(0.1,0.3,0.5)
3	Low influence (L)	2	(0.3,0.5,0.7)
4	High influence (HL)	3	(0.5,0.7,0.9)
5	Very high influence (VH)	4	(0.7,0.9,1.0)

Source: Tyagi et al. (2015a)

Table 11.3 Average direct-relation matrix

	DP1	DP2	DP3	DP4	DP5	DP6	DP7	DP8	DP9
DP1	0.000	0.548	0.548	0.595	0.690	0.595	0.643	0.595	0.643
DP2	0.359	0.000	0.311	0.263	0.265	0.217	0.217	0.359	0.217
DP3	0.405	0.357	0.000	0.263	0.125	0.125	0.125	0.217	0.310
DP4	0.595	0.548	0.690	0.000	0.690	0.548	0.690	0.690	0.690
DP5	0.452	0.263	0.405	0.452	0.000	0.405	0.311	0.405	0.452
DP6	0.690	0.643	0.643	0.643	0.641	0.000	0.548	0.643	0.783
DP7	0.783	0.829	0.783	0.875	0.875	0.783	0.000	0.829	0.829
DP8	0.359	0.500	0.500	0.500	0.500	0.500	0.406	0.000	0.452
DP9	0.737	0.737	0.737	0.690	0.690	0.783	0.737	0.737	0.000

Table 11.4 Normalized direct-relation matrix

	DP1	DP2	DP3	DP4	DP5	DP6	DP7	DP8	DP9
DP1	0.000	0.083	0.083	0.090	0.105	0.090	0.098	0.090	0.098
DP2	0.054	0.000	0.047	0.040	0.040	0.033	0.033	0.054	0.033
DP3	0.061	0.054	0.000	0.040	0.019	0.019	0.019	0.033	0.047
DP4	0.090	0.083	0.105	0.000	0.105	0.083	0.105	0.105	0.105
DP5	0.069	0.040	0.061	0.069	0.000	0.061	0.047	0.061	0.069
DP6	0.105	0.098	0.098	0.098	0.097	0.000	0.083	0.098	0.119
DP7	0.119	0.126	0.119	0.133	0.133	0.119	0.000	0.126	0.126
DP8	0.054	0.076	0.076	0.076	0.076	0.076	0.062	0.000	0.069
DP9	0.112	0.112	0.112	0.105	0.105	0.119	0.112	0.112	0.000

Table 11.5 Total relation matrix

	DP1	DP2	DP3	DP4	DP5	DP6	DP7	DP8	DP9
DP1	0.159	0.236	0.243	0.238	0.257	0.227	0.224	0.245	0.247
DP2	0.123	0.073	0.121	0.109	0.112	0.098	0.094	0.125	0.104
DP3	0.121	0.116	0.067	0.100	0.084	0.078	0.075	0.097	0.108
DP4	0.248	0.243	0.270	0.162	0.264	0.227	0.236	0.264	0.260
DP5	0.170	0.146	0.170	0.168	0.108	0.155	0.137	0.166	0.170
DP6	0.263	0.258	0.266	0.253	0.260	0.153	0.221	0.260	0.274
DP7	0.312	0.320	0.323	0.319	0.328	0.294	0.176	0.322	0.317
DP8	0.172	0.192	0.197	0.188	0.193	0.179	0.160	0.122	0.184
DP9	0.286	0.288	0.296	0.276	0.284	0.275	0.259	0.290	0.185

11.4 Results and Discussion

Based on the (M + N) values shown in Table 11.6, the digital practices under consideration can be prioritized, whereby a higher (M + N) value corresponds to a greater importance rating. The observed importance rating sequence of each factor is as follows: DP7 > DP9 > DP6 > DP4 > DP1 > DP8 > DP5 > DP2 > DP3. Therefore, it is evident that digital practice, specifically 'Intelligent equipment for increased

Table 11.6 The score of digital practices and respective values for cause and effect groups

	M	N	M + N	M − N
DP1	2.077	1.856	3.933	0.221
DP2	0.961	1.873	2.834	−0.912
DP3	0.847	1.954	2.800	−1.107
DP4	2.174	1.814	3.988	0.360
DP5	1.390	1.890	3.279	−0.500
DP6	2.208	1.686	3.894	0.521
DP7	2.711	1.582	4.294	1.129
DP8	1.588	1.891	3.480	−0.303
DP9	2.440	1.849	4.289	0.591

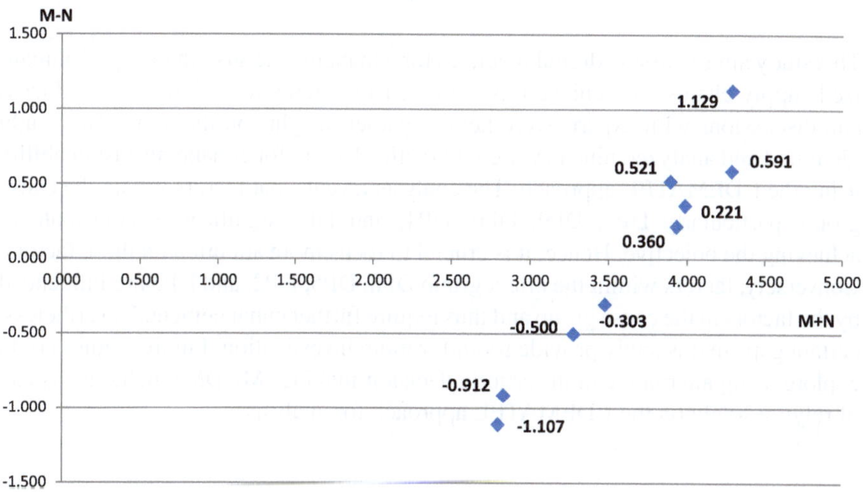

Fig. 11.1 Cause and effect diagram

adaptability and large scale customization (DP7)' is rated as having a higher importance rating (4.294) than the other practices. Consequently, '(DP7)' becomes essential digital practice for the enhancement of humanized SC.

Using a cause and effect diagram (Fig. 11.1) & depicting the (M − N) values, these practices are considered within their respective cause and effect groups namely as Intelligent equipment for increased adaptability and large-scale customization (DP7), Using augmented reality, provide advice and support remotely (DP9), Digitally updated procedures and quality control (DP6), Information technology proficiency (DP4), and Production powered by robotics and material recovery (DP1) have positive (M – N) values and falls in the cause category group. Conversely, practices such as Interoperability and system integration (DP8), System for making decisions based on big data (DP5), supply chain visibility, traceability and transparency (DP2), and social sustainability and legitimacy (DP3) have negative (M – N) values and belong in an effect category group.

Key factors in a cause group exert influence over factors within the effect group and among all factors, "Intelligent equipment for increased adaptability and large scale customization (DP7)" emerges as significant having the highest (M + N) value of 4.294. Additionally, it is identified as a particularly influential factor due to the highest (M − N) value of 1.129. Hence DP7 is considered as the most influential factor for enhancing the sustainability of HSC. While on the other side, "social sustainability and legitimacy (DP9)" factor is characterized by the lowest (M + N) and (M − N) values as 2.800 and −1.107 and is considered as the least influential factor respectively.

11.5 Conclusion and Future Scope

This study aims to assess digital practices for enhancing the sustainability of humanized supply chains. To achieve this, an extensive literature review was conducted and discussions with experts were held to gather insights on the topic. This study identified and analyzed nine key digital practice factors for enhancing sustainability using the f-DEMATEL approach. The analysis reveals that factors within the cause group, particularly DP7, DP9, DP6, DP4, and DP1 significantly contribute to achieving the objective. Hence, it is crucial to focus more attention on these factors. Conversely, factors within the effect group DP8, DP5, DP2, and DP3 are influenced by the factors in the cause group and thus require further enhancement. Nevertheless, certain gaps in this study provide room for more investigation. Future studies could explore using alternative multi-criteria decision-making (MCDM) methods instead of relying solely on the f-DEMATEL approach for analysis.

References

Agarwal V, Mathiyazhagan K, Malhotra S, Saikouk T (2022) Analysis of challenges in sustainable human resource management due to disruptions by Industry 4.0: an emerging economy perspective. Int J Manpow 43(2):513–541. https://doi.org/10.1108/IJM-03-2021-0192

Agarwal S, Tyagi M, Garg RK (2023) Conception of circular economy obstacles in context of supply chain: a case of rubber industry. Int J Product Perform Manag 72(4):1111–1153. https://doi.org/10.1108/IJPPM-12-2020-0686

Antikainen M, Uusitalo T, Kivikytö-Reponen P (2018) Digitalisation as an enabler of circular economy. Procedia CIRP 73:45–49. https://doi.org/10.1016/j.procir.2018.04.027

Asees Awan M, Ali Y (2019) Sustainable modeling in reverse logistics strategies using fuzzy MCDM: case of China Pakistan Economic Corridor. Manag Environ Qual 30(5):1132–1151. https://doi.org/10.1108/MEQ-01-2019-0024

Awan U, Sroufe R, Shahbaz M (2021) Industry 4.0 and the circular economy: a literature review and recommendations for future research. Bus Strateg Environ 30(4):2038–2060. https://doi.org/10.1002/bse.2731

Bag S, Telukdarie A, Pretorius JHC, Gupta S (2018) Industry 4.0 and supply chain sustainability: framework and future research directions. Benchmarking 28(5):1410–1450. https://doi.org/10.1108/BIJ-03-2018-0056

Bag S, Rahman MS, Srivastava G, Giannakis M, Foropon C (2023) Data-driven digital transformation and the implications for antifragility in the humanitarian supply chain. Int J Prod Econ 266:109059. https://doi.org/10.1016/j.ijpe.2023.109059

Bernstein ES (2017) Making transparency transparent: the evolution of observation in management theory. Acad Manag Ann 11(1):217–266. https://doi.org/10.5465/annals.2014.0076

Bressanelli G, Adrodegari F, Perona M, Saccani N (2018) Exploring how usage-focused business models enable circular economy through digital technologies. Sustainability 10(3):639. https://doi.org/10.3390/su10030639

Dubey R, Luo Z, Gunasekaran A, Akter S, Hazen BT, Douglas MA (2018) Big data and predictive analytics in humanitarian supply chains: enabling visibility and coordination in the presence of swift trust. Int J Logist Manag 29(2):485–512. https://doi.org/10.1108/IJLM-02-2017-0039

Dubey R, Bryde DJ, Foropon C, Tiwari M, Dwivedi Y, Schiffling S, Luo Z, Gunasekaran A, Akter S, Hazen BT, Douglas MA, Gupta S, Modgil S, Kumar A, Sivarajah U, Irani Z (2021) An investigation of information alignment and collaboration as complements to supply chain agility in humanitarian supply chain. Int J Prod Econ 59:1586–1605. https://doi.org/10.1016/j.ijpe.2022.108642

Elkington AR (1987) Book review. Eye 1(5):636–636. https://doi.org/10.1038/eye.1987.103

Fera M, De Padova V, Di Pasquale V, Caputo F, Caterino M, Macchiaroli R (2020) Workers' aging management—human fatigue at work: an experimental offices study. Appl Sci (Switzerland) 10(21):1–21. https://doi.org/10.3390/app10217693

Fragapane G, Ivanov D, Peron M, Sgarbossa F, Strandhagen JO (2022) Increasing flexibility and productivity in Industry 4.0 production networks with autonomous mobile robots and smart intralogistics. Ann Oper Res 308(1–2):125–143. https://doi.org/10.1007/s10479-020-03526-7

Gupta S, Modgil S, Gunasekaran A, Bag S (2020) Dynamic capabilities and institutional theories for Industry 4.0 and digital supply chain. Supply Chain Forum 21(3):139–157. https://doi.org/10.1080/16258312.2020.1757369

Gupta S, Modgil S, Kumar A, Sivarajah U, Irani Z (2022) Artificial intelligence and cloud-based collaborative platforms for managing disaster, extreme weather and emergency operations. Int J Prod Econ 254:108642. https://doi.org/10.1016/j.ijpe.2022.108642

Halvarsson Lundkvist A, Gustavsson M (2018) Conditions for employee learning and innovation—interweaving competence development activities provided by a workplace development programme with everyday work activities in SMEs. Vocat Learn 11(1):45–63. https://doi.org/10.1007/s12186-017-9179-6

Jaouadi MHO (2022) Investigating the influence of big data analytics capabilities and human resource factors in achieving supply chain innovativeness. Comput Ind Eng 168:108055. https://doi.org/10.1016/j.cie.2022.108055

Javaid M, Haleem A, Singh RP, Rab S, Suman R, Khan S (2022) Exploring relationships between Lean 4.0 and manufacturing industry. Ind Robot 49(3):402–414. https://doi.org/10.1108/IR-08-2021-0184

Jazdi N (2014) Cyber physical systems in the context of Industry 4.0. In: Proceedings of 2014 IEEE international conference on automation, quality and testing, robotics, AQTR 2014, pp 9–11. https://doi.org/10.1109/AQTR.2014.6857843

Kayikci Y, Kazancoglu Y, Gozacan-Chase N, Lafci C (2022) Analyzing the drivers of smart sustainable circular supply chain for sustainable development goals through stakeholder theory. Bus Strateg Environ 31(7):3335–3353. https://doi.org/10.1002/bse.3087

Koca G, Mathiyazhagan K (2020) An assessment of the barriers to social dimension of SSCM practice implementation using fuzzy DEMATEL approach: a case from Turkey. Ege Akademik Bakis (Ege Academic Review) 20(2):75–89. https://doi.org/10.21121/eab.729526

Kumar R, Singh SP, Lamba K (2018) Sustainable robust layout using big data approach: a key towards Industry 4.0. J Clean Prod 204:643–659. https://doi.org/10.1016/j.jclepro.2018.08.327

Lai Y, Sun H, Ren J (2018) Understanding the determinants of big data analytics (BDA) adoption in logistics and supply chain management: an empirical investigation. Int J Logist Manag 29(2):676–703. https://doi.org/10.1108/IJLM-06-2017-0153

Nascimento DLM, Alencastro V, Quelhas OLG, Caiado RGG, Garza-Reyes JA, Lona LR, Tortorella G (2019) Exploring Industry 4.0 technologies to enable circular economy practices in a manufacturing context: a business model proposal. J Manuf Technol Manag 30(3):607–627. https://doi.org/10.1108/JMTM-03-2018-0071

Nobre GC, Tavares E (2017) Scientific literature analysis on big data and internet of things applications on circular economy: a bibliometric study. Scientometrics 111:463–492. https://doi.org/10.1007/s11192-017-2281-6

Panghal A, Pan S, Vern P, Mor RS, Jagtap S (2024) Blockchain technology for enhancing sustainable food systems: a consumer perspective. Int J Food Sci Technol 59(5):3461–3468. https://doi.org/10.1111/ijfs.16835

Pham TT, Kuo TC, Tseng ML, Tan RR, Tan K, Ika DS, Lin CJ (2019) Industry 4.0 to accelerate the circular economy: a case study of electric scooter sharing. Sustainability (Switzerland) 11(23):1–16. https://doi.org/10.3390/su11236661

Rajput S, Singh SP (2020) Industry 4.0 model for circular economy and cleaner production. J Clean Prod 277:123853. https://doi.org/10.1016/j.jclepro.2020.123853

Rosa P, Sassanelli C, Urbinati A, Chiaroni D, Terzi S (2020) Assessing relations between Circular Economy and Industry 4.0: a systematic literature review. Int J Prod Res 58(6):1662–1687. https://doi.org/10.1080/00207543.2019.1680896

Shah N, Irani Z, Sharif AM (2017) Big data in an HR context: exploring organizational change readiness, employee attitudes and behaviors. J Bus Res 70:366–378. https://doi.org/10.1016/j.jbusres.2016.08.010

Sharma J, Tyagi M, Kazançoğlu Y (2024) Impact of digital technologies on the risk assessment in food supply chain: a wake towards digitalisation. Int J Food Sci Technol 59(5):3491–3504. https://doi.org/10.1111/ijfs.17035

Shayganmehr M, Kumar A, Garza-Reyes JA, Moktadir MA (2021) Industry 4.0 enablers for a cleaner production and circular economy within the context of business ethics: a study in a developing country. J Clean Prod 281:125280. https://doi.org/10.1016/j.jclepro.2020.125280

Tsang YP, Yang T, Chen ZS, Wu CH, Tan KH (2022) How is extended reality bridging human and cyber-physical systems in the IoT-empowered logistics and supply chain management? Internet of Things (Netherlands) 20:100623. https://doi.org/10.1016/j.iot.2022.100623

Tyagi M, Kumar P, Kumar D (2015a) Analysis of interactions among the drivers of green supply chain management. Int J Bus Perform Supply Chain Model 7(1):92–108. https://doi.org/10.1504/IJBPSCM.2015.068137

Tyagi M, Kumar P, Kumar D (2015b) Parametric selection of alternatives to improve performance of green supply chain management system. Procedia Soc Behav Sci 189:449–457. https://doi.org/10.1016/j.sbspro.2015.03.197

Yu W, Zhao G, Liu Q, Song Y (2021) Role of big data analytics capability in developing integrated hospital supply chains and operational flexibility: an organizational information processing theory perspective. Technol Forecast Soc Chang 163:120417. https://doi.org/10.1016/j.techfore.2020.120417

Chapter 12
Revolutionizing Supply Chain Resilience: Life Cycle Assessment and Sustainable Decision-Making Strategies in Beverage Industry 5.0

Meenu Singh, Millie Pant, and Václav Snášel

Abstract The significance of packaging extends beyond mere aesthetic appeal, serving as a protective measure for products during transit and storage. The choice of packaging materials, ranging from glass and paper to tin and plastic, is product-specific. Nonetheless, the ecological footprint left by packaging waste has emerged as a pressing issue. In response to increasingly stringent environmental regulations, there is a mounting demand for eco-friendly packaging solutions. This study introduces a novel method for selecting sustainable packaging materials, integrating Life Cycle Assessment (LCA) with Multi-criteria Decision Making (MCDM) techniques. The methodology involves a dual-phase analysis: first, evaluating eight environmental impact categories through the Impact 2002++ method, and subsequently, utilizing MCDM to assess the performance of nine different packaging materials. The findings highlight the utility of this integrated approach in fostering sustainable practices and bolstering supply chain robustness within the framework of Industry 5.0.

Keywords Sustainable packaging · LCA · AHP · TOPSIS · Green packaging material selection · Supply chain resilience · Industry 5.0

M. Singh (✉) · V. Snášel
Faculty of Electrical Engineering and Computer Science, VŠB-Technical University of Ostrava, Ostrava, Czech Republic
e-mail: meenu.singh@vsb.cz; vaclav.snasel@vsb.cz

M. Pant
Department of Applied Mathematics and Scientific Computing, Indian Institute of Technology, Roorkee, Uttarakhand, India

Mehta Family School of Data Science and Artificial Intelligence, Indian Institute of Technology, Roorkee, Uttarakhand, India
e-mail: pant.milli@as.iitr.ac.in

K. Mathiyazhagan et al. (eds.), *Humanizing Circular Supply Chain Management*, Circular Economy and Sustainability,
https://doi.org/10.1007/978-3-031-86562-6_12

12.1 Introduction

Over the last few decades, the impact of human activities on the environment has significantly increased, leading to numerous environmental problems on local, regional, and global scales leading to a focus towards environmentally friendly products (Saleh 2016). According to the Central Pollution Control Board (CPCB) report, India produces approximately 707 million metric tons of plastic every year, and most of it is used toward packaging purposes. While India's per capita consumption of plastic is 9.7 kg, which is lower than the global average. However, the demand for plastic in packaging keeps on increasing.

New regulations and environmental standards are being set up by the governments worldwide to minimize carbon footprints (CPCB 2018). This has put pressure on industries to design their products in accordance with minimum customers' needs and environmental damage. Emphasis is given to the designing of goods, which have provisions for refurbishment, remanufacturing, or recycling to generate less waste and pollution (Aryan et al. 2019). One of the effective approaches to curb the environmental threats is to evaluate the total life cycle of a product, from the time of raw material extraction to transportation and disposal (Saleh 2016).

Life Cycle Assessment (LCA) is a valuable tool for analyzing the environmental impact of products or services (Aryan et al. 2019; Ye et al. 2017). LCA follows a very thorough, all-embracing approach of "system boundary" in assessing the full life cycle of a product, as is illustrated in Fig. 12.1, from raw material extraction to the process of disposal. LCA thus became relevant on the horizon as a useful technique for assessing a product in terms of its environmental impacts. Most packaging materials have devastating negative environmental impacts during their production and treatment in terms of depleting natural resources, generating hazardous wastes,

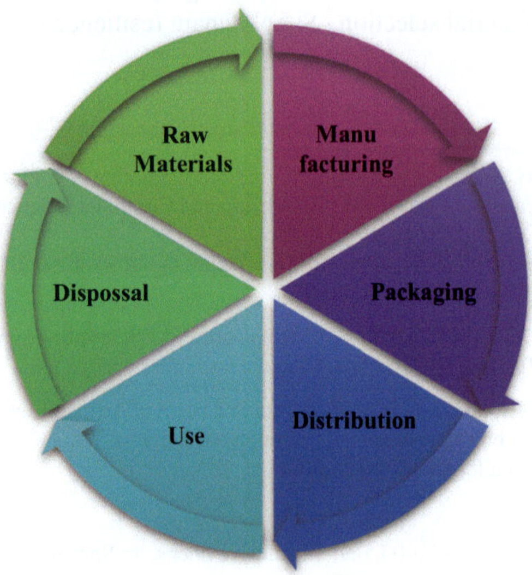

Fig. 12.1 Life cycle stages: packaging material

and emitting greenhouse gases to the atmosphere (Amienyo and Azapagic 2016; Saleh 2016; Ye et al. 2017). This calls for genuine solutions to create a sustainable packaging material that reduces environmental impacts throughout the life cycles of the materials, considering the scarcity and price rise of natural resources in India. The method had an accurate evaluation on the environmental impacts of different packaging materials serving the same function but having various environmental effects at the end of their life cycle period. Particularly, beverage packaging materials are amongst the very vital waste management concerns, due to large volumes of their generation.

Packaging substitutes for milk, juices, and water are made by four types of packaging: glass, metal cans, PET, and cartons. Each one has different environmental impacts during manufacture, processing, transport, storage, and use (Aryan et al. 2019; Biona et al. 2015; Marathe et al. 2019; Yadav et al. 2024). Beverage containers help in protecting salvation quality, preventing contamination, and facilitating transportation, yet imposing significant environmental challenges if not managed in a sustainable way. Packaging materials such as PET, LDPE, HDPE, and PP are dominant and thus, contribute significantly towards environment burdens such as greenhouse gas emissions and energy consumption (CPCB 2018).

This study aims to compare the environmental footprints of various beverage packaging materials prevalent in India by using LCA. Recognizing that LCA might not provide a definitive ranking on its own, the research proposes a comprehensive evaluation method that integrates LCA with Multi-criteria Decision Making (MCDM) techniques. The methodology is applied to nine types of packaging materials commonly used in the Indian beverage industry: PET Bottles, Glass Bottles, Paper Bags, Jute Bags, HDPE Bags, PP Cups, Paper Cups, PP Woven Sacks, and LDPE Pouches. The assessment adheres to the standards set by the International Organization for Standardization (ISO) (CPCB 2018). The environmental impact categories identified through the Impact 2002+ method is further prioritized using the MCDM approach, specifically the Analytic Hierarchy Process (AHP) coupled with the Technique for Order of Preference by Similarity to Ideal Solution (TOPSIS). This integrated method facilitates a more accurate and holistic ranking of the packaging materials based on their sustainability. Its rankings are also compared with the other six MCDM methods in order to validate the obtained ranking results.

The synergy of LCA with MCDM enhances the analysis by incorporating both qualitative and quantitative perspectives, accommodating diverse client needs, and substantially diminishing uncertainty in the outcomes for sustainable decision-making. This integrated approach supports the identification of eco-friendly packaging solutions, fostering resilience in the supply chain. The study's findings, grounded in actual CPCB data from Indian beverage industries, offer stakeholders actionable insights for making environmentally conscious choices that reflect the current ecological landscape.

The chapter proceeds as follows: Sect. 12.2 reviews pertinent literature on the application of LCA across various packaging types and products. The methodology employed is described in Sect. 12.3. The results derived from the Life Cycle Impact Assessment (LCIA) and the combined LCA-MCDM analysis, are compared and discussed in Sect. 12.4. Managerial implications are presented in Sect. 12.5. Finally, the conclusion is encapsulated in Sect. 12.6.

12.2 Literature Review

LCA has demonstrated its usefulness in various industrial applications by evaluating the sustainability of different products. The company also considered recycling of their returnable bottles with the disposable cans and was doing study on the new packaging materials for the bottles of their soft drinks. Due to the increasing concern of the environment in the mind of the consumer Coca-Cola is also looking for the new packaging options (Coca-Cola Sustainable Packaging 2018). In a life cycle study, plastic bottles utilize fewer hydrocarbons than glass bottles, despite plastics being made up from petroleum products (Yadav et al. 2024). This unexpected finding provides confidence to the companies to move toward the use of plastic bottles over glass ones.

Mata and Costa (2001) conducted an LCA study on glass to measure the environmental impacts of the returnable and non-returnable glass beer bottles, using the data from a Portuguese beer manufacturer. They assessed potential ecological effects of emissions across various impact categories. According to a study by Banar and Cokaygil (2008), the environmental impact of glass bottles is higher than that of cartons, as determined using SimaPro LCA software (Speck et al. 2016).

Packaging in the Indian Context In India, packaging is crucial for product distribution and preservation. According to a CAGR forecasting report, packaged water bottles are projected to become the most packed product unit by 2030, surpassing milk, baked goods, and cigarettes. PET is the most efficient material for packaging, with its production rising significantly from 982 KT in 2014–15 to 1458 KT in 2015–16. Major PET manufacturers in India include Reliance Industries Limited and Dhunseri Petrochem and Tea Limited. In 2015–16, approximately 650 KT of PET was exported to countries such as the UAE, Bangladesh, Israel, the United States, Ukraine, and Italy. The volume of exports has increased in recent years, with production levels closely monitored. PET bottles are widely used for packaging various products, including milk, juice, water, and food (Marathe et al. 2019).

12.3 Case Study

This study evaluates the environmental impacts of nine packaging materials (PET Bottle, Glass Bottle, Paper Bag, Jute Bag, HDPE Bag, PP Cups, Paper Cups, PP woven Sacks, and LDPE Pouches) across their entire life cycle, including production, transportation, use, and disposal (CPCB 2018). The environmental impact categories considered include water consumption, energy consumption, human toxicity, and global warming potential, among others, as identified according to the Impact 2002++ method (Owsianiak et al. 2014, p. 200).

The study applies a functional unit of one Ton of packaging material as the common factor for all packaging materials. Due to unavailability data related to energy requirement, some assumptions are considered in this study align with relevant literature and industry reports in the context of India. These assumptions were made based on standard practices found in CPCB's LCA report of plastic products (CPCB 2018), that underscores the necessity of considering the full lifecycle of plastic products for estimating energy consumption:

- Recycling energy is estimated at 10–20% of the cumulative energy demand (CED).
- Feed stock energy of the product is estimates at 20–50% of CED.
- Process energy is estimated at 50–80% of the gross energy demand (GED), which is the addition of process energy and feed stock.
- Energy for use, including transport energy, is assumed at 5–10% of the feedstock energy.

These assumptions are highly critical as the accuracy for the percentage of energy required for different products differs at different stages of production and varies with size and type of products (CPCB 2018). By adopting these assumptions rooted in the report, the present study shall be able to efficiently estimate the amount of energy used by various packaging materials. It is imperative to note that these assumptions would have some degree of uncertainty, which could affect the results. Additionally, as indicated in the CPCB report, plastic packaging materials, especially its CO_2 emissions, depend on the type and recyclable character of plastic. For instance, LCA values for LDPE and PP are lower than those materials like paper or jute, which further strengthens the objective set in this paper to be adequately consumed with comprehensive energy consumption analysis.

Here, in this study, nine packaging materials are ranked based on their environmental impact scores, calculated using a hybrid approach combining LCA and MCDM. The study's scope and system boundary are defined in Fig. 12.2, and the framework for selecting the most sustainable packaging material is presented in Fig. 12.3.

12.4 Results and Discussion

12.4.1 Life Cycle Inventory (LCI) Analysis

The LCI results show the environmental burdens associated with 1 ton of different packaging materials over their entire life cycle. The data was collected and analyzed using EXCEL software, and the life cycle model was developed based on estimated and measured values of input and output data.

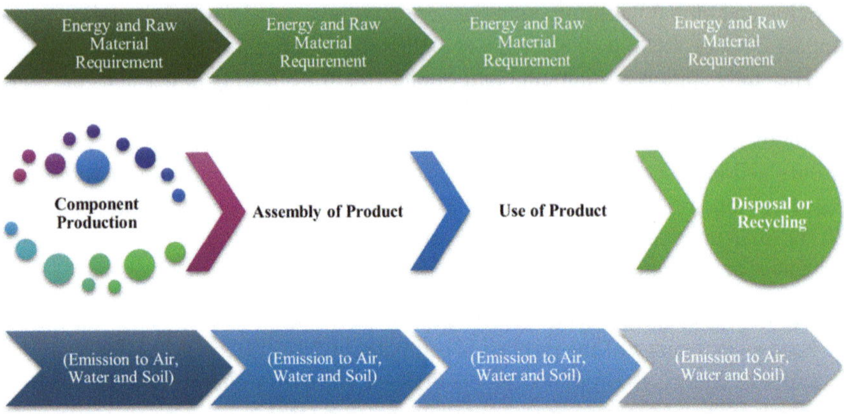

Fig. 12.2 Environmental impact of packaging material across its life cycle (CPCB 2018)

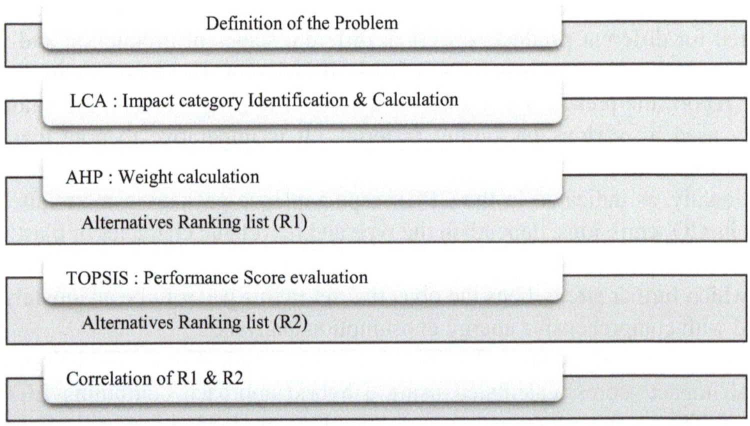

Fig. 12.3 Framework for selection of sustainable packaging materials

Table 12.1 includes water and energy consumption, material usage, and emissions & wastes generated during the production, transportation, use, and disposal stages. The results of different packaging materials can be used to compare their environmental impacts and to identify opportunities for reducing the environmental footprint of packaging (CPCB 2018).

Table 12.1 LCI results for 1 ton of packaging materials

S. no.	Environmental burden	Unit	PET bottle	Glass bottle	Paper bags	Jute bag	HDPE bags	PP cups	Paper cups	PP woven sacks	LDPE pouches
1	Energy consumed	MJ	31,651.2	37,465.2	98,388	26,366.4	16,214.4	31,230	93,924	17,326.8	15,285.6
2	Water consumed	L	7529	46,728	381,000	101,520	3069	287,917	381,000	113,250	3005
Atmospheric emissions released											
3	Dust (silica particles)	kg	0.71	1.353	5.727	0.067	177*10^(−6)	73.55*10^(−6)	8.71	177*10^(−6)	0.0257
4	CO	kg	20.18	0.45	0.36	0.054	0.0029*10^(−6)	0.0029*10^(−6)	0.36	0.002946*10^(−6)	0.497
5	CO_2	kg	1001	1580	4989	6.61	702.64	689.1	4989	702.64	702.64
6	SO_2	kg	4.5553	13.47	27.85	0.134	0.0244*10^(−6)	0.0138*10^(−6)	37.78	0.024403*10^(−6)	0.1091
7	NO_2	kg	0.5604	8.4	13.397	0.068	0.2978*10^(−6)	0.3067*10^(−6)	13.397	0.2976*10^(−6)	0.639

12.4.2 Life Cycle Impact Assessment (LCIA)

The goal of LCIA is to evaluate the environmental impact of packaging materials. The selected impact categories include global warming potential, respiratory effects, human toxicity, photochemical oxidation, terrestrial acidification, eco cost, water consumption, solid waste, and energy consumption. Eco-cost is not a part of the IMPACT 2002+ categories, but it was chosen due to its significance and the availability of its LCI data. Using SimaPro and GaBi software, LCA studies were conducted to model 1 ton of four materials to determine their environmental impact (Speck et al. 2016).

- **Classification:** Here, the midpoint impact categories are defined and the LCI exchanges are classified into these categories. Table 12.2 below presents the classification of the LCI results to the midpoint impact categories.
- **Characterization and Normalization:** Here, the inventory data is quantitatively modeled and integrated with the impact categories to express their environmental significance. By definition given by Huang et al. (2009), the characterization stage can be expressed as the product of inventory data with the characterization factors of each impact category. The contribution of each inventory data is quantified at this step, to the respective impact categories. The normalization phase determines the relative contribution of each inventory data to each impact category, bringing all categories to a common reference point. This step enables a better understanding of the relative significance of each impact category. For the nine selected materials, the summary of characterization of the impact categories is presented in Table 12.3.

12.4.2.1 Discussions

The study reveals significant disparities in the environmental impact of various packaging materials. Key findings from Table 12.3 are discussed below, and the corresponding values are highlighted in bold in the table:

- **Respiratory Effects**: CO_2 and dust emissions are highest for PET bottles.
- **Human Toxicity**: Primarily attributed to HDPE, PP cups, and PP woven sacks, with PET bottles contributing the most.
- **Terrestrial Acidification**: High for paper bags and cups, negligible for PP cups, HDPE, and PP woven sacks.
- **Photo-oxidant Formation**: Paper cups are the highest contributors, while PP cups, HDPE, PP woven sacks, glass bottles, and jute bags contribute minimally.
- **Eco-cost**: Paper has the highest eco-cost; HDPE the lowest, followed by LDPE and PP woven sacks.
- **Water Intensity**: Paper bags (0.281 gal/$) are the highest, LDPE pouches the lowest (0.033 gal/$).

Table 12.2 LCI results classified to midpoint impact categories[a]

	Midpoint impact category							
	Global warming potential (kg CO_2 equiv.)	**Respiratory effects (kg)**	**Human toxicity (kg 1,4-dichlorobenzene eq.)**	**Photo chemical oxidation (kg ethylene equiv.)**	**Terrestrial acidification (kg SO_2-equiv.)**	**Eco-cost**	**Water consumption (L)**	**Energy consumption (MJ)**
Energy								X
Water							X	
Atmospheric emissions released								
Damage category	**Climate change (life support system)**	**Human health**	**Human health**	**Human health**	**Ecosystem quality**	**Ecosystem quality, resources**	**Ecosystem quality, resources, human health**	**Resources**
Dust		X						
CO		X	X	X				
CO_2	X					X		
SO_2			X	X	X			
NO_2	X		X	X	X			

[a] (X) presents the connection along with damage categories for each category

Table 12.3 Characterized impact categories

Alternatives	Impact category: Global warming potential	Respiratory effects	Photo oxidant formation	Terrestrial acidification	Human toxicity	Eco-cost	Water consumption	Energy consumption
PET bottle	1167.9992	**20.89**	0.7792056	4.94758	**49.5417888**	996.8798088	7529	31,651.2
Glass bottle	26.874	0.121	0.009794	0.1816	**0.224064**	996.9803028	46,728	37,465.2
Paper bag	**8981.306**	6.087	1.721636	**37.2279**	19.614	3821.135742	381,000	**98,388**
Jute bag	26.874	0.121	0.009794	0.1816	**0.224064**	702.7067196	101,520	26,366.4
HDPE bag	702.6400887	**0.0001770029**	0.0000000095879	**0.00000023286**	**0.0000003666624**	**524.7650136**	3069	16,214.4
PP cups	689.1000914	**0.0000735529**	0.0000000093283	**0.00000022849**	**0.0000003763248**	921.0607323	287,917	31,230
Paper cups	**8981.306**	9.07	**2.198276**	**47.1579**	20.56728	**3702.777246**	381,000	93,924
PP woven sacks	702.6400887	**0.000177002946**	0.000000009583686	**0.000000232723**	**0.000000366533088**	**554.2591872**	113,250	17,326.8
LDPE pouches	893.062	0.5227	0.0365478	0.5564	1.9700736	**525.8457684**	3005	15,285.6
Total	22,171.80147	36.81212756	4.755253428	90.25298069	92.14127151	12,746.41052	1,325,018	367,851.6

- **Energy Intensity**: Paper bags are the most energy-intensive; LDPE pouches the least.

Overall, HDPE emerges as the most environmentally sustainable packaging material, closely followed by LDPE. This underscores the need to select packaging materials based on comprehensive LCA impacts to ensure functional and environmental sustainability.

12.4.3 MCDM Methods

In this research, MCDM methods are utilized with a dual focus: (1) to assess the significance of various criteria and (2) to evaluate the performance of different alternatives (Singh and Pant 2021). The said objectives will lead to selecting the most sustainable packaging material. The AHP is initially employed to weigh the alternatives based on their relative importance. These weights are, therefore, fed into TOPSIS to compute performance scores for each of the packaging materials. A comparison between the two results is then analyzed in view of the decision-making purposes.

12.4.3.1 With AHP

AHP is utilized to assign weights to each category of environmental impact (Singh et al. 2018). The overall environmental impact scores for the packaging materials are depicted in Fig. 12.4, which has been validated through discussions with other experts and researchers. The pairwise comparison matrix is studied for consistency,

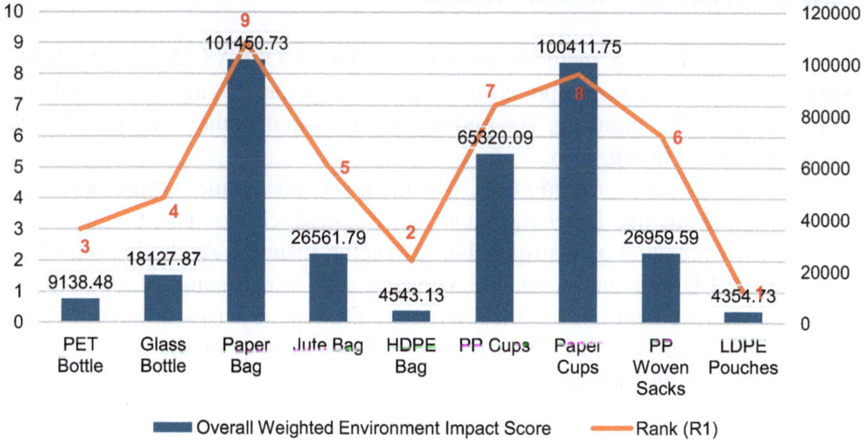

Fig. 12.4 Rank (R1) and Overall Environmental Impact Score of packaging materials by LCA-AHP

yielding a Consistency Ratio (CR) of 0.0178, which falls below the acceptable threshold of 0.10, indicating a reliable set of weights.

The findings indicate that LDPE pouches and HDPE bags have the lowest impact scores, with values of 4354.73 and 4543.13, respectively, placing them first and second in terms of environmental performance. Following these, PET bottles and glass bottles are ranked with scores of 9138.48 and 18,127.87, respectively. Conversely, paper bags and paper cups exhibit the highest impact scores at 101,450.73 and 100,411.75, respectively, making them the most environmentally detrimental options and thus ranked lowest for preference in sustainable packaging.

12.4.3.2 With AHP-TOPSIS

These weights obtained from AHP serve as input to TOPSIS for evaluating the performance scores of individual packaging materials (Singh et al. 2020a, b). The decision matrix considered for TOPSIS method presents the characterized impact categories of packaging materials. The values undergo normalization and are subsequently multiplied by the AHP weights for the creation of a weighted, normalized decision matrix. Table 12.4 expresses the performance scores and ranks (R2) of each packaging material. Performance scores for the different packaging materials, along with their ranks against the scores, are shown in Fig. 12.5.

Notably, Paper Bags demonstrate the lowest efficiency at 2.3%, while HDPE exhibits the highest at 15.5% among all materials. Both Paper Bag and Paper Cups exhibit similarly low efficiency due to their high-water consumption. Further comparisons of each packaging material are visualized in Fig. 12.5, illustrating their numerical efficiency.

12.4.3.3 With Other MCDM Methods

The other six MCDM methods such as VIKOR, COPRAS, MOORA, WSM, MABAC and EDAS were applied to compare with the ranking results obtained from AHP-TOPSIS approach. The ranking results from the other six MCDM methods are presented in Fig. 12.6.

The rankings across different MCDM methods show some consistency, especially for materials like Paper Bags and Jute Bags, which are ranked higher, highlighting their potential as a sustainable packaging material. At the bottom of the rank are the LDPE Pouches and PP Woven Sacks, which could further indicate scope for improvement on sustainability ends. The results across different MCDM methods effectively validate the AHP-TOPSIS approach, reinforcing the accuracy and reliability of the ranking results.

Table 12.4 The ideal worst (S^-), ideal best (S^+), overall performance score (P_i) and rank of packaging materials

Alternatives	Global warming potential (kg CO_2 equiv.)	Respiratory effects (kg)	Photo oxidant formation (kg ethylene equiv.)	Human toxicity (kg 1, 4-dichlorobenzene eq.)	Terrestrial acidification (kg SO_2-equiv.)	Water consumption (m^3)	Eco-cost ($)	S^-	S^+	Performance score (P_i)	Rank (R2)
PET bottle	0.01717885138	0.04429622603	0.01413726553	0.05079898834	0.003299045365	0.002398147959	0.03156270175	0.2161525602	0.07652972115	0.7385	**6**
Glass bottle	0.0003952609316	0.0002565745979	0.0001776942806	0.0002297499707	0.0001210908441	0.01488387008	0.03156588354	0.2280200908	0.03889279502	0.8543	**5**
Paper bag	0.1320964269	0.01290718659	0.03123594757	0.0201117356	0.02482355636	0.1213566705	0.1209828574	0.04522836508	0.2454974141	0.1556	**8**
Jute bag	0.0003952609316	0.0002565745979	0.0001776942806	0.0002297499707	0.0001210908441	0.03233629709	0.02224874294	0.2315571857	0.03591463959	0.8657	**3**
HDPE bag	0.01033438178	0.0000003753260156	0.0000000001739549717	0.0000000003759670257	0.0000000001552710019	0.0009775423145	0.0166148431	0.2497245679	0.01003527556	0.9614	**1**
PP cups	0.01013523644	0.0000001559653367	0.0000000001692450029	0.000000000385874624	0.0000000001523570868	0.09170773886	0.0291621567	0.2033372897	0.09515170782	0.6812	**7**
Paper cups	0.1320964269	0.01923249258	0.03988371171	0.02108920655	0.03144487839	0.1213566705	0.1172354509	0.03961390046	0.2431560021	0.1401	**9**
LDPE pouches	0.01313509407	0.001108359854	0.0006630932234	0.002020067266	0.0003710074099	0.000957156942	0.01664906141	0.2483020159	0.01296883357	0.9504	**2**
PP woven sacks	0.01033438178	0.0000003753261132	0.0000000001738785164	0.0000000003758344322	0.0000000001551796504	0.03607255364	0.01754867263	0.2335516423	0.03663363832	0.8644	**4**

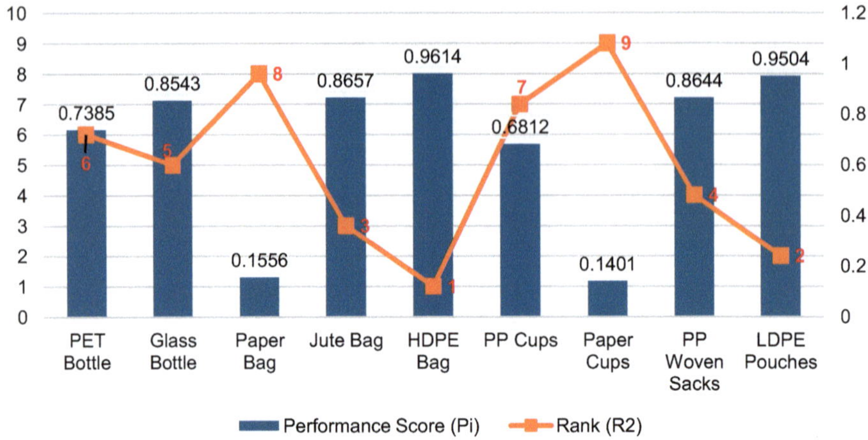

Fig. 12.5 Performance scores and Rank (R2) of packaging materials by AHP-TOPSIS

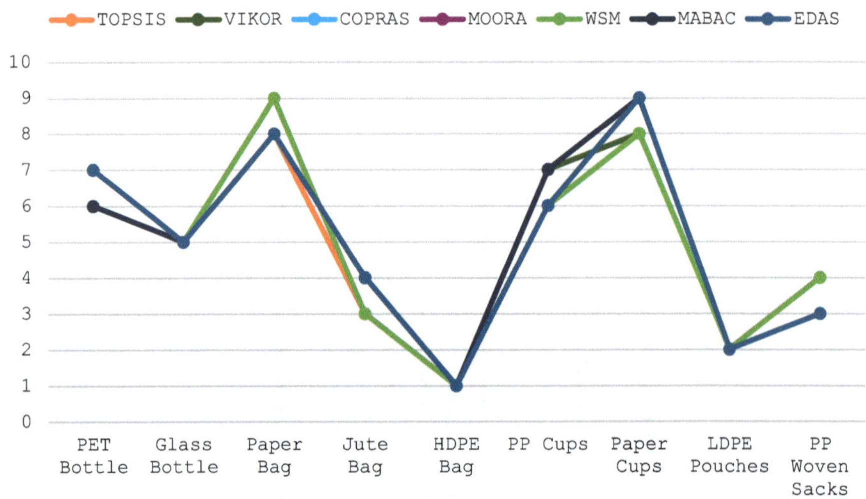

Fig. 12.6 Ranking results by other MCDM methods

12.4.4 Statistical Analysis

The ranking results of the packaging materials were then compared using Spearman's correlation coefficient for different MCDM methods (Singh et al. 2020a). This statistical method was employed to investigate the association between the overall performance score and validate AHP-TOPSIS results. The heatmap, depicted in Fig. 12.7, reveals a significant positive correlation between the rankings of all the MCDM methods, with a highest coefficient of 1.000.

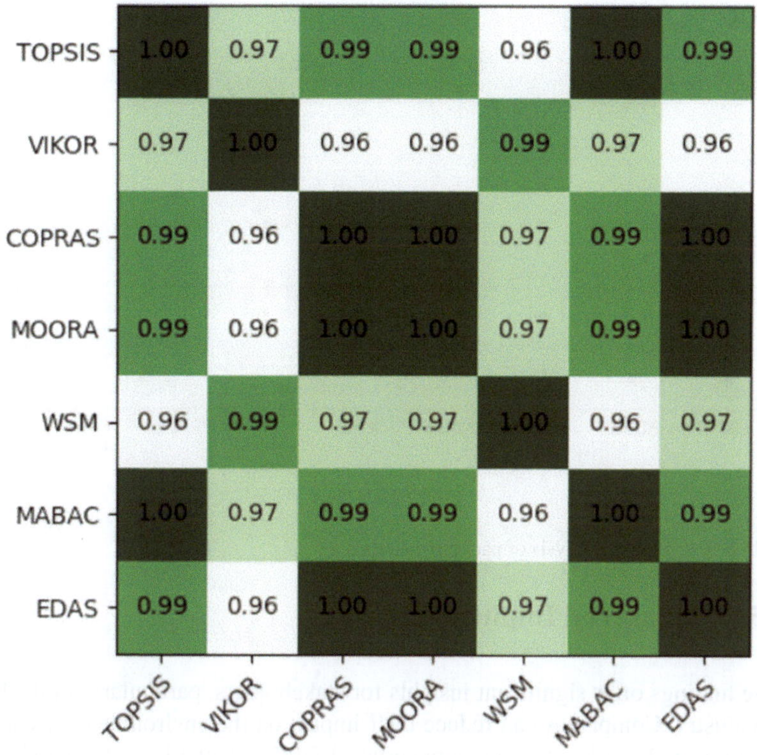

Fig. 12.7 Correlation analysis MCDM methods

The rankings derived from LCA and those from AHP-TOPSIS, denoted as R1 and R2, were also statistically compared to investigate the association between the overall performance score and the overall weighted environmental impact score. The scatter chart, depicted in Fig. 12.8, reveals a significant positive correlation, with a coefficient of 0.81667. This finding suggests a robust concordance between both ranking approaches, implying that packaging materials with greater environmental impact scores generally correspond to lower performance scores.

While LCA and MCDM methods offer comprehensive evaluations, validation of these results through real-world data is yet a challenge due to limited data availability in the Indian context. Some preliminary form of validation could be cross-validation with data of international case studies or similar industries. Additionally, expert opinion could also assist as an indirect validation approach for the MCDM process. Direct validation through real-world data is not feasible, this limitation should be acknowledged. The outcomes of these LCAs depict the environmental burdens in 1 ton of different packaging materials throughout their life cycles. It can be used to compare the results obtained with the industry benchmarks to contextualize the findings. These comparisons then validate the model while demonstrating how alternative packaging materials perform relative to industry norms.

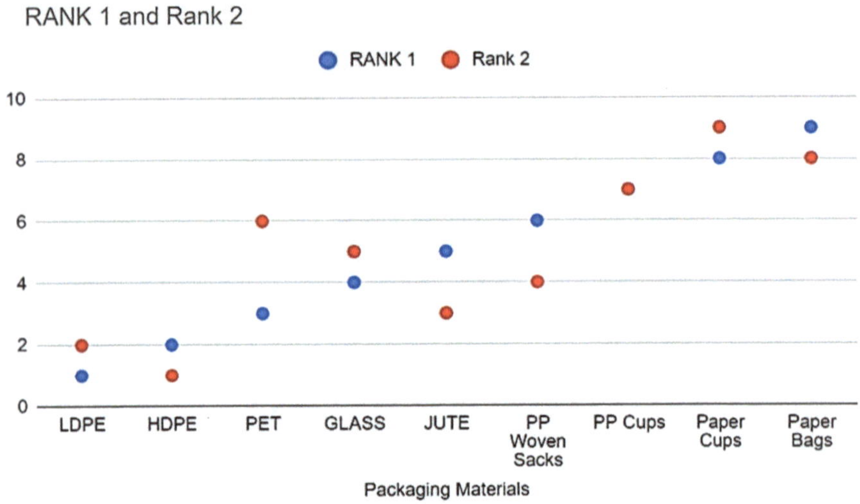

Fig. 12.8 Correlation analysis of packing materials

12.5 Managerial Implications

These findings offer significant insights for stakeholders, particularly in the beverage industry. Companies can reduce their impact on the environment by choosing more impacts-lowering packaging materials. LDPE and HDPE are found within the range of the lowest environmental impacts, while others form a wide range of impacts. Stakeholders are encouraged to transition to these materials, which are more amendable with cost-effective and are environmentally friendly alternatives. In addition, the results do indeed highlight the appropriateness of making sustainability a consideration in the purchase decision, not just on environmental grounds but also on long-term cost saving and compliance with increasingly stringent environmental regulations. These insights, therefore, can be leveraged by companies in their strategies on supply chains to help in enhancing resilience and future-proof their operations against evolving environmental standards.

12.6 Conclusion and Future Considerations

This study evaluates the energy input/output data of nine packaging materials used in the Indian beverage industry, considering environmental impacts. Here, LCA is integrated with MCDM for total sustainability of packaging decisions, covering environmental, economic, and social concerns. Some concluding remarks for the above study are:

1. The integration of LCA with AHP-TOPSIS provides a robust framework for sustainable packaging decisions.
2. Sustainability encompasses not only environmental but also socio-economic aspects.
3. The proposed approach ensures consideration of these factors, accommodating diverse client perspectives and reducing decision-making uncertainties.
4. This approach supports business preferences related to eco-packaging and material selection.
5. The study demonstrates a strong correlation (0.816) between overall weighted environmental impact scores and performance scores.
6. Decision-makers can utilize either ranking method confidently.
7. The findings offer valuable insights for stakeholders in the beverage packaging industry to enhance sustainability in material choices.

Future research could expand to include more social dimensions, ensuring a comprehensive evaluation in varying environmental conditions. Exploration of fuzzy environments and incorporation of longitudinal data analysis could be considered in future studies. Fine-tuning of the data regarding energy consumptions could be considered to observe the effect on overall environmental impact results. Sensitivity analysis could be performed to assess the robustness of the model under different scenarios.

References

Amienyo D, Azapagic A (2016) Life cycle environmental impacts and costs of beer production and consumption in the UK. Int J Life Cycle Assess 21(4):492–509. https://doi.org/10.1007/s11367-016-1028-6

Aryan Y, Yadav P, Samadder SR (2019) Life cycle assessment of the existing and proposed plastic waste management options in India: a case study. J Clean Prod 211:1268–1283. https://doi.org/10.1016/j.jclepro.2018.11.236

Banar M, Cokaygil Z (2008) A comparative life cycle analysis of two different juice packages. Environ Eng Sci. https://www.liebertpub.com/doi/abs/10.1089/ees.2007.0079

Biona JBMM, Gonzaga JA, Ubando AT, Tan HC (2015) A comparative life cycle analysis of plastic and paper packaging bags in the Philippines. In: 2015 International conference on humanoid, nanotechnology, information technology, communication and control, environment and management (HNICEM), pp 1–6. https://doi.org/10.1109/HNICEM.2015.7393237

Coca-Cola Sustainable Packaging (2018) Packaging solutions for a world without waste. https://www.coca-colacompany.com/sustainability/packaging-sustainability

CPCB (2018) Report of 1. Central Pollution Control Board (CPCB) study of life cycle assessment of plastics packaging products, March 2018. https://cpcb.nic.in/uploads/plasticwaste/LCA_Report_15.05.2018.pdf

Huang Y, Bird R, Heidrich O (2009) Development of a life cycle assessment tool for construction and maintenance of asphalt pavements. J Clean Prod 17(2):283–296. https://doi.org/10.1016/j.jclepro.2008.06.005

Marathe, K. V., Chavan, K. R., & Nakhate, P. (2019). 8—Life cycle assessment (LCA) of PET bottles. In S. Thomas, A. Rane, K. Kanny, Abitha VK, & M. G. Thomas (Eds.), Recycling

of polyethylene terephthalate bottles (pp. 149–168). William Andrew Publishing. https://doi.org/10.1016/B978-0-12-811361-5.00008-0
Mata TM, Costa CAV (2001) Life cycle assessment of different reuse percentages for glass beer bottles—ProQuest. https://www.proquest.com/docview/664647982?pq-origsite=gscholar&fromopenview=true
Owsianiak M, Laurent A, Bjørn A, Hauschild MZ (2014) IMPACT 2002+, ReCiPe 2008 and ILCD's recommended practice for characterization modelling in life cycle impact assessment: a case study-based comparison. Int J Life Cycle Assess 19(5):1007–1021. https://doi.org/10.1007/s11367-014-0708-3
Saleh Y (2016) Comparative life cycle assessment of beverages packages in Palestine. J Clean Prod 131:28–42. https://doi.org/10.1016/j.jclepro.2016.05.080
Singh M, Pant M (2021) A review of selected weighing methods in MCDM with a case study. Int J Syst Assur Eng Manag 12(1):126–144. https://doi.org/10.1007/s13198-020-01033-3
Singh M, Pant M, Kaul A, Jha PC (2018) Online academic social networking sites (ASNSs) selection through AHP for placement of advertisement of E-learning website. In: Abraham A, Kr. Muhuri P, Muda AK, Gandhi N (eds) Intelligent systems design and applications. Springer, pp 1008–1017. https://doi.org/10.1007/978-3-319-76348-4_97
Singh M, Pant M, Godiyal RD, Kumar Sharma A (2020a) MCDM approach for selection of raw material in pulp and papermaking industry. Mater Manuf Process 35(3):241–249. https://doi.org/10.1080/10426914.2020.1711917
Singh M, Verma S, Pant M (2020b) Integrating customer preference rating approach with TOPSIS to evaluate the quality of pulp. In: Pant M, Sharma TK, Verma OP, Singla R, Sikander A (eds) Soft computing: theories and applications. Springer, pp 1391–1401. https://doi.org/10.1007/978-981-15-0751-9_127
Speck R, Selke S, Auras R, Fitzsimmons J (2016) Life cycle assessment software: selection can impact results. J Ind Ecol 20(1):18–28. https://doi.org/10.1111/jiec.12245
Yadav P, Silvenius F, Katajajuuri J-M, Leinonen I (2024) Life cycle assessment of reusable plastic food packaging. J Clean Prod 448:141529. https://doi.org/10.1016/j.jclepro.2024.141529
Ye L, Qi C, Hong J, Ma X (2017) Life cycle assessment of polyvinyl chloride production and its recyclability in China. J Clean Prod 142:2965–2972. https://doi.org/10.1016/j.jclepro.2016.10.171

Chapter 13
Entrepreneurial Strategies for Pioneering Sustainable Practices in Circular Supply Chains: Insights from India

Madhusudan Narayan, M. Dileep Kumar, Pooja Shukla, and Bhavna Taneja

Abstract Purpose: This study investigates how entrepreneurial strategies foster sustainable practices in India's circular supply chains, addressing challenges such as fragmented supply networks, reliance on informal sectors, and regulatory gaps. It bridges the research gap in understanding context-specific strategies for circular economy adoption. **Design/Methodology/Approach**: This study employs qualitative research methods, including a systematic literature review of peer-reviewed articles (2015–2023) and thematic analysis of policy documents and industry reports. By synthesizing theoretical frameworks and practical examples, the research explores how Indian entrepreneurs integrate circular economy principles into supply chains. Data triangulation ensures alignment with entrepreneurship, sustainability, and Sustainable Development Goals (SDGs). **Findings**: Indian entrepreneurs capitalize product innovation exemplified by Banyan Nation's plastic recycling, collaborative partnerships such as Goonj's waste reuse networks, and digital tools including blockchain for traceability to align with Sustainable Development Goals 9 and 12. Key barriers include inconsistent policy enforcement, consumer price sensitivity, and infrastructural deficits in waste management. **Practical Implications**: Policymakers should harmonize regulations such as tax incentives for circular economy research and development and invest in waste infrastructure. Entrepreneurs must prioritize affordability-optimized circular models and stakeholder education. **Originality/Value**: This study provides a framework for integrating circular economy principles into India's supply chains, emphasizing the role of informal sectors, digital innovation, and Sustainable Development Goal-aligned entrepreneurship.

M. Narayan (✉) · P. Shukla · B. Taneja
Amity University Jharkhand, Ranchi, India

M. Dileep Kumar
Hensard University, Toru-Orua, Bayelsa, Nigeria

K. Mathiyazhagan et al. (eds.), *Humanizing Circular Supply Chain Management*, Circular Economy and Sustainability,
https://doi.org/10.1007/978-3-031-86562-6_13

Keywords Entrepreneurial strategies · Sustainable practices · Circular supply chains · Circular economy · Environmental stewardship · Sustainable Development Goals (SDGs)

13.1 Background

The linear "take-make-dispose" supply chain model, characterized by resource depletion and environmental degradation (McDonough and Braungart 2002; RBS 2023), is increasingly being replaced by circular supply chains (CSCs). CSCs emphasize resource optimization through reuse, refurbishment, and recycling, aligning with circular economy (CE) principles to minimize waste and regenerate natural systems (D'Amato and Korhonen 2021; Kumar et al. 2021; Alhawari et al. 2021). Integrating sustainability into supply chains not only reduces costs and enhances operational efficiency (Genovese et al. 2017) but also addresses urgent global challenges like climate change and pollution (Kumar et al. 2022; Kopnina and Padfield 2021).

The United Nations Sustainable Development Goals (SDGs), particularly SDG 12 (responsible consumption and production), provide a framework for aligning business practices with global sustainability targets (Kusi-Sarpong et al. 2021; Liu et al. 2023). CSCs directly contribute to SDGs such as clean energy (SDG 7), sustainable cities (SDG 11), and climate action (SDG 13) by reducing environmental impacts and fostering inclusive economic growth (Kurniawan and Fernando 2023; D'Amato and Korhonen 2021; Goyal et al. 2021). For instance, collaborative models in CSCs, such as industrial symbiosis and co-learning frameworks, have demonstrated success in formalizing waste recovery and driving systemic change (Sanchez et al. 2020; Bakırlıoğlu and McMahon 2021).

13.1.1 Entrepreneurial Strategies for Circular Supply Chains in India

India's rapidly growing economy, supported by initiatives like Swachh Bharat Abhiyan and Make in India, offers a unique context for CE adoption. However, challenges such as inadequate waste management infrastructure, reliance on informal sectors for material recovery, and consumer preferences for affordability over sustainability hinder widespread CSC implementation (Yin et al. 2023; Knäble et al. 2022). For example, informal waste collectors, who handle ~80% of India's recyclables, often operate outside formal frameworks, leading to inefficiencies in material recovery (Yang et al. 2023; Manshoven and Gillabel 2021; Knäble et al. 2022).

Entrepreneurship plays a pivotal role in overcoming these barriers. India's thriving startup ecosystem, bolstered by digital technologies like IoT and blockchain, drives innovations in recycling, upcycling, and resource optimization (World Economic Forum 2019; Schulz-Mönninghoff and Evans 2023). Firms like Banyan Nation capitalize on advanced recycling to transform plastic waste into high-quality materials, while platforms like Rentomojo adopt product-as-a-service models to extend product lifecycles (Venugopal 2021; Rentomojo 2019; Voulvoulis et al. 2022). Such strategies align with India's policy goals and SDGs, creating pathways for SMEs to integrate CE principles into business models (Del Vecchio et al. 2021).

This study explores how entrepreneurial strategies foster sustainable practices in India's Circular Supply Chains (CSCs), emphasizing Industry 4.0 technologies, collaborative models, and evolving policies. While global research highlights circular practices, there is limited insight into how Indian entrepreneurs address fragmented supply networks, informal sectors, and regulatory complexity. By examining context-specific challenges and innovations, this study fills a critical gap in emerging economy literature. It assesses the influence of technology, policy, and consumer behavior on sustainable transitions, offering actionable insights for business leaders and policymakers aligned with SDG targets.

13.2 Literature Review

Sustainable development, defined as meeting present needs without compromising future generations, gained global traction with the UN's adoption of the Sustainable Development Goals (SDGs) in 2015. These 17 goals address interconnected social, economic, and environmental challenges (UN 2015), extending the Millennium Development Goals (MDGs) to emphasize systemic change (Biermann et al. 2017; Goyal et al. 2021; Torre-Marin Rando et al. 2021). While criticized for lacking corporate accountability (Sachs 2012), the SDGs provide a shared agenda for aligning business practices with sustainability, particularly through circular economy (CE) principles (D'Amato and Korhonen 2021; Voulvoulis et al. 2022).

The CE framework, which replaces the linear "take-make-dispose" model, promotes closed-loop systems via remanufacturing, recycling, and waste reduction (EMF 2015; Geissdoerfer et al. 2017; Knäble et al. 2022). CE adoption in supply chains enhances resource efficiency and reduces environmental impacts while creating economic value (Murray et al. 2017; Bocken et al. 2021). For instance, reverse logistics and durable product design extend lifecycle utility and align with SDG 12 (Kristoffersen et al. 2020; Sumter et al. 2020). However, challenges such as product disassembly complexity and fragmented supply chain partnerships hinder implementation (Loomba and Nakashima 2012).

Entrepreneurship is critical for advancing CE practices. Entrepreneurs drive innovation through circular business models (CBMs) like product-as-a-service and industrial symbiosis, which prioritize resource regeneration. Startups capitalize on digital tools such as IoT and blockchain to enhance supply chain transparency and efficiency. Yet, financial barriers persist: SMEs often lack capital for circular transitions, while consumer resistance to novel models complicates market entry (Kopnina and Padfield 2021).

In India, CE adoption is shaped by socio-economic dynamics. The informal sector, responsible for 80% of waste collection, remains underintegrated into formal supply chains, limiting scalability (Gupta et al. 2021; Manshoven and Gillabel 2021). Policy initiatives like Extended Producer Responsibility (EPR) and *Swachh Bharat Abhiyan* aim to formalize waste management but face inconsistent enforcement (Sinha et al. 2021; Lagrasta et al. 2021). Entrepreneurial ventures, such as agritech firms using AI for resource optimization (Kumar et al. 2021) and manufacturers adopting remanufacturing (Bansal and Kumar 2021), demonstrate CE's potential in emerging economies. Collaborative frameworks, like co-learning platforms and simulation-based training, further bridge knowledge gaps (Bakırlıoğlu and McMahon 2021; Del Vecchio et al. 2021).

13.3 Research Gap

Existing studies emphasize the global role of circular supply chains (CSCs) in advancing sustainability goals (Korhonen et al. 2018). However, there is a conspicuous gap in understanding entrepreneurial strategies tailored to India's unique socio-economic, technological, and regulatory context (Lagrasta et al. 2021). While prior research highlights CE adoption in developed economies (Sorin and Sivarajah 2021), few studies address how Indian entrepreneurs navigate fragmented supply chains, informal sectors, and evolving policies like Extended Producer Responsibility (EPR) to implement sustainable practices (Gupta et al. 2021; Kopnina and Padfield 2021). This gap limits actionable insights for stakeholders seeking scalable, context-specific CSC frameworks in emerging economies (Fig. 13.1).

This matrix synthesizes literature themes (green) with unresolved gaps (red), demonstrating how the study addresses India-specific CE challenges through entrepreneurial strategies. Dashed lines highlight the transition from global theoretical foundations to context-driven solutions for SDG-aligned supply chains.

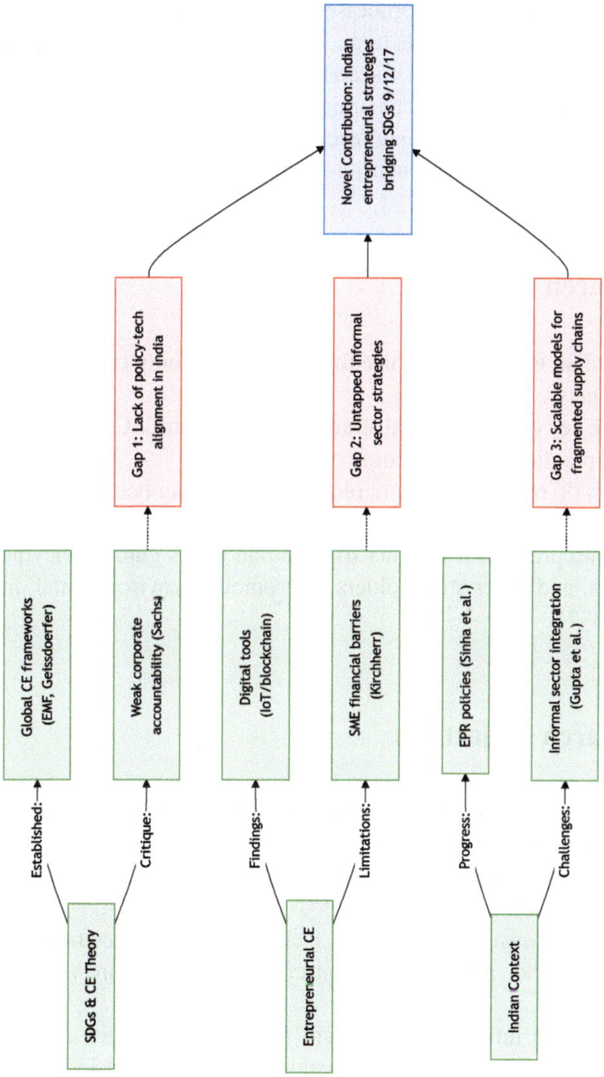

Fig. 13.1 Literature-research gap bridging matrix for circular supply chains in emerging economies

13.4 Problem Statement

Despite the global momentum toward circular supply chains (CSCs), developing nations like India face systemic barriers to integrating CE principles. Fragmented regulations, infrastructural deficits, and consumer resistance to sustainable alternatives hinder cohesive circular processes. While entrepreneurial ventures show promise in leveraging digital tools (e.g., IoT, blockchain) and collaborative models (Bakırlıoğlu and McMahon 2021), scalability remains constrained by inconsistent policy enforcement, financial constraints, and gaps in stakeholder education (Manshoven and Gillabel 2021; Del Vecchio et al. 2021). A structured framework is urgently needed to align entrepreneurial strategies with India's sustainability goals.

13.5 Research Questions

1. How can Indian entrepreneurs use the circular economy concept to their supply chain strategies?
2. What are the key entrepreneurial strategies that Indian CSCs are using to improve sustainable practices and overcome obstacles?
3. What impact do regulations, technology, and customer behaviour have on Indian entrepreneurs' adoption of circular supply chain practices?
4. How can entrepreneurial insights from Indian CSCs guide policymakers, business owners, and other stakeholders in promoting environmental and financial sustainability?

13.6 Research Objectives

1. To investigate how circular economy principles are incorporated into supply chain operations by Indian entrepreneurs.
2. To explore what obstacles entrepreneurs face when trying to adopt and accept sustainable practices.
3. To inspect how consumer behaviour, technological advancements, and governmental frameworks affect Indian entrepreneurs' adoption of circular supply chain strategies.
4. To offer valuable information that will assist policymakers, company owners, and stakeholders in implementing sustainable practices in Indian CSCs, hence assisting in the achievement of sustainability goals related to the environment and finances (Fig. 13.2).

This framework systematically maps research questions to objectives, thematic insights, and SDG targets, illustrating how entrepreneurial strategies transition from barrier diagnosis (RQ1-RO1) to scalable solutions (RQ4-RO4). It bridges local

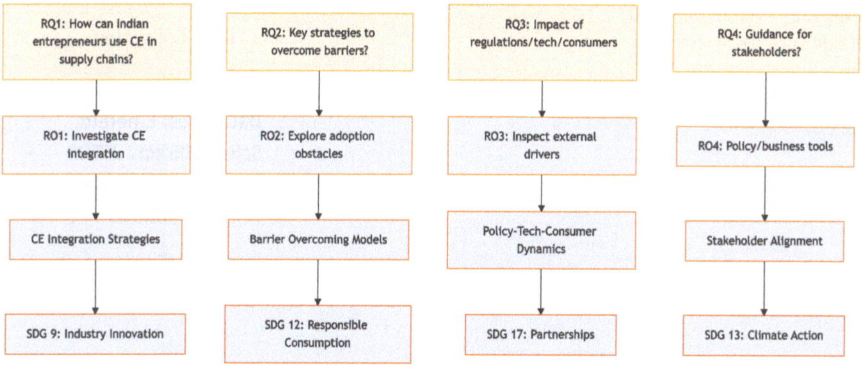

Fig. 13.2 Research alignment matrix for circular supply chain strategies

circular supply chain challenges with global sustainability goals through policy-tech-consumer dynamics, offering actionable pathways for stakeholders.

13.7 Research Methodology

This study employs a systematic literature review (SLR) grounded in qualitative research principles to analyze entrepreneurial strategies for integrating circular economy (CE) principles into India's supply chains. Aligned with PRISMA guidelines, the methodology synthesizes peer-reviewed articles, policy documents, and industry reports from databases including Emerald Insight, ScienceDirect, JSTOR, Web of Science, and SpringerLink, alongside Indian government portals (e.g., NITI Aayog) and industry publications. Keywords such as "circular economy," "sustainable supply chains," "entrepreneurial strategies," and "Sustainable Development Goals (SDGs)" were used to identify relevant sources (2015–2023), prioritizing empirical studies, India-specific analyses, and policy evaluations for contextual relevance. Thematic analysis categorized recurring patterns, including infrastructural gaps, funding limitations, and policy incentives like India's Extended Producer Responsibility framework, while credibility was assessed through methodological rigor, contextual alignment, and source diversity (e.g., case studies on SMEs). Emerging trends, such as digital platforms enabling resource recovery and the impact of India's 2023 Circular Economy Action Plan, were identified through multidisciplinary integration of operations management and socio-political governance perspectives. Limitations, such as reliance on existing literature, were mitigated by triangulating findings with recent policy drafts (e.g., Plastic Waste Management Rules) and global benchmarks (e.g., EU frameworks). Future research could validate theoretical frameworks through empirical studies or quantitative modeling, particularly in informal sectors like textiles. This methodology bridges theoretical and practical insights, offering stakeholders a roadmap to align CE innovation with India's net-zero and SDG targets while fostering equitable growth (Fig. 13.3).

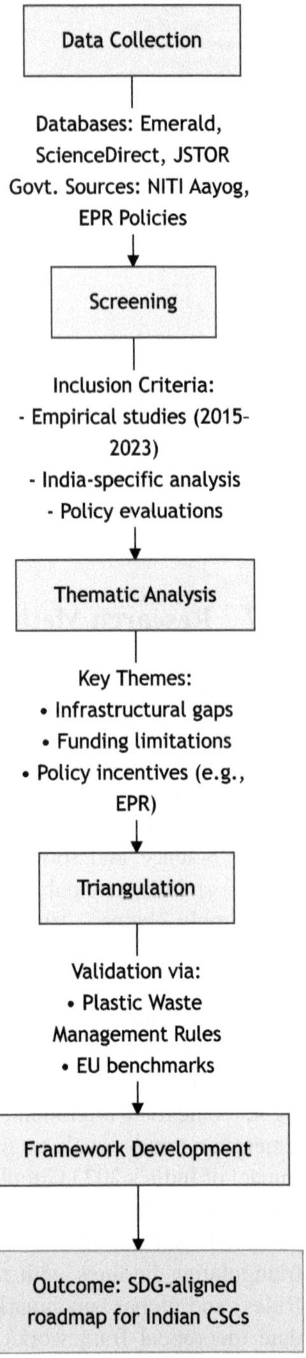

Fig. 13.3 Methodological framework for systematic literature review on circular supply chains

This flowchart visualizes the PRISMA-aligned systematic review process, from multidisciplinary database searches (2015–2023) to thematic synthesis and policy triangulation. It highlights how infrastructural gaps, SDG alignment, and India's EPR framework are analyzed to bridge theoretical CE models with entrepreneurial implementation strategies.

13.8 Results and Discussion

13.8.1 Entrepreneurial Strategies for Circular Economy Integration

Indian entrepreneurs are advancing circular economy (CE) principles through product innovation and collaborative partnerships. Startups like Banyan Nation transform plastic waste into high-quality materials using advanced recycling, while Tata Motors extends product lifecycles through refurbishment programs. Access-based models, such as Rentomojo's product-as-a-service (PaaS) framework, promote resource sharing and reduce waste (Rentomojo 2019). Collaborative efforts, like Goonj's urban waste reuse networks, formalize informal sectors and align with global CE frameworks (Schulz-Mönninghoff and Evans 2023). These strategies mirror successful models in Tanzania's coffee sector, where partnerships enhance resource efficiency (Lagrasta et al. 2021). Additionally, digital platforms like TraceX Technologies employ blockchain to enhance traceability in agricultural supply chains (TraceX Technologies 2021), demonstrating how technology amplifies circularity. Such innovations not only reduce environmental footprints but also create economic value, aligning with SDG 12 (Responsible Consumption) and SDG 9 (Industry Innovation) (D'Amato and Korhonen 2021; Torre-Marin Rando et al. 2021).

13.8.2 Challenges in Scaling Circular Practices

Despite progress, regulatory gaps and financial constraints hinder scalability. Policies like Extended Producer Responsibility (EPR) lack consistent enforcement, creating compliance hurdles for SMEs (Yin et al. 2023). High upfront costs for technologies like blockchain or IoT limit adoption among small firms (Knäble et al. 2022). For example, TraceX Technologies' blockchain solutions for agricultural traceability (TraceX Technologies 2021) remain inaccessible to resource-constrained entrepreneurs. Consumer behavior adds complexity: while demand for sustainability grows, price sensitivity and mistrust of eco-friendly products slow market adoption (Yang et al. 2023; Manshoven and Gillabel 2021). These challenges reflect global barriers observed in African and European contexts (Kopnina and Padfield

2021; Del Vecchio et al. 2021). Skill gaps further exacerbate issues, as SMEs often lack expertise to implement advanced CE practices (Schulz-Mönninghoff and Evans 2023). For instance, only 30% of Indian manufacturers have adopted lean-circular hybrids due to technical knowledge deficits (Afum et al. 2022), underscoring the need for targeted training programs.

13.8.3 Policy, Technology, and Consumer Dynamics

India's regulatory landscape, though progressive, struggles with implementation gaps. The Swachh Bharat Abhiyan and Plastic Waste Management Rules (2016) aim to reduce waste but face uneven enforcement (Khan et al. 2020; Gupta et al. 2021). Entrepreneurs advocate for harmonized policies and fiscal incentives, such as tax breaks for circular innovations (Lacy et al. 2015). Digitally, AI and IoT enhance efficiency TATA Steel's AI-driven waste reduction (Fraccascia et al. 2019) and blockchain-enabled transparency exemplify this potential. However, a digital divide excludes rural SMEs, perpetuating reliance on linear models (Voulvoulis et al. 2022; Bocken et al. 2021). Consumer engagement remains pivotal: brands like Fabindia integrate sustainability into value propositions (Manshoven and Gillabel 2021), yet affordability barriers persist. Educational campaigns, such as Mahindra Group's "Rise for Good" (Mahindra and Mahindra Limited 2023), are critical for shifting cultural norms. Lessons from Scandinavia's hospitality sector, where eco-certifications boosted consumer trust (Sinha et al. 2021), suggest India could adopt similar labeling schemes to bridge the awareness gap.

13.8.4 Actionable Insights for Stakeholders

To accelerate CE adoption, collaborative ecosystems and policy alignment are essential. Public-private partnerships, like those in the Quintuple Helix Model, can subsidize technology access for SMEs. Harmonizing policies (e.g., EPR compliance) and incentivizing circular practices (e.g., tax breaks) will bridge regulatory gaps (Lacy et al. 2015). Stakeholder education, as seen in European CE programs (Voulvoulis et al. 2022) can address consumer skepticism. Initiatives like the ADKAR model for waste management emphasize systemic collaboration, while industrial symbiosis frameworks offer scalable blueprints. Governments should prioritize grants for circular R&D and establish monitoring bodies to track policy efficacy, as seen in the EU's Circular Economy Action Plan. By integrating these measures, India can achieve SDG 12 targets while fostering inclusive growth, particularly for informal workers contributing to material recovery (Gupta et al. 2021) (Fig. 13.4).

This figure depicts the progression from entrepreneurial strategies to SDG-aligned outcomes, highlighting how regulatory, financial, and consumer challenges

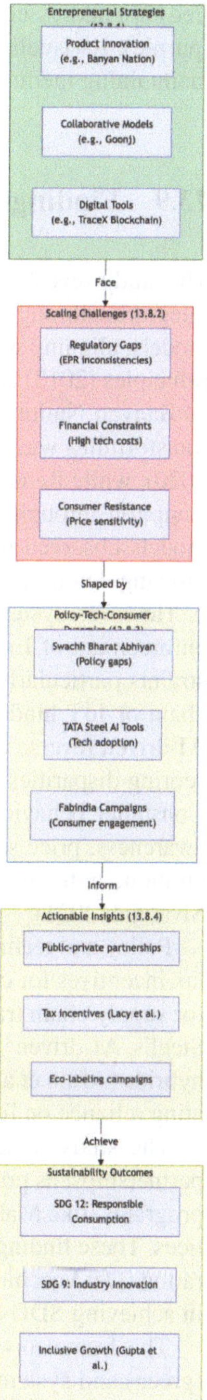

Fig. 13.4 Circular economy transition framework: strategies, challenges, and outcomes

(red) are addressed through policy-tech synergies (blue) and collaborative insights (purple). It visually anchors your case studies (Banyan Nation, TATA Steel) while maintaining methodological alignment with cited literature.

13.9 Findings of Study

The study reveals that Indian entrepreneurs are pioneering circular economy (CE) integration through product innovation, collaborative partnerships, and access-based models, aligning with global frameworks like the Ellen MacArthur Foundation's CE principles (2013) and SDG 12 targets (D'Amato and Korhonen 2021). Startups such as Banyan Nation and Attero Recycling exemplify resource-efficient practices by transforming waste into high-value materials (Bocken et al. 2021; Sumter et al. 2020), while Rentomojo's product-as-a-service (PaaS) model reduces material consumption through shared ownership. These strategies mirror industrial symbiosis models observed in Tanzania's coffee sector (Lagrasta et al. 2021) and reflect the growing emphasis on circular design competencies (Sumter et al. 2020).

However, systemic barriers persist. Regulatory ambiguities, such as inconsistent enforcement of Extended Producer Responsibility (EPR) rules, and financial constraints particularly for SMEs lacking access to advanced technologies like blockchain or IoT hinder scalability. Only 30% of Indian manufacturers have adopted AI-driven reverse logistics due to high costs and skill gaps (Liu et al. 2023), underscoring disparities highlighted in African CE studies (Kopnina and Padfield 2021). Consumer behavior further complicates adoption: despite growing environmental awareness, price sensitivity and mistrust of sustainable products limit market penetration, a trend also observed in Scandinavian hospitality sectors (Sorin and Sivarajah 2021).

Policy and technology emerge as dual enablers. Harmonized regulations, such as tax incentives for circular R&D (Lacy et al. 2015), and digital tools like blockchain for supply chain transparency could bridge gaps. Successful cases, including TATA Steel's AI-driven waste reduction, demonstrate the potential of "lean-circular" hybrids (Afum et al. 2022). However, a digital divide excludes rural SMEs, perpetuating reliance on linear models.

The study underscores the need for collaborative ecosystems. Public-private partnerships, as proposed in the Quintuple Helix Model, and stakeholder education programs, like Mahindra Group's "Rise for Good", are critical for scaling CE practices. These findings align with European frameworks advocating systemic collaboration (Del Vecchio et al. 2021) and emphasize the role of circular entrepreneurship in achieving SDG-aligned economic growth (Fig. 13.5).

This framework synthesizes study findings, mapping entrepreneurial innovations (green) and systemic barriers (red) to policy-tech enablers (blue) and collaborative solutions (purple). It culminates in SDG-aligned outcomes (gold), emphasizing how Indian CE models bridge global sustainability principles with local socio-economic realities.

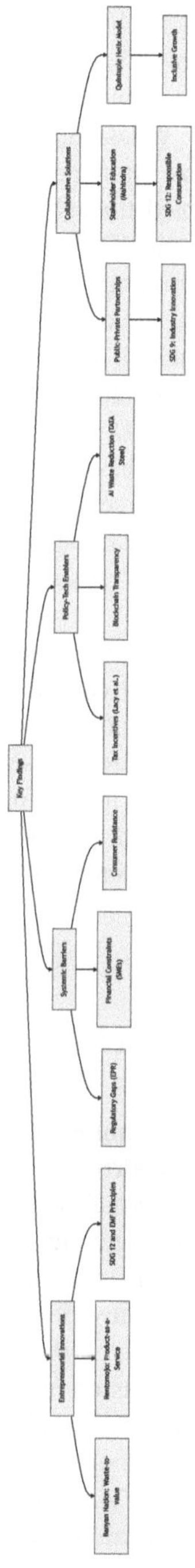

Fig. 13.5 Findings

13.10 Conclusion and Policy Recommendations

This study establishes circular supply chains (CSCs) as transformative tools for achieving SDG-aligned sustainability in India, driven by entrepreneurial innovation in resource efficiency, product redesign, and cross-sector collaboration. Entrepreneurs have demonstrated how strategies like industrial symbiosis and digital platforms (e.g., blockchain-enabled traceability) can mitigate waste and align with SDGs 9 (Industry Innovation) and 12 (Responsible Consumption). However, scaling these practices requires addressing systemic barriers, including regulatory fragmentation, technological divides, and consumer price sensitivity.

To accelerate progress, policymakers must prioritize harmonized regulatory frameworks, such as stringent enforcement of Extended Producer Responsibility (EPR) and tax incentives for circular R&D. Public-private partnerships, modeled on the Quintuple Helix approach should subsidize SMEs' adoption of IoT and blockchain to enhance transparency and resource optimization. Concurrently, nationwide campaigns inspired by the EU's Circular Economy Action Plan are critical to educate consumers and foster demand for sustainable products, bridging the gap between awareness and action.

By integrating these measures, India can transition fragmented linear systems into resilient, low-carbon CSCs, driving equitable growth while fulfilling global climate commitments. Collaboration among governments, financial institutions, and civil society is essential to support informal sectors and SMEs, ensuring inclusive participation in circular transitions. These efforts, coupled with robust monitoring frameworks, will position India as a leader in sustainable supply chain innovation, advancing both environmental stewardship and socio-economic resilience.

13.11 Scope for Future Research

Future studies should prioritize technology-driven circular models, such as blockchain-enabled traceability and AI-driven reverse logistics (Liu et al. 2023), to assess their scalability in India's informal sectors like textiles and agriculture. Comparative analyses with emerging economies, such as Tanzania's circular coffee sector (Lagrasta et al. 2021) or Brazil's industrial symbiosis frameworks, could uncover context-specific barriers and innovations. Longitudinal research is critical to evaluate the long-term efficacy of policies like EPR and Swachh Bharat Abhiyan, particularly their socio-economic impacts on informal waste workers (Gupta et al. 2021; Kopnina and Padfield 2021). Additionally, interdisciplinary work should explore socio-cultural dynamics, including caste and regional disparities in CE adoption, to tailor inclusive strategies (D'Amato and Korhonen 2021). Finally, integrating circular entrepreneurship education and multi-stakeholder collaboration models into empirical studies could bridge theory-practice gaps, advancing SDG-aligned frameworks for Global South economies.

References

Afum E, Li Y, Han P, Sun Z (2022) Interplay between lean management and circular production system: implications for zero-waste performance, green value competitiveness, and social reputation. J Manuf Technol Manag 33(8):1213–1231. https://doi.org/10.1108/JMTM-02-2021-0192

Alhawari O, Awan U, Bhutta MKS, Ülkü MA (2021) Insights from circular economy literature: a review of extant definitions and unravelling paths to future research. Sustainability 13(2):859. https://doi.org/10.3390/su13020859

Bakırlıoğlu Y, McMahon M (2021) Co-learning for sustainable design: the case of a circular design collaborative project in Ireland. J Clean Prod 279:123474. https://doi.org/10.1016/j.jclepro.2020.123474

Bansal M, Kumar V (2021) Forcing responsibility? Examining earnings management induced by mandatory corporate social responsibility: evidence from India. Rev Acc Financ 20(2):194–216. https://doi.org/10.1108/RAF-06-2020-0151

Biermann F, Kanie N, Kim RE (2017) Global governance by goal-setting: the novel approach of the UN Sustainable Development Goals. Curr Opin Environ Sustain 26–27:26–31. https://doi.org/10.1016/j.cosust.2017.01.010

Bocken NMP, Weissbrod I, Antikainen M (2021) Business model experimentation for the circular economy: definition and approaches. Circ Econ Sustain 1:49–81. https://doi.org/10.1007/s43615-021-00026-z

D'Amato D, Korhonen J (2021) Integrating the green economy, circular economy and bioeconomy in a strategic sustainability framework. Ecol Econ 188:107143. https://doi.org/10.1016/j.ecolecon.2021.107143

Del Vecchio P, Secundo G, Mele G, Passiante G (2021) Sustainable entrepreneurship education for circular economy: emerging perspectives in Europe. Int J Entrep Behav Res 27(8):2096–2124. https://doi.org/10.1108/IJEBR-03-2021-0210

Ellen MacArthur Foundation (2015) What is a circular economy? https://www.ellenmacarthurfoundation.org/topics/circular-economy-introduction/overview. Accessed 6 May 2024

Fraccascia L, Giannoccaro I, Albino V (2019) Business models for industrial symbiosis: a taxonomy focused on the form of governance. Resour Conserv Recycl 146:114–126. https://doi.org/10.1016/j.resconrec.2019.03.017

Geissdoerfer M, Savaget P, Bocken NMP, Hultink EJ (2017) The circular economy—a new sustainability paradigm? J Clean Prod 143:757–768. https://doi.org/10.1016/j.jclepro.2016.12.048

Genovese A, Acquaye AA, Figueroa A, Koh SCL (2017) Sustainable supply chain management and the transition towards a circular economy: evidence and some applications. Omega 66(Part B):344–357. https://doi.org/10.1016/j.omega.2015.05.015

Goyal S, Chauhan S, Mishra P (2021) Circular economy research: a bibliometric analysis (2000–2019) and future research insights. J Clean Prod 287:125011. https://doi.org/10.1016/j.jclepro.2020.125011

Gupta H, Kumar A, Wasan P (2021) Industry 4.0, cleaner production and circular economy: an integrative framework for evaluating ethical and sustainable business performance of manufacturing organizations. J Clean Prod 295:126253. https://doi.org/10.1016/j.jclepro.2021.126253

Khan HU, Ahmad M, Khan S (2020) The impact of corporate social responsibility on firm performance: evidence from Pakistan. Asian J Sustain Soc Responsib 5:1–19

Knäble D, de Quevedo Puente E, Pérez-Cornejo C, Baumgärtler T (2022) The impact of the circular economy on sustainable development: a European panel data approach. Sustain Prod Consum 34:233–243. https://doi.org/10.1016/j.spc.2022.09.016

Kopnina H, Padfield R (2021) (Im)possibilities of "circular" production: learning from corporate case studies of (un)sustainability. Environ Sustain Indic 12:100161. https://doi.org/10.1016/j.indic.2021.100161

Korhonen J, Nuur C, Feldmann A, Birkie SE (2018) Circular economy as an essentially contested concept. J Clean Prod 175:544–552. https://doi.org/10.1016/j.jclepro.2017.12.111

Kristoffersen E, Blomsma F, Mikalef P, Li J (2020) The smart circular economy: a digital-enabled circular strategies framework for manufacturing companies. J Bus Res 120:241–261. https://doi.org/10.1016/j.jbusres.2020.07.044

Kumar P, Singh RK, Kumar V (2021) Managing supply chains for sustainable operations in the era of industry 4.0 and circular economy: analysis of barriers. Resour Conserv Recycl 164:105215. https://doi.org/10.1016/j.resconrec.2020.105215

Kumar R, Mishra JS, Mali SS, Mondal S, Meena RS, Lal R, Jha BK, Naik SK, Biswas AK, Hans H, Sundaram PK, Choudhary AK, Monobrullah M, Kumar S, Kumar S, Raman RK, Bhatt BP, Kumar U (2022) Comprehensive environmental impact assessment for designing carbon-cum-energy efficient, cleaner and eco-friendly production system for rice-fallow agro-ecosystems of South Asia. J Clean Prod 331:129973. https://doi.org/10.1016/j.jclepro.2021.129973

Kurniawan I, Fernando Y (2023) Circular economy supply chain for Sustainable Development Goals (SDGs): a review and future opportunities. Int J Ind Manag 17(1):9211. https://doi.org/10.15282/ijim.17.1.2023.9211

Kusi-Sarpong S, Gupta H, Khan SA, Jabbour CJC, Rehman ST, Kusi-Sarpong H (2021) Sustainable supplier selection based on Industry 4.0 initiatives within the context of circular economy implementation in supply chain operations. Prod Plan Control 34(10):999–1019. https://doi.org/10.1080/09537287.2021.1980906

Lacy P, Long D, Spindler T (2015) Waste to wealth: the circular economy advantage. Palgrave Macmillan. https://doi.org/10.1057/9781137530707

Lagrasta FP, Pontrandolfo P, Scozzi B (2021) Circular economy business models for the Tanzanian coffee sector: a teaching case study. Sustainability 13(24):13931. https://doi.org/10.3390/su132413931

Liu Y, Farooque M, Lee C-H, Gong Y, Zhang A (2023) Antecedents of circular manufacturing and its effect on environmental and financial performance: a practice-based view. Int J Prod Econ 259:108866. https://doi.org/10.1016/j.ijpe.2023.108866

Loomba APS, Nakashima K (2012) Enhancing value in reverse supply chains by sorting before product recovery. Prod Plan Control 23(2–3):205–215. https://doi.org/10.1080/09537287.2011.591652

Mahindra & Mahindra Limited (2023) Sustainability report 2021–22. https://www.mahindra.com/sites/default/files/2023-01/Mahindra-Sustainability-Report-2021-22_0.pdf

Manshoven S, Gillabel J (2021) Learning through play: a serious game as a tool to support circular economy education and business model innovation. Sustainability 13(23):13277. https://doi.org/10.3390/su132313277

McDonough W, Braungart M (2002) Cradle to cradle: remaking the way we make things. North Point Press

Murray A, Skene K, Haynes K (2017) The circular economy: an interdisciplinary exploration of the concept and application in a global context. J Bus Ethics 140:369–380. https://doi.org/10.1007/s10551-015-2693-2

RBS—Rome Business School (2023, November 30) Why is the linear economic model no longer sustainable? https://romebusinessschool.com/blog/linear-economy/#:~:text=What%20is%20a%20Linear%20Economy,the%20depletion%20of%20natural%20resources

Rentomojo (2019) Rent furniture, appliances, and electronics online. https://www.rentomojo.com/

Sachs JD (2012) From millennium development goals to sustainable development goals. Lancet 379(9832):2206–2211. https://doi.org/10.1016/S0140-6736(12)60685-0

Sanchez B, Ballinas-Gonzalez R, Rodriguez-Paz MX, Nolazco-Flores JA (2020) Integration of circular economy principles for developing sustainable development competences in higher education: an analysis of bachelor construction management courses. In: 2020 IEEE Global Engineering Education Conference (EDUCON). IEEE, pp 988–996. https://doi.org/10.1109/EDUCON45650.2020.9125307

Schulz-Mönninghoff M, Evans S (2023) Key tasks for ensuring economic viability of circular projects: learnings from a real-world project on repurposing electric vehicle batteries. Sustain Prod Consum 35:559–575. https://doi.org/10.1016/j.spc.2022.11.025

Sinha A, Mishra S, Sharif A, Yarovaya L (2021) Does green financing help to improve environmental & social responsibility? Designing SDG framework through advanced quantile modelling. J Environ Manag 292:112751. https://doi.org/10.1016/j.jenvman.2021.112751

Sorin F, Sivarajah U (2021) Exploring circular economy in the hospitality industry: empirical evidence from Scandinavian hotel operators. Scand J Hosp Tour 21(3):265–285. https://doi.org/10.1080/15022250.2021.1921021

Sumter D, de Koning J, Bakker C, Balkenende R (2020) Circular economy competencies for design. Sustainability 12(4):1561. https://doi.org/10.3390/su12041561

Torre-Marin Rando A, González-Rojí SJ, Hurni K, Mathez-Stiefel SL, Messmer M, Raible CC, Snethlage M, Stocker TF, Fischer M (2021) Towards sustainable futures for nature and people: an appraisal report for Madre de Dios, Peru (Wyss Academy Report 1). Wyss Academy for Nature at the University of Bern. https://boris.unibe.ch/161385/

TraceX Technologies (2021) TRACEX blockchain powered platform. https://tracextech.com/technology/#:~:text=TRACEX%20Blockchain%20Powered%20Platform,traceability%20at%20any%20given%20time

United Nations (2015) Goal 12: ensure sustainable consumption and production patterns. United Nations Department of Economic and Social Affairs. https://sdgs.un.org/goals/goal12. Accessed 6 May 2025

Venugopal S (2021, June 17) Hyderabad startup Banyan Nation recognised as TechPioneers 2021 by WEF. The Hans India. https://www.thehansindia.com/startups/hyderabad-startup-banyan-nation-recognised-as-techpioneers-2021-by-wef-691432

Voulvoulis N, Giakoumis T, Hunt C, Kioupi V, Petrou N, Souliotis I, Vaghela C, binti Wan Rosely WIH (2022) Systems thinking as a paradigm shift for sustainability transformation. Glob Environ Chang 75:102544. https://doi.org/10.1016/j.gloenvcha.2022.102544

World Economic Forum (2019, October 8) How to end a decade of lost productivity growth. https://www.weforum.org/publications/how-to-end-a-decade-of-lost-productivity-growth/

Yang M, Chen L, Wang J et al (2023) Circular economy strategies for combating climate change and other environmental issues. Environ Chem Lett 21:55–80. https://doi.org/10.1007/s10311-022-01499-6

Yin S, Jia F, Chen L, Wang Q (2023) Circular economy practices and sustainable performance: a meta-analysis. Resour Conserv Recycl 190:106838. https://doi.org/10.1016/j.resconrec.2022.106838

Chapter 14
Essential Skills for Employees in Effective Circular Supply Chains

Danny Ngo Lung Yao, Farhad Nadi, Suan See Tan, and Yap Bee Wah

Abstract Driven by globalization, complex global supply chains have traditionally focused on efficiency, speed, and cost-effectiveness in a linear model. However, the growing focus on sustainability demands a shift towards circular supply chains that extend product lifecycles and minimize waste. This transition requires a skilled workforce equipped with new knowledge and competencies. This chapter explores the challenges and opportunities of developing such a workforce, highlighting the emergence of new roles like green procurement specialists and circular leadership. It proposes a multi-faceted approach to bridge the circular economy skills gap. This strategy emphasizes integrating circular economy principles into company culture, developing targeted training programs, leveraging existing expertise, fostering collaboration with educational institutions, and utilizing digital technology for learning and collaboration. Organizations can empower their workforce by implementing these strategies to catalyze the transformation to a sustainable future.

Keywords Supply chain management · Digital technology · Circular economy · Skills

D. N. L. Yao · F. Nadi (✉) · Y. B. Wah
School of Information Technology, UNITAR International University, Petaling Jaya, Selangor, Malaysia

Center for Innovation and Technology Adoption (CITA), UNITAR International University, Petaling Jaya, Selangor, Malaysia
e-mail: farhad.nadi@unitar.my

S. S. Tan
Eagle Horizon Sdn. Bhd., Klang, Selangor, Malaysia

K. Mathiyazhagan et al. (eds.), *Humanizing Circular Supply Chain Management*, Circular Economy and Sustainability,
https://doi.org/10.1007/978-3-031-86562-6_14

14.1 Introduction

The preceding decades have been marked by a remarkable transformation in global trade. National economies have become intricately linked, fostering a free flow of goods, services, and people. This phenomenon, known as globalization (Sato 2018), has fueled a significant rise in international trade. A complex system known as the global supply chain has emerged to manage this movement of products and information efficiently. Traditionally, supply chains have focused on optimizing costs, speed, and agility to meet the demands of global trade. This linear model prioritizes finding the most efficient path from raw materials to finished products, delivered to consumers as quickly and cheaply as possible, as illustrated by the Supply Chain Operations Reference (SCOR) model (Huan et al. 2004). While this approach has undoubtedly powered global trade, recent years have increasingly emphasized supply chain sustainability.

The widely accepted linear model with its "take-make-dispose" approach as shown in Fig. 14.1, is increasingly unsustainable. One of the very reasons is the environmental impact due to the heavy reliance on virgin resource extraction for product creation. The production and disposal processes also contribute to climate change with excessive greenhouse gas emissions. The World Economic Forum reported that eight key industries, including food, construction, and electronics, generate over half of global greenhouse gas emissions (Forum 2021). Furthermore, Carbon intensity in supply chains is more than five times higher than in other business operations (Scott 2019).

Besides, waste management has become a burden as the linear models prioritize production and disposal over reuse and recycling. Malaysia's daily waste generation of 39,078 tonnes (Ravimalar 2024) illustrates the strain traditional waste management systems face. Landfill overflow and incineration might be the easy solution but would not be sustainable as more pollution appears.

Circular supply chains offer a transformative alternative by focusing on resource efficiency through extended product use (Amir et al. 2023; Lahane et al. 2023;

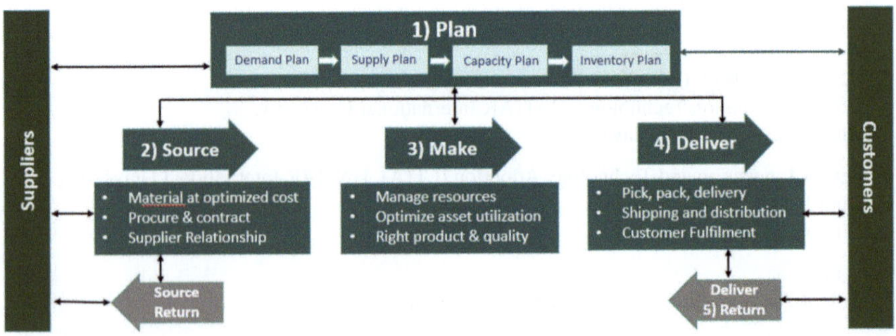

Fig. 14.1 Five-stage depiction of the linear supply chain model. Note. Figure is created by Tan Suan See, Eagle Horizon

Zhang et al. 2023). This approach prioritizes strategies like product design for longevity and recyclability, repair and refurbishment of used products, and closed-loop materials recycling. By increasing the useful life of products and materials, circular supply chains can significantly reduce waste generation and resource consumption.

The adoption of circular practices is significantly influenced by regulatory frameworks in different regions. These frameworks play a crucial role in shaping the transition to a circular economy by either encouraging or discouraging the implementation of circular strategies. The effectiveness of these regulations can vary depending on various factors such as regional governance structures, economic conditions, and sector-specific policies (Do et al. 2022; Liu et al. 2023; Mhatre et al. 2021; Varjú et al. 2022; Zarbà et al. 2021).

Businesses are increasingly adopting circular supply chain models due to economic, regulatory, environmental, and resource-related factors. Economically, these models offer opportunities to reduce waste, boost future profits, and build a sustainable business strategy (Alavi et al. 2021; Orji and Ojadi 2022). Regulatory pressures, including environmental and social regulations, as well as the need for alternative resource sources, are also driving adoption of circular supply chain. Environmental benefits like reduced impact, increased sustainability, and addressing climate change are significant motivators. Additionally, businesses face resource scarcity, price fluctuations, and the need for supply chain stability, all of which are mitigated by circular practices. Furthermore, consumer demand for sustainable products, stakeholder pressures from various groups, and the need to redefine growth by focusing on societal benefits and eliminating waste, all contribute to the shift towards circular supply chains. This shift towards circularity necessitates new skills like circular product design, remanufacturing and refurbishment, reverse logistics management, and supply chain data analytics by supply chain professionals.

14.2 The Emergence of Circular Economy

Despite the absence of a strict definition, the Circular Economy (CE) is widely regarded as an industrial paradigm centered on restorative and regenerative principles (Julian Kirchherr et al. 2023; Kirchherr et al. 2017). A Circular Economy (CE) seeks to minimize waste and optimize resource use, particularly for scarce and energy-intensive materials, throughout the supply chain. By prioritizing reuse, refurbishment, and recycling, CE aims to improve environmental conditions while generating profits.

In short, CE presents an innovative and forward approach with three core principles (Fig. 14.2) for economic growth and aims to achieve sustainability by establishing systems that continuously reuse resources instead of discarding them as waste. This involves creating closed-loop systems for materials, enhancing product

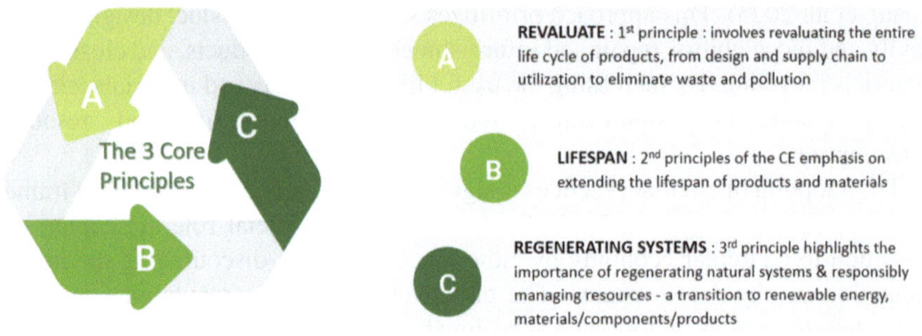

Fig. 14.2 The three core principles of circular economy. Note. The figure is created by Tan Suan See, Eagle Horizon

durability through repairability, remanufacturing, recycling, and redesigning supply chain stages for sustainability.

Despite being only 8.6% circular, the global economy holds the potential to slash emissions by 39%, according to a 2021 report (Circle Economy 2021). The 2023 Circularity Gap Report indicates growing interest in the circular economy but insufficient action. Global circularity is declining. Secondary material use dropped 21% from 9.1% in 2018 to 7.2% in 2023.

14.2.1 How Does Circular Economy Strengthen Supply Chains?

Supply chains within a circular economy framework are commonly designated as either closed-loop or circular supply chains (Geissdoerfer et al. 2018). Representing a departure from the traditional "take-make-dispose" system, the circular economy emphasizes a regenerative model centered on restoring and renewing products, materials, and energy. It alters conventional performance metrics of value creation and encourages manufacturers to design products with durability, repairability, and recyclability in mind.

Circular supply chains (CSCs) represent a paradigm shift towards sustainable and regenerative practices by applying circular economy concepts to supply chain management, thereby minimizing the business's ecological footprint.

Here are three key tenets where CE principles strengthen supply chain durability:

1. Reduce waste, complexity, toxicity, and harmful substances within products to enhance end-of-life management across the entire supply chain.
2. Maximize the lifespan and quality of raw materials, components, and products.
3. Recycle materials effectively to replenish resource pools.

Hence, integrating CE principles offers a solution to build supply chain resilience as they help supply chains adapt to and maintain operations amidst unforeseen

Fig. 14.3 Traditional linear efficient supply chains. Note. The figure is created by Tan Suan See, Eagle Horizon

disruptions, fostering sustainable growth and resilience in an interconnected world. By implementing circular economy practices such as recycling, remanufacturing, part recovery, repair, recommerce, and refurbishment, organizations can decrease reliance on limited resources and components, fostering adaptable and resilient supply chains.

14.2.2 Making the Transition to Circular Supply Chain (CSC) Models

CSC will fully materialize when all waste and by-products become valuable inputs for other supply chain processes. This transformation will significantly expand the repair, refurbishment, remanufacturing, and recycling industries. Ultimately, CSCs will replace the traditional linear supply chain model depicted in Fig. 14.3.

Transitioning from the prevalent linear supply chain models (Figs. 14.3 and 14.4) of today to CSC requires a shift in thinking, organization capabilities, new employee skills, and strategic ecosystem collaboration to overcome the barriers that lead to the falling rate of CE principles adaptations globally as reported in Circular Gap Report 2023 (Circle Economy 2023).

14.2.3 Fostering Circular Thinking and Performance

We know supply chains have been the core enabler for organizations to achieve every aspect of economic value target. To successfully transition to a circular economy, organizations must establish clear, measurable goals and track their progress. Circular thinking necessitates a shift in performance metrics for traditional linear supply chains. These new metrics are essential for aligning stakeholders and partners around shared strategies and objectives. Transitioning to a CSC requires overcoming deep-rooted resistance at multiple levels of the supply chain network. Internally, in general, organizations face inertia from established linear efficient supply chain processes which have been long developed, with the prevailing "this is how we've always done it" mentality posing a significant barrier to change or transformation.

Externally, at the upstream supply chain, the transition may disrupt the economic value of some suppliers and service providers at the initiation stage, leading to resistance to transformation. While the downstream of supply chain, changing customer

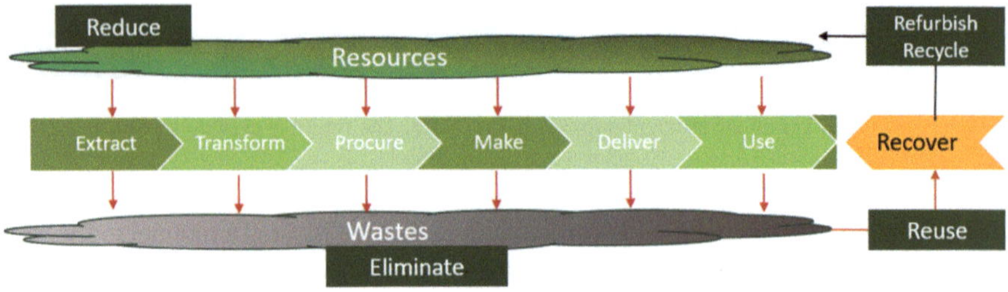

Fig. 14.4 Integrate circular economy in SCM: circular supply chain. Note. The figure is created by Tan Suan See, Eagle Horizon

preferences to align with circular practices requires concerted effort and engagement, as it challenges traditional ways, patterns, and behaviors.

14.2.4 Supply Chain Network Re-Design

Today's supply chain networks focus on optimizing linear material and product flows seamlessly. Circular supply chains require cost-environmentally effective ways to recapture products to extract value. Network design needs to evolve to handle the distribution, collection, and sorting process, allowing product recovery to eliminate waste and reuse products. Consider new partnership models like competitors pooling resources for material reprocessing.

14.2.4.1 Collaboration with Supply Chain Network Partners

Suppliers are vital in guaranteeing consistent supply, high quality, and clear origin of these circular resources. Supply chain leaders need to set up guidelines, criteria, and expectations for their suppliers—and provide support to building suppliers' capability by incorporating circular requirements into supplier qualification processes and agreements.

Customers need to be engaged, too. The Nike or H&M projects don't work unless consumers are willing to send back their shoes or clothes. Getting customers to participate in a closed-loop system in the circular supply chain will require education resources to encourage sustainable behavior.

A McKinsey study shows that 78% of consumers say a sustainable lifestyle is important to them and companies that deliver on that promise can see gains in brand loyalty. As such, a circular supply chain can improve customer retention. The right rate or trade-in program can encourage additional purchases.

Circular supply chains need to ensure return logistics align to maximize product capture and reprocessing as well as multidirectional communication and innovation, so the technology must provide easy ways to accomplish this. It takes a proactive and diligent approach to make the change.

Successful circular supply chain practices require a pragmatic, practical, and full-scale supply-chain-sustainability strategy with circularity in practice. This approach can significantly reduce carbon emissions while accelerating the shift towards a circular economy and enhancing societal well-being.

Organizations must take a systemic approach that transforms everything from strategic goals, policy key performance metrics, incentives, processes, systems, and infrastructure to product design and stakeholder engagement. Breakthroughs on these fronts, powered by advancements in technology for transparency, will be key to realizing supply chains that maximize the value and lifespan of products and materials.

Existing supply chain technologies optimize for linear processes and transactions, so organizations need a fresh approach to redesign the closed-loop system and five stages of supply chain processes, beyond organization boundaries to manage the total life cycle of products and services effectively from tracking and creating visibility with real-time data analytics for an informed decision. For example, leveraging big data analytics, IoT, and AI within supply chain management can significantly enhance visibility into operations. By fostering effective information sharing among network partners and implementing robust tracking, monitoring, and auditing systems, organizations can accurately trace the journey of circular materials, achieve CSC objectives, and ensure regulatory compliance.

14.2.4.2 Supply Chain Network, Model and Process Re-Design

Today, the entire supply chain network is designed to optimize material and product flows at speed. CSCs require cost-effective ways to recapture products to extract value. Therefore, supply chain network design needs to evolve to handle the distribution, collection, and sorting process, allowing product recovery to eliminate waste and reuse products. Consider new partnership models like competitors pooling resources for material reprocessing. The circular economy demands a holistic approach, encompassing all supply chain stages from product development to end-of-life management, including procurement, production, logistics, distribution, after-sales service, return, reuse, remanufacturing, and recycling, as illustrated in Fig. 14.5.

Product Design and Development Stage

Circular supply chains (CSCs) necessitate product lifecycle reorientation, prioritizing end-of-life considerations. This involves incorporating circular economy principles to minimize resource use, extend product lifespan, and optimize product design for easy repair, refurbishment, repurposing, and recycling. The transition requires exploring product-as-a-service models, implementing comprehensive

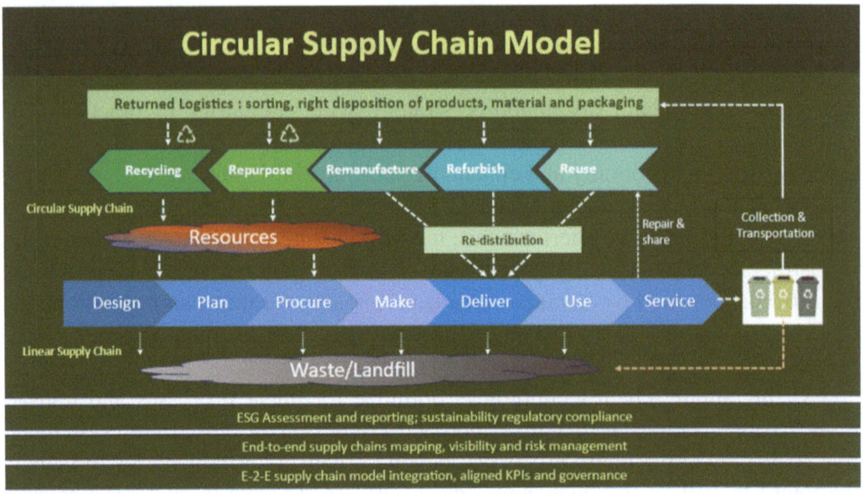

Fig. 14.5 Circular supply chain process. Note. The figure is created by Tan Suan See, Eagle Horizon

product lifecycle management, leveraging digital technologies like digital twins, blockchain, and IoT for data transparency, and establishing recycling programs throughout the supply chain, including take-back initiatives. Key questions arise regarding the organization's capability to utilize less scarce, environmentally friendly materials, extend product lifecycles, and implement cost-effective repair or recycling processes on a large scale.

Source and Procure Stage

Traditionally, procurement focused primarily on economic factors such as price, quality, cost reduction, and delivery speed, often neglecting ESG considerations. Consequently, supplier evaluations rarely included critical criteria like emissions, pollution, energy usage, raw material sourcing, and waste management.

To successfully implement circular supply chains (CSCs), procurement must evolve into an orchestrator of internal functions and a collaborator with external value chain partners. This transformation demands new skill sets to lead and manage an extended network, including second-tier suppliers.

To drive circularity, a robust governance framework is essential for defining, implementing, monitoring, and measuring procurement sustainability. This includes increasing transparency through the sourcing of secondary materials and refurbished components, promoting renewable energy and recycled material usage across products and services, and applying lean principles to minimize waste.

Make or Manufacturing Stage

Manufacturing has historically prioritized efficiency, incorporating energy and resource conservation. However, the focus has traditionally been on economic gains, neglecting scarce resource preservation and broader sustainability goals. By integrating circularity objectives, manufacturers can optimize energy use, transition to renewable sources, minimize waste, and maximize resource recovery, ultimately reducing disposal costs.

Deliver Stage

Logistics, a cornerstone of supply chains, consumes significant energy and resources throughout its processes. While traditional logistics focuses on cost and speed, sustainability is paramount. To optimize efficiency and reduce environmental impact, innovative planning, transportation, and warehousing strategies are essential. Leveraging human-machine collaboration and digitalization, advancements such as electric vehicles, 3D printing, and reusable packaging can transform logistics operations. Implementing a comprehensive roadmap centered on the 4R principles (Reduce, Reuse, Recycle, Repair) and incorporating reverse logistics is vital. By adopting these strategies, businesses can create a circular logistics system, minimizing resource consumption, reducing emissions, and driving sustainable growth.

Use Stage

While circular practices often focus on product design, sourcing, manufacturing, and logistics, the after-sales service phase is frequently overlooked. To maximize value recovery, organizations should implement end-of-life management strategies that incentivize product returns for remanufacturing and recycling. Additionally, product-as-a-service models can extend product lifecycles and generate valuable data on product performance and material flows.

By adopting circular principles, businesses can significantly reduce waste, lower raw material costs, and enhance regulatory compliance.

14.2.5 Key Performance Indicators for Circular Supply Chains

Key Performance Indicators (KPIs) are essential tools for measuring and improving sustainability, resource efficiency, and economic viability in circular economy supply chains. They provide a framework for evaluating the effectiveness of circular strategies like recycling, reuse, and remanufacturing.

One major family of KPIs focuses on circularity metrics. These metrics assess the extent to which resources are reused and recycled, minimizing waste and environmental impact. For instance, the Product Circularity Indicator (PCI) measures the circularity of products at a micro-level, identifying areas for improvement (Bracquené et al. 2022).

Another family of KPIs encompasses the economic, environmental, and social dimensions of circular supply chain management (CSCM). Many independent KPIs have been proposed in literature for each dimension. For example, metrics such as lifecycle greenhouse gas emissions and Total Logistics Costs evaluate the economic feasibility and environmental impact of reusable packaging systems (Betts et al. 2022).

To facilitate the measurement of these KPIs, several frameworks have been developed. The Triple Bottom Line (TBL) Metrics incorporate environmental, social, and economic sustainability metrics for a comprehensive assessment of circular supply chains. Additionally, the Modified Balanced Scorecard adapts traditional performance measurement frameworks to include circular economy principles (Saroha et al. 2022; Kazançoğlu et al. 2024).

14.3 Workforce and Skills Needed for Transitioning to CSCs

The integration of circular economy principles into supply chains is being driven by environmental concerns and economic potential. This section investigates the workforce development needs arising from this transition to circular supply chains. An analysis of the evolving skills required at various industry levels is presented. Additionally, developing effective training programs to equip the workforce with the necessary skills and knowledge to thrive in this new paradigm is essential.

14.3.1 New Employment Effect

There are some significant changing qualification needs in jobs observed for climate change mitigation compliance, green and renewable energies and employee skill profiles in environment and eco-innovations across all sectors. Many jobs will need greening, CE practices skills with a spectrum from sourcing, making, and delivering green goods or services (Kudtarkar and Srivalli 2023; Yun et al. 2023). Employees from all levels of an organization need to expand their skill sets so they can help safeguard the sustainability of the supply chain and the environment.

This section explores the evolving job landscape and categorizes key roles with their corresponding skill sets.

14.3.1.1 Green Supply Chain Expertise

Green Supply Chain Learning Specialists: These specialists develop and deliver training programs that foster employees' understanding of sustainable practices and how they contribute to CE principles. They indirectly impact performance by driving environmentally conscious behavior and innovation.

Environmental Specialists: These professionals possess knowledge of environmental regulations and best practices. They play a crucial role in implementing resource-efficient processes, minimizing environmental impact, and ensuring compliance with regulations (Agyabeng-Mensah et al. 2022; Bassi and Guidolin 2021).

14.3.1.2 Green Procurement and Logistics Professionals

Green Procurement Specialists: These specialists source materials responsibly and sustainably, aligning with CE's focus on resource recovery. They may collaborate with suppliers to identify opportunities for using recycled materials or minimizing packaging waste.

Eco-Design Engineers: These engineers design products for disassembly, reuse, or remanufacture. Their expertise is essential for extending product lifecycles and minimizing waste generation.

Green Logistics Coordinators: These coordinators optimize transportation networks and logistics processes to minimize environmental impact during product movement. They may implement strategies such as route optimization, consolidation of shipments, and utilizing fuel-efficient vehicle (Sarkis and Dou 2017).

14.3.1.3 Circular Leadership and Data Analysts

Sustainable Supply Chain Executives: These leaders possess a strong understanding of environmental regulations, supply chain management principles, and data analytics. They foster a corporate culture that prioritizes sustainability and leverages data to optimize circular practices within the supply chain.

Digital Supply Chain Analysts: These analysts utilize AI-driven big data analytics to strengthen decision-making and facilitate efficient resource management within a circular supply chain. They translate complex data sets into actionable insights for optimizing resource utilization and minimizing waste (Bag and Rahman 2023).

14.3.1.4 E-commerce and Green Operations Managers

Sustainable E-commerce Operations Managers: These managers oversee e-commerce supply chains with a focus on minimizing packaging waste and maximizing recycling. They may implement strategies such as using recycled packaging materials, offering reusable packaging options, and collaborating with partners to improve recycling infrastructure (Kudtarkar and Srivalli 2023).

14.3.1.5 Green Manufacturing and Industry Collaboration Specialists

Green Manufacturing Engineers: These engineers possess the specialized skills and knowledge required to implement CE principles within the manufacturing sector. This may involve optimizing production processes to minimize waste, designing products for disassembly, or developing systems for remanufacturing or refurbishing used products.

Industry Collaboration Specialists: These specialists facilitate collaboration between different players within the industry, such as manufacturers, recyclers, and retailers. This collaboration is crucial for creating efficient resource-sharing networks and infrastructure development to support a circular economy (Schöggl et al. 2023).

14.3.1.6 Environmental Goods and Services Sector Professionals

Environmental Goods and Services Sector (EGSS) specialists is a broad category encompasses a diverse range of professionals focused on delivering environmental goods and services that support sustainable development within the CE framework. Job titles can vary depending on the specific good or service, but may include roles such as renewable energy engineers, waste management specialists, or specialists in developing and implementing eco-technologies. These diverse skills and job roles are crucial for driving the transition towards circular supply chains. They ensure environmental sustainability, resource efficiency, and economic resilience within this evolving economic model (Sulich and Sołoducho-Pelc 2021).

14.3.1.7 Developing Competencies for a Circular Economy: Skills Matrix for Supply Chain Transformation

The rapid evolution of supply chains necessitates continuous advancement in the field of supply chain management, demanding ongoing development from professionals and their organizations. Effective execution and leadership are crucial, requiring managers and leaders with expertise in global standards and practices that demonstrably improve supply chain performance (Cooper et al., 2016).

Table 14.1 The knowledge, skills, and competencies necessary for circular supply chain planning and procurement

CSC skills	Knowledge	Technical skills	Soft skill	Competency
Leadership	Awareness of regulations and standards for sustainable procurement and zero waste	Map SC network, perform supplier assessments, and build frameworks to minimize waste with KPIs aligned to sustainability goals	Advocate for circular supply chains, motivate suppliers towards innovation, and lead change across the SC network	Create and deliver value through implementing circular practices
Management	Embrace environmental issues, climate crisis impact, and deep knowledge of CSC and carbon footprint	Perform supplier assessments, analyze sustainable products, monitor waste reduction, and integrate network processes	Manage supplier relationships, promote transparency and collaboration across the value chain	Drive innovation towards circular supply chains at multiple levels
System, Operations Process	Understand organization's sustainability vision, align KPIs with strategic goals	Collect and analyze data on resource use and carbon impact, comply with best circular practices	Monitor and report on circularity of products and materials	Competent in tracing, monitoring, and reporting within circular supply chains

This section proposes two circular supply chain skills matrix tables (Tables 14.1 and 14.2) developed through insights gathered from interviews, training, and consultations with a broad spectrum of supply chain professionals, CEOs across various industries, and private sector entities. These skills matrices aim to equip organizations with practical tools to assess their unique training needs in the context of transitioning to a circular supply chain.

Furthermore, these matrices serve as a reference for companies committed to transforming their linear supply chains into circular models. They can be utilized for:

1. **Workforce Competency Assessment**: Identifying knowledge and skill gaps (technical and soft) within the existing workforce regarding circular supply chain implementation.
2. **Environmental Challenge Evaluation**: Gauging the organization's current capacity to address environmental challenges such as climate change, waste, pollution, supply chain disruptions, material shortages, biodiversity loss, and rising material costs.
3. **Employee Development**: Supporting employees with existing sustainability and circularity skills to leverage their expertise to unlock new opportunities and develop strategies that generate regenerative value for the company, people, and the planet.

Table 14.2 Supply chain operations: manufacturing and logistics

CSC skills	Knowledge	Technical skills	Soft skills	Capabilities
Leadership	Awareness of sustainability with closed-loop systems; understand ESG, cycle management, CE models, and principles	Use data-driven scenario planning and digital twin modelling for informed decisions on production and green logistics	Promote continuous improvement and circular thinking; communicate vision and share commitment with stakeholders	Drive organization and partners to operate sustainably, aiming for growth with a triple bottom line
Management	Master sustainability and closed-loop systems, Scope 1, 2, 3, ESG reporting, and SHE compliance	Implement circular projects, monitor and report circularity risks and opportunities across the supply chain	Engage stakeholders with circular SC change management, establish guidelines, frameworks, and targets	Adapt to change, anticipate, and spot opportunities for waste reduction and efficient resource use
System, Operations Process	Understand new materials, recyclable packaging, process techniques, and technologies for circular economy solutions	Manufacturing refurbishment, de-cycle/ repurpose, reverse logistics, digitalize or automate operations	Manufacturing refurbishment, de-cycle/ repurpose, reverse logistics, digitalize or automate operations	Build circular SC operations processes and systems with KPIs for operational efficiency and sustainable manufacturing and logistics solutions

4. **Role Enhancement Analysis**: Utilizing the matrix to evaluate how embedded circular economy skills can strengthen key roles within the organization and champion the transition towards a circular model.

Table 14.1 details the knowledge, skills, and competencies necessary for circular supply chain planning and procurement.

14.3.1.8 Sector-Specific Considerations

The transition to a circular supply chain presents unique challenges and opportunities across different sectors.

- **Construction and Infrastructure**: These sectors often utilize materials with limited regenerative potential, such as concrete and cement. Consequently, their transition to net-zero emissions and circularity may be slower. Close collaboration with suppliers to define requirements and drive innovation towards sustainable solutions will be crucial for these sectors.

- **Retail and Manufacturing**: Sectors that deal in physical materials or products will see growing emphasis on material science, retrofitting, repurposing, and remanufacturing. Additionally, global supply chain management will necessitate increasingly complex data analysis, tracking, and reporting of carbon footprint, waste generation, and biodiversity impact throughout all stages of the supply chain.
- **Heavy Asset Sectors**: Exploring new business models, such as product-as-a-service systems, will be critical for sectors that rely on heavy assets.
- **Finance and Digital Sectors**: These sectors, which focus on selling services or solutions, will prioritize innovation in sustainable service design, customer experience, and strategies to encourage consumer behaviour that reduces their carbon footprint and embraces lower-carbon alternatives.

14.3.1.9 Skills Shifts and Restructuring: Enablers and Obstacles for Transition

This section explores the challenges and enablers associated with adopting circular economy skills within the context of transitioning to a circular supply chain (see Table 14.3).

A key driver of this technical shift is the transformation of required skillsets. Certain efficient supply chain skills optimized for linear systems may become less valuable in a circular model. Conversely, implementing circular economy (CE) strategies necessitates that businesses extend their activities throughout the product lifecycle and foster deeper integration with business partners and the broader ecosystem to stimulate circular innovation.

Performance-related skills become even more crucial in this context. Organizations must develop the ability to connect performance metrics with their network supply chain strategy and establish sustainable key performance indicators (KPIs) that effectively balance cost, quality, time, and the circularity of the entire supply chain network, moving beyond a focus on a single company's performance.

Table 14.3 Challenges and enablers in each stage of implementing CE and CSC

	Challenges	Enablers	References
Understanding CE	• Leaders have limited understanding of CE and sustainable practices could add value in the business and unclear direction for change • Employees have limited idea of CE and sustainable practice, and how to be engaged to affect CE	• Commit and communicate the company's commitment to CE • Create awareness with employee engagement on CE and sustainability competency-based training	Palombi et al. (2024), Fernandez et al. (2020)
Developing CSC	• Lack of clear strategy and plan actions in CE • Rigid processes and structure to adapt to the change • Sustainable with CE practice seems less important	• Forming task forms to focus on adopting CE practices with agility • Prioritize of work with CE values • On-job learning of CSC	Abu-Bakar and Charnley (2024), Glomann (2021)
Extending CSC	• Leaders are unwilling to invest and pursue CE Innovation amidst uncertainty in business • CE not embedded for the long term, limited focus on sustainability, and CE with difficulty in tracking and seeing the value of CE	• Clear policy and direction with sustainable goals • Continuous sustainability and CE learning and development	Stingl et al. (2023), Ibrahim (2024), Palombi et al. (2024)
Leading Network Partner	• Network partners show limited commitment to the sustainable CE journey, misalignment with suppliers • Limited specialist talent with CE technical skill	• Active supplier collaborative program to improve circularity • Supplier upskilling workforce with CE and certify themselves to green circularity • Partner with educational institution to develop new talent to lead the change	Walker et al. (2022)

14.4 Bridging the Circular Economy Skills Gap: Strategies for Workforce Development

Addressing the skills and competency gaps within the existing workforce, particularly supply chain personnel necessitate a multi-pronged strategic approach. This section outlines key steps that organizations can implement to bridge the circular economic gap through comprehensive strategies.

14.4.1 Integrating Circular Economy into Corporate Culture

The foundation for a successful transition lies in embedding circular economy principles into the very core of the organization. This can be achieved by revising the company vision, mission, and values to explicitly include CE principles. Clear communication of these revised values to all employee levels is essential to ensure understanding and buy-in. Additionally, strong leadership commitment is crucial. Leaders must champion the circular economy by setting a positive example and demonstrating their dedication through visible and tangible actions.

14.4.2 Implementing Training and Development Programs

14.4.2.1 Internal Development

Tailored Training Programs

Develop training programs specifically focused on circular economy principles and practices. This may include workshops, seminars, and online learning courses.

Competency-Based Learning

Utilize competency-based learning as a core element of workforce development. This approach involves assessing the current skills and knowledge of the supply chain workforce against the competencies outlined in the skills matrix. The goal is to identify skill gaps and provide targeted training that addresses these specific needs, focusing on developing practical skills rather than solely theoretical knowledge. The scope of training can encompass:

- Cognitive Skills: The ability to think critically and creatively within a circular economy framework, including problem-solving approaches that prioritize sustainability.

- Technical Skills: Expertise relevant to specific job functions, products, policies, and processes within the circular economy context.
- Soft Skills: Emotional intelligence to foster effective communication and collaboration, alongside behavioural aspects like positive attitudes and adaptable work habits.
- Knowledge-Based Skills: Understanding of relevant regulations, certifications, rules, laws, compliance requirements, and core circular economy concepts.

To prepare for competency-based training, a five-step process is recommended:

1. **Identify Competencies**: Define the specific competencies required for each job role at every organizational level (leader, manager, execution employee), with desired outcomes as a guiding principle.
2. **Develop Job Competency Standards and Assessment Criteria**: Establish organizational job competency standards and assessment criteria. Clearly communicate these standards to the workforce involved in the circular supply chain initiative.
3. **Performance Appraisal**: Evaluate employee performance against the established competency standards. Record identified skill gaps and areas for improvement, along with employee strengths.
4. **Training and Development Opportunities**: Provide training and development opportunities to address the identified competency gaps. It's crucial to frame any identified shortcomings in a positive light.
5. **Continuous Monitoring and Assessment**: Continuously monitor and evaluate employee performance and competency progress to ensure ongoing development. Implementing regular reinforcement training courses is a valuable approach to keep employees on track.

External Training

To bridge the skills gap in supply chain sustainability, organizations can leverage comprehensive training programs offered by reputable providers. Some notable options include:

- **APICS**: A leading industry authority that offer comprehensive education, certification, benchmarking, and best practices to develop supply chain talent and enhance overall supply chain performance.
- **Cambridge Institute for Sustainability Leadership**: Offers specialized programs designed to equip professionals with the knowledge and skills to build sustainable supply chains.
- **International Supply Chain Education Alliance**: Provides a Certified Sustainable Supply Chain Professional program.
- **Sustainability Academy**: Offers a Diploma in Sustainable Supply Chain Management.

- **PECB Certified ISO 20400 Lead Manager from Abilene Academy**: Provides training on the ISO 20400 standard for sustainable procurement.
- **Sustainable Procurement Institute**: Offers training programs in Sustainable Procurement for Professionals.

14.4.3 Leveraging Existing Expertise

- **Identifying Champions:** Recognize employees who demonstrate passion and strong knowledge/experience in circular economy practices. Appoint them to lead, mentor, and train other employees throughout the transition journey.
- **Forming Cross-Functional Teams:** To foster innovation and collaboration, establish cross-functional teams comprising employees from diverse departments and external partners to drive circular economy initiatives. This facilitates knowledge sharing, expertise exchange, and collaborative problem-solving.

14.4.4 Collaboration with Educational Institutions

- **Partnerships**: Develop partnerships with universities and colleges to create curriculum and programs focused on circular economy and its integration with supply chain management. This collaboration might also involve joint research projects.
- **Expert Speakers**: Invite industry experts and academics with circular economy knowledge to speak to employees about the latest trends, technologies, CE business case studies, and best practices.

14.4.5 Utilizing Digital Technology

- **Digital Platforms**: Leverage digital platforms and tools to facilitate learning and collaboration across the supply chain network. These may include e-learning platforms, knowledge management systems, and collaboration tools for planning, communication, and reporting within the organization and with external network partners.
- **Simulations and Models**: Implement simulation tools and models to help employees understand and practice circular economy scenarios in a risk-free environment.

14.4.6 Digital Competency

The main technologies that can be leveraged as digital competencies in adopting CE practices include the IoT, Big Data, Blockchain, AI, Building Information Modelling (BIM), Material Passports (MPs), and digital twins. IoT and Big Data are instrumental in enhancing resource monitoring, traceability, and data management, thereby improving recycling, recovery, and reuse processes (Carlos et al. 2024). Blockchain technology provides security, reliability, and traceability, which are crucial for implementing CE practices effectively (Carlos et al. 2024; Trubina et al. 2024). AI, along with machine learning, aids in material classification, predictive maintenance, and optimizing resource allocation, thus facilitating efficient decision-making and reducing waste (Aponte 2024; De Wolf et al. 2024). BIM and MPs are particularly significant in the construction industry, where they help in assessing resource availability, optimizing waste reduction, and enabling smart deconstruction planning (Trubina et al. 2024). Digital twins and augmented reality also play a role in simulating and visualizing processes, which can lead to more efficient resource utilization and process optimization (Aponte 2024). Additionally, digital product passports and AI-assisted material classification enhance the reuse of construction materials by providing detailed information on material properties and facilitating their cataloguing and inventory management (De Wolf et al. 2024). Interorganizational collaboration, supported by these digital technologies, is essential for the successful implementation of CE practices, as it enhances sustainability and economic performance at the firm level (Schöggl et al. 2023). By integrating these digital technologies, organizations can accelerate the transition to a circular economy while driving innovation and efficiency across industries, positioning them as essential tools for sustainable growth.

Figure 14.6 highlights the main barriers to integrating CE principles in the Supply Chain, the diversity of workforce required with different expertise and workforce development needed are also included.

14.5 Conclusion

This paper demonstrated that a strategic approach to workforce development is crucial for organizations transitioning to circular supply chains. The strategic approach emphasizes leveraging the existing talent pool by providing upskilling and knowledge acquisition opportunities. By employing a multi-faceted learning and development strategy that incorporates both top-down and bottom-up initiatives, organizations can effectively bridge the circular economy skills gap. This, in turn, empowers the workforce to drive sustainable and circular practices not only within the organization itself, but also throughout all stages of the supply chain ecosystem.

To successfully integrate circular economy principles into supply chain management, organizations must develop a strategic roadmap. Investing in employee

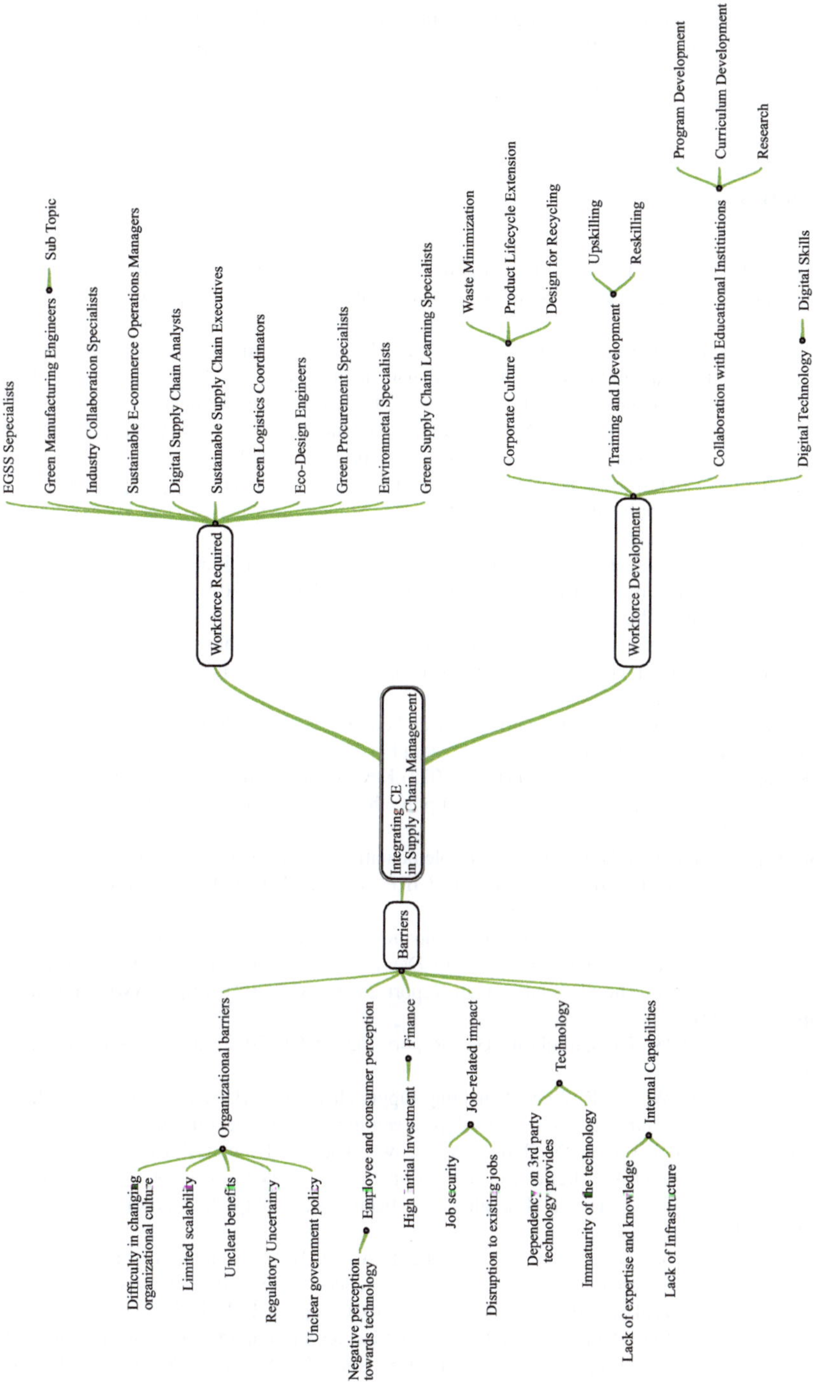

Fig. 14.6 Integrating CE in supply chain management

reskilling and upskilling is essential to acquire the necessary expertise for implementing circular supply chain practices and maximizing the potential of emerging digital technologies. A holistic approach that aligns people, processes, and technology is crucial for achieving CE objectives.

References

Abu-Bakar H, Charnley F (2024) Developing a strategic methodology for circular economy roadmapping: a theoretical framework. Sustainability 16(15):6682. https://doi.org/10.3390/su16156682

Agyabeng-Mensah Y, Afum E, Baah C, Essel D (2022) Exploring the role of external pressure, environmental sustainability commitment, engagement, alliance and circular supply chain capability in circular economy performance. Int J Phys Distrib Logist Manag 52(5/6):431–455

Alavi B, Tavana M, Mina H (2021) A dynamic decision support system for sustainable supplier selection in circular economy. Sustain Prod Consumpt 27:905–920

Amir S, Salehi N, Roci M, Sweet S, Rashid A (2023) Towards circular economy: a guiding framework for circular supply chain implementation. Bus Strateg Environ 32(6):2684–2701. https://doi.org/10.1002/bse.3264

Aponte GM (2024) Tendencias en la producción científica en el área de tecnologías digitales y economía circular. Rev Gestión Las Personas Tecnol 17(49):156–176. https://doi.org/10.35588/6436n243

Bag S, Rahman MS (2023) The role of capabilities in shaping sustainable supply chain flexibility and enhancing circular economy-target performance: an empirical study. Supply Chain Manag Int J 28(1):162–178

Bassi F, Guidolin M (2021) Resource efficiency and circular economy in European SMEs: investigating the role of green jobs and skills. Sustain For 13(21):12136

Betts K, Gutierrez-Franco E, Ponce-Cueto E (2022) Key metrics to measure the performance and impact of reusable packaging in circular supply chains. Front Sustain 3:910215. https://doi.org/10.3389/frsus.2022.910215

Bracquené E, Lindemann J, Duflou J (2022) Implementation of circularity indicators in a household product manufacturing company. Procedia CIRP 105(3):660–665. https://doi.org/10.1016/j.procir.2022.02.110

Carlos RL, De Souza EB, Mattos CA (2024) Enhancing circular economy practices through the adoption of digital technologies. Bus Strategy Dev 7(1):e330. https://doi.org/10.1002/bsd2.330

Circle Economy (2021) The Circularity Gap Report 2021. CGR 2021. https://www.circularity-gap.world/2021

Circle Economy (2023) The Circularity Gap Report 2023. CGR 2023. https://www.circularity-gap.world/2023

Cooper S, Watson D, Worrall R (2016) Managing supply chain networks: a framework for achieving superior performance through leadership capabilities development in supply chain node

De Wolf C, Byers BS, Raghu D, Gordon M, Schwarzkopf V, Triantafyllidis E (2024) A 5D digital circular workflow: digital transformation towards matchmaking of environmentally sustainable building materials through reuse from disassembly. https://doi.org/10.21203/rs.3.rs-4349460/v1

Do Q, Mishra N, Colicchia C, Creazza A, Ramudhin A (2022) An extended institutional theory perspective on the adoption of circular economy practices: insights from the seafood industry. Int J Prod Econ 247(4):108400. https://doi.org/10.1016/j.ijpe.2021.108400

Fernandez A, Kullu FD, Shankar R (2020) A grounded research approach to sustainable leadership practices and competencies. In: Vanka S, Rao MB, Singh S, Pulaparthi MR (eds)

Sustainable human resource management. Springer, Singapore, pp 71–86. https://doi.org/10.1007/978-981-15-5656-2_5
Fraser M, Haigh L, Soria A (2023) Circularity Gap Report 2023: towards a circular economy. Circle Economy, Netherlands. Retrieved from https://coilink.org/20.500.12592/wr5sc0. On 06 May 2025. 20.500.12592/wr5sc0
Geissdoerfer M, Morioka SN, de Carvalho MM, Evans S (2018) Business models and supply chains for the circular economy. J Clean Prod 190:712–721
Glomann L (2021) Agile circular design. In: Ahram TZ, Karwowski W, Kalra J (eds) Advances in artificial intelligence, software and systems engineering, vol 271. Springer International Publishing, pp 279–286. https://doi.org/10.1007/978-3-030-80624-8_35
Huan SH, Sheoran SK, Wang G (2004) A review and analysis of supply chain operations reference (SCOR) model. Supply Chain Manage Int J 9(1):23–29
Ibrahim M (2024) Integrating circular economy principles for sustainable development. In: Ordóñez De Pablos P (ed) Advances in finance, accounting, and economics. IGI Global, pp 142–165. https://doi.org/10.4018/979-8-3693-2750-0.ch007
Kazançoğlu Y, Ekinci E, Mangla SK, Sezer MD, Choudhary S, Dora M (2024) Smart closed-loop food supply chain in circular economy. IEEE Trans Eng Manag 71:12346–12363. https://doi.org/10.1109/TEM.2023.3303272
Kirchherr J, Reike D, Hekkert M (2017) Conceptualizing the circular economy: An analysis of 114 definitions. Resour Conserv Recycl 127:221–232
Kirchherr J, Yang NHN, Schulze-Spüntrup F, Heerink MJ, Hartley K (2023) Conceptualizing the circular economy (revisited): an analysis of 221 definitions. Resour Conserv Recycl 194:107001
Kudtarkar SG, Srivalli B (2023) A circular supply chain in E-commerce businesses in India. In: Handbook of research on designing sustainable supply chains to achieve a circular economy. IGI Global, pp 82–96
Lahane S, Kant R, Shankar R, Patil SK (2023) Circular supply chain implementation performance measurement framework: a comparative case analysis. Prod Plan Control 35(11):1–20. https://doi.org/10.1080/09537287.2023.2180684
Liu Z, James S, Walpole G, White GR (2023) A communities of practice approach to promoting regional circular economy innovation: evidence from East Wales. Eur Plan Stud 31(5):988–1006
Mhatre P, Panchal R, Singh A, Bibyan S (2021) A systematic literature review on the circular economy initiatives in the European Union. Sustain Prod Consum 26:187–202. https://doi.org/10.1016/j.spc.2020.09.008
Orji IJ, Ojadi F (2022) The circular supply chain: basic principles and techniques. CRC Press
Palombi G, Schiaroli V, Fraccascia L, Nonino F (2024) Identifying barriers and good practices for implementing circular economy principles in small and medium enterprises. Bus Strateg Environ. https://doi.org/10.1002/bse.3835
Ravimalar R (2024) Is it a good idea to use waste to generate energy? https://www.tnb.com.my/assets/newsclip/20052024g1.pdf
Sarkis J, Dou Y (2017) Green supply chain management: a concise introduction. Routledge
Saroha M, Garg D, Luthra S (2022) Analyzing the circular supply chain management performance measurement framework: the modified balanced scorecard technique. Int J Syst Assur Eng Manag 13(Suppl 2):951–960
Sato E (2018) A dinâmica da integração internacional das economias e o Brasil: Um exercício teórico. Revista de Estudos e Pesquisas Avançadas Do Terceiro Setor 5(1):635. https://doi.org/10.31501/repats.v5i1.9805
Schöggl JP, Stumpf L, Baumgarter RJ (2023) The role of interorganizational collaboration and digital technologies in the implementation of circular economy practices: empirical evidence from manufacturing firms. Bus Strategy Env 33(3):1–25. https://doi.org/10.1002/bse.3593
Scott M (2019) Companies look to supply chains for sustainability gains. Forbes
Stingl V, Fuglsig LV, Hoveling C (2023) Uncertainty management in circular business model innovation: the case of circular plastics. Proc Des Soc 3:3671–3680. https://doi.org/10.1017/pds.2023.368

Sulich A, Sołoducho-Pelc L (2021) Renewable energy producers' strategies in the Visegrád group countries. Energies 14(11):3048

Trubina N, Leindecker G, Askar R, Karanafti A, Gómez-Gil M, Blázquez T, Güngör B, Bragança L (2024) Digital technologies and material passports for circularity in buildings: an in-depth analysis of current practices and emerging trends. In: Ungureanu V, Bragança L, Baniotopoulos C, Abdalla KM (eds) 4th International conference "coordinating engineering for sustainability and resilience" & midterm conference of circularb "implementation of circular economy in the built environment", vol 489. Springer Nature, pp 690–699. https://doi.org/10.1007/978-3-031-57800-7_64

Varjú V, Óvári Á, Mezei C, Suvák A, Vér C (2022) Efforts and barriers shifting a city region towards circular transition: lessons from a living lab from Pécs, Hungary. Future Cities Environ 8:10. https://doi.org/10.5334/fce.157

Walker AM, Opferkuch K, Lindgreen ER, Simboli A, Vermeulen WJV, Raggi A (2022) Assessing sustainability across circular inter-firm networks. In: Wojnarowska M, Ćwiklicki M, Ingrao C (eds) Sustainable products in the circular economy, 1st edn. Routledge, pp 154–171. https://doi.org/10.4324/9781003179788-10

World Economic Forum (2021) Net-zero challenge: the supply chain opportunity. Insight Rep 46

Zarbà C, Chinnici G, La Via G, Bracco S, Pecorino B, D'Amico M (2021) Regulatory elements on the circular economy: driving into the agri-food system. Sustainability 13(15):8350. https://doi.org/10.3390/su13158350

Zhang A, Duong L, Seuring S, Hartley JL (2023) Circular supply chain management: a bibliometric analysis-based literature review. Int J Logist Manage 34(3):847–872. https://doi.org/10.1108/IJLM-04-2022-0199

Chapter 15
A Comparative Analysis of Sustainability Frameworks for Organizational Decision-Making

Khaoula Khlie, Zoubida Benmamoun, and M. Aruna

Abstract In an era where sustainability has become a paramount concern for organizations worldwide, the proliferation of sustainability frameworks has provided both opportunities and challenges. Various entities, from multinational corporations to educational institutions, are increasingly expected to measure, report, and improve sustainability performance using frameworks such as the UN Green Metrics, THE Impact, QS Stars, STARS, and QS Sustainability. However, selecting the most appropriate framework amidst this multitude of options remains daunting for many organizations. To close the gap, this work provides a thorough framework for comparison that aids businesses in choosing the most appropriate sustainability measures.

Keywords The impact · QS stars · QS sustainability · UN green · Sustainability frameworks · Comparison · Decision making

15.1 Introduction

Numerous sustainability metrics frameworks have been developed in response to the pressing need to address global sustainability issues and assist firms in implementing more ethical practices. These frameworks, which range from sector-specific projects like QS Stars for Higher Education to worldwide initiatives like the United

K. Khlie · M. Aruna
Department of Industrial Management, College of Business, Liwa University, Abu Dhabi, UAE

Z. Benmamoun (✉)
Department of Mechanical and Industrial Engineering, College of Engineering, Liwa University, Abu Dhabi, UAE
e-mail: zoubida.benmamoun@lc.ac.ae

K. Mathiyazhagan et al. (eds.), *Humanizing Circular Supply Chain Management*, Circular Economy and Sustainability,
https://doi.org/10.1007/978-3-031-86562-6_15

Nations Sustainable Development Goals (UN SDGs), provide many ways to assess and analyze sustainability performance.

Our study compares sustainability frameworks, including UN Green Metrics, THE Impact Rankings, QS Stars, STARS, and QS Sustainability. Organizations can, however, sort through the various frameworks available and choose the one that best fits their objectives, core values, and practical requirements.

Through a comprehensive evaluation of each framework's advantages, disadvantages, and suitability, our goal is to develop a new comparative framework that assists companies in making well-informed decisions regarding their sustainability reporting and measuring plans. Furthermore, using our suggested comparison methodology on a real-world case study, Liwa College, this work goes beyond theoretical analysis. By examining the unique requirements, concerns, and obstacles that Liwa College encounters, we demonstrate how our framework may direct organizational choices toward more sustainable practices. Taking this all-encompassing approach, we hope to further sustainability initiatives in practical settings and add to scholarly conversations.

The following sections of this document will be organized as outlined below: Sect. 15.2 furnishes a comprehensive outline of the prevailing sustainability frameworks under scrutiny, delineating their principal characteristics and objectives. Section 15.3 introduces the proposed comparison framework and discusses its applicability in guiding organizational decision-making. Section 15.4 offers insights from the application of the framework to the case study of Liwa College. Section 15.5 provides the paper's concluding remarks by offering insights into the research's contributions and implications and presenting recommendations for prospective studies. Organizational sustainability encompasses a range of dimensions, including environmental, social, and economic aspects. Utilization of sustainability frameworks helps an organization to evaluate and improve its sustainability performance by using structured methodologies for measurement and assessment.

This comparative analysis examines multiple widely used sustainability frameworks like THE Impact Rankings, UI Green Metrics, STARS, QS Stars, and QS Sustainability with a focus on their strengths, weaknesses and suitability in various organizational contexts. Choosing a sustainability framework depends on the goals, priorities, and resources available to an organization. By weighing the pros and cons of each framework thoroughly, organizations can choose the one that best suits their sustainability targets as well as strategic objectives. Such methodology allows decision-makers to make informed decisions that enhance sustainability performance and influence future outcomes of an organization.

15.2 Literature Review

Assess the sustainability framework for higher education institutions (HEIs) to be measured by considering how well each level aligns with the current curriculum through its transformative potential and in light of contribution towards sustainable development goals. This has included assessing the strengths and limitations of different frameworks, including but not limited to the Sustainable Development Goals (SDGs), three-pillar framework, possibility frontiers analysis, and extended accounting sandwiches technique balance sheets incorporated into a capital approach (Simon and Angela 2023). The Times Higher Education (THE) Impact Rankings are more comfortable for assessing how well universities in the world are doing in promoting the UN Sustainable Development Goals (SDGs) (Sheila Mae et al. 2023). In addition, the research above focusing on humanities and social sciences schools of Wuhan University in China advocated that establishing a complete evaluation framework can help to systematically evaluate school/college performance at HEIs concerning their sustainable development (James and Ballinger 2023). These tools and frameworks used by HEIs might help develop Sustainability in their curricula, thus contributing to the establishment of a sustainable world.

How higher education is taking on the challenge of Sustainability has attracted interest from all over the world, and different types of frameworks have been developed to support efforts for assessing and enhancing institutions' activities related to Sustainability. They are designed to evaluate various environmental, social, and economic sustainability attributes. The predominant evaluation tool in this area is the Sustainability Tracking, Assessment and Rating System (STARS), which represents a well-thought-out structure of indicators for assessing institutional sustainability (Kosta 2019). Furthermore, methodologies, including the Campus Sustainability Assessment Framework (CSAF) and over-arching measures such as those used in the case of Governance for Sustainable Universities: Developing a Delphi Process, sought to explore approaches employed by universities that initiate sustainability schemes (Sayed and Asmuss 2013).

Numerous research studies have found that developing a comprehensive evaluation tool for higher education establishments such as universities and liberal arts colleges is essential for Sustainability. Another proposal suggested a methodology for ranking university sustainability, including measurements from STARS, the American Colleges and Universities Presidents' Climate Commitment (ACUPCC), and the Green Report Card (Yturzaeta 2023). It was also highlighted that at a higher education institution level, there is a demand for cradle-to-grave assessment tools for Sustainability reporting with respect to proficiency in Education/reflections/research/outreach (Sepasi et al. 2018). The literature also underscores the necessity of including Sustainability in education and calls for skill development toward sustainability education at higher levels.

A single study examined the essential abilities and instructional techniques for sustainability education applications in better schooling. They have emphasized the importance of institutional sustainability administrators and coordinators in advancing sustainability projects (Kosta 2019). The authors analyzed the development and alertness of a framework for evaluating students' sustainability capabilities. They determined that the DAB framework as it should be depicts college students' sustainability skills (Besong and Holland 2015).

The research highlights the importance of identifying and integrating sustainability skills into better schooling curricula to equip college students to cope with problematic sustainability-demanding situations in their destiny personal and expert endeavours. The study stresses the need for creative and efficient coaching techniques to train approximately sustainability. Active knowledge of palms-on studying, problem-primarily based mastering, and interdisciplinary strategies are recommended to beautify scholar engagement and information on sustainability problems (Evans 2019).

The look at outcomes suggests that a complete and integrated technique to better schooling is necessary to expand vital talents for sustainable development. The studies emphasize the significance of updating the curriculum, promoting critical thinking, and fostering interdisciplinary mastering. It highlights the involvement of all stakeholders, together with students, school, and directors, in developing getting-to-know surroundings that help sustainability. The authors emphasize the importance of continuous expert improvement for educators to successfully equip them with the necessary abilities to educate about sustainability (Barth et al. 2007). They argue that sustainability training needs to be conscious of developing abilities consisting of structure questioning and anticipatory, normative, strategic, and interpersonal competencies.

In addition, the research proposes a framework for improving instructional programs, highlighting the significance of experiential and problem-based learning processes. The authors additionally stress the need for collaboration across disciplines. They advise that better education establishments implement supportive systems and policies to integrate these competencies into their curricula (Wiek et al. 2011).

The authors argue that sustainability declarations and charters are essential for guiding higher training establishments closer to sustainable improvement. These declarations offer a structure for universities to set up sustainability objectives, engage with stakeholders, and tune progress. The authors also advocate the significance of management dedication, institutional recommendations, and the combination of sustainability in coaching, studies, and operations. The look underscores the need for ongoing assessment and adjustment of sustainability techniques to address changing and emerging demanding situations (Lozano et al. 2013). The observation indicates that including sustainability in college guides is elaborate because more human beings need to study it, schools need more assistance, and teachers want extra training. To repair these problems, they want to make more human beings aware of sustainability, assist instructors in coaching it, and encompass sustainability within the school's plans and actions. Thomas emphasizes the function of management in driving the sustainability agenda and creating supportive surroundings for curricular alternatives (Thomas 2004).

In precis, organizing and imposing effective sustainability evaluation frameworks in higher training is essential for nurturing a culture of sustainability within academic institutions. The frameworks above play an essential function in evaluating and improving sustainability practices across various parameters and, in the long run, selling sustainable development targets within higher training.

15.3 Methodology

To facilitate the evaluation and selection manner of sustainability metrics frameworks, we have evolved a comprehensive evaluation framework encompassing important criteria essential for organizational selection-making. The numerous criteria function as evaluative measures for assessing the appropriateness and concordance of sustainability frameworks with organizational goals.

15.3.1 Factor Identification

The choice of comparison criteria becomes decided based totally on the company objectives outlined in the sustainability projects. The following criteria have been selected to compare our frameworks:

15.3.1.1 Operational Excellence

Operational excellence encompasses evaluating the operational framework, focusing on the efficiency, effectiveness, and consistency of processes and activities. This criterion emphasizes optimizing resource utilization and productivity to achieve high performance and continuous improvement in operations. The chosen framework must facilitate streamlined processes and effective resource management, ultimately contributing to sustainable business practices. A comprehensive assessment of various lean production frameworks indicates that while they enhance operational efficiency, surpassing traditional lean paradigms is essential for sustainable operational excellence. Integrating lean principles with other management strategies is recommended to elevate productivity and resource utilization. Noteworthy findings underscore the significance of continuous improvement, waste reduction, and active employee engagement in achieving operational excellence (Rich and Bateman 2003).

15.3.1.2 Focus on Net Zero Carbon Emissions

Given the institution's paramount long-term objective of achieving net zero carbon emissions, it has been integrated as a pivotal criterion. This criterion assesses the extent to which each framework prioritizes the reduction of greenhouse gas emissions and the transition to carbon neutrality or net-zero emissions. Frameworks must endorse robust strategies for diminishing carbon footprints in alignment with global climate goals. The Paris Agreement underscores the international dedication to capping global warming below 2 °C above pre-industrial levels, striving to limit the increase to 1.5°. The agreement delineates specific measures to mitigate greenhouse gas emissions, including advancing the adoption of renewable energy, enhancing energy efficiency, and promoting sustainable land use practices (Focus on Net Zero Carbon Emissions: United Nations 2015). It also stresses the necessity of international collaboration and the mobilization of financial resources to support climate action.

15.3.1.3 International Benchmarking

A critical criterion is the framework's incorporation of global sustainability standards and benchmarks to facilitate international comparison and alignment. This aspect is pivotal in ensuring global recognition and compliance with international standards, promoting broader acceptance and implementation. By embracing internationally benchmarked frameworks, an organization can bolster its credibility and cultivate global partnerships. The research emphasizes that national prosperity is not inherited but created. A nation's competitive advantage hinges on its industries' ability to innovate and upgrade. International benchmarking is instrumental in identifying best practices and nurturing a culture of continuous improvement. The research asserts that nations can attain a competitive edge through robust domestic competition, strong supplier networks, and the presence of sophisticated and discerning local customers (Porter 1990). Consequently, the significance of international benchmarking in learning from global leaders and aligning with international standards is underscored.

15.3.1.4 Local Benchmarking

The assessment focuses on the framework's ability to integrate local contextual factors and benchmarks, enhancing its relevance and suitability for addressing regional sustainability challenges and priorities. The framework's adaptability to local environmental, economic, and social conditions is evaluated through local benchmarking to promote practical and feasible sustainability initiatives. The work highlights the significance of sustainable business models in addressing local sustainability challenges. Organizations can tailor global sustainability practices to local contexts by leveraging local benchmarking, ensuring their practicality and relevance. Noteworthy findings of the study include the identification of diverse, sustainable business model archetypes, such as valorizing waste, transitioning to renewable resources, and emphasizing functionality over ownership. Furthermore, the study underscores the importance of businesses engaging with local stakeholders and comprehending regional sustainability priorities to formulate efficient strategies (Bocken et al. 2014).

15.3.1.5 Alignment with the National Sustainability Strategy

Assessing the framework's alignment with the objectives and goals mentioned within the National Sustainability Strategy is essential to ensure compliance with countrywide sustainability priorities and agendas. The National Sustainability Strategy defines the United States lengthy period as imaginative and prescient and targets sustainable development. Ensuring organizational frameworks are coherent with this approach enables the success of broader sustainability dreams and countrywide priorities (National Sustainability Strategy Document n.d.). The plan addresses critical consciousness regions, which include renewable power adoption, waste management, water conservation, and sustainable agriculture, presenting a roadmap for achieving sustainability targets and emphasizing the significance of multi-stakeholder collaboration and policy harmony.

15.3.1.6 Alignment with the institution's strategy

The evaluation criterion assesses the quantity to which a framework aligns with a group's mounted sustainability strategy and dreams, making sure of coherence and synergy with institutional priorities and aspirations. It emphasizes various aspects, including the framework's comprehensive and worldwide visibility and thinking about partnerships with stakeholders, including industry collaborators, governmental organizations, and non-income businesses. By systematically comparing sustainability frameworks with those criteria, agencies can gain insights into every framework's strengths, boundaries, and relevance within their specific contexts and goals. This method empowers decision-makers to make informed alternatives, optimizing the effectiveness and effect of sustainability tasks within their businesses. Throughout the following analysis, those criteria will serve as the muse for assessing the suitability of sustainability frameworks and guiding the choice system. Kaplan and Norton's findings emphasize the importance of efficiently aligning an organization's method with its operations to translate strategic goals into actionable plans. The observation identifies overall performance management structures, such as the Balanced Scorecard, as practical equipment for aligning approaches with operations. These systems enable corporations to monitor progress toward strategic objectives, promote duty, and facilitate continuous improvement. The look additionally highlights the significance of enticing stakeholders and cultivating partnerships to reap strategic goals (Kaplan and Norton 2008).

Table 15.1 Ratings for sustainability frameworks

Rating	Meaning
2	Completely inadequate, does not address the criterion at all
4	Inadequate, addresses the criterion to a limited extent, but significant improvements are needed
6	Moderate, partially addresses the criterion but could be enhanced for better effectiveness
8	Good, meets the criterion satisfactorily with minor areas for improvement
10	Excellent, fully meets the criterion and demonstrates exceptional performance

15.3.2 Ratings for Frameworks Ranking

We will verify the five sustainability frameworks using six criteria aligning with institutional priorities, each weighted. Each framework will be evaluated on a scale from 2 to 10, indicating how well it meets the necessities of every criterion, as precisely in Table 15.1:

For this research, the evaluation panel accommodates the sustainability committee, which includes three sustainability specialists, three college contributors, three directors, and one scholar. Each member will examine the five frameworks using the standards and score scale.

15.4 Case Study

During the evaluation, Liwa College aimed to harmonize its sustainability practices with its strategic priorities. The standards, tailor-made to the college's particular dreams and desires, have been meticulously assigned weights to replicate their significance as they should be. Subsequently, each framework became a score based on this weighted assessment. Evaluating sustainability frameworks involves assigning weights to each framework and computing ratings based on precise criteria to ensure a comprehensive assessment. Liwa College evaluated the most suitable sustainability framework to enhance sustainability practices.

When we pick the best sustainability frameworks, we're looking at established ideas about sustainability that show how important it is to think about sustainability in different ways, depending on the situation and for the long term. We ensure that our decisions are well-supported by solid theories like the Triple Bottom Line, Ecological Modernization, Sustainability Transitions, Institutional Theory, Natural Capitalism, and Social Learning Theory. These theories help us pick the right frameworks and emphasize how important they are for dealing with difficult sustainability problems and making big changes in how things work. We've also created a chart, Table 15.2, that clearly shows how these theories connect to specific frameworks and why we chose them. Figure 15.1 represents the Triple Bottom Line.

Table 15.2 Theories connect to specific frameworks

Sustainability theory	Frameworks aligned	Justification
Triple Bottom Line (Elkington 1994)	STARS, QS Stars, THE Impact Rankings, UN Green Metrics	Addresses the three pillars of sustainability (environmental, social, economic)
Ecological Modernization (Mol and Spaargaren 2000)	STARS, QS Sustainability, UN Green Metrics	Encourages innovation and efficiency for both economic and environmental improvements
Sustainability Transitions (Geels 2002)	THE Impact Rankings, QS Stars, STARS	Promotes long-term, systemic transformation across various levels of sustainability
Sustainability Transitions (Geels 2002)	QS Stars, UN Green Metrics, THE Impact Rankings	Enhances legitimacy and reputation through widely accepted global frameworks
Natural Capitalism (Hawken et al. 1999)	THE Impact Rankings, UN Green Metrics	Focuses on reducing environmental impacts and regenerating natural capital
Social Learning Theory (Bandura 1977)	QS Stars, QS Sustainability	Facilitates benchmarking and learning from global best practices to improve performance

The assessment scrutinized inner and outside criteria, cautiously allocating weights to deliver their relative importance. Frameworks assessed in this system covered THE Impact Rankings, UI Green Metrics, STARS, QS Stars, and QS Sustainability. Based on the assigned weights and the evaluation of sustainability frameworks in opposition to the specified standards, the subsequent ratings were calculated for every framework (Table 15.3):

The assessment standards consisted of inner and external categories, every with particular weights.

15.4.1 Internal Criteria

1. Alignment with Liwa Strategy (Weight: 0.3): This diploma evaluates the quantity to which every framework aligns with the university's strategic dreams and targets.
2. Best for our Operations (Weight: zero.25): This degree measures the framework's effectiveness in improving operational excellence and specializes in efficiency, effectiveness, and consistency of techniques. Three.
3. Focus on Net Zero Carbon Emissions (Weight: 0.2): This phase assesses the framework's emphasis on attaining 0 carbon emissions, an essential prolonged-time period objective for Liwa College.

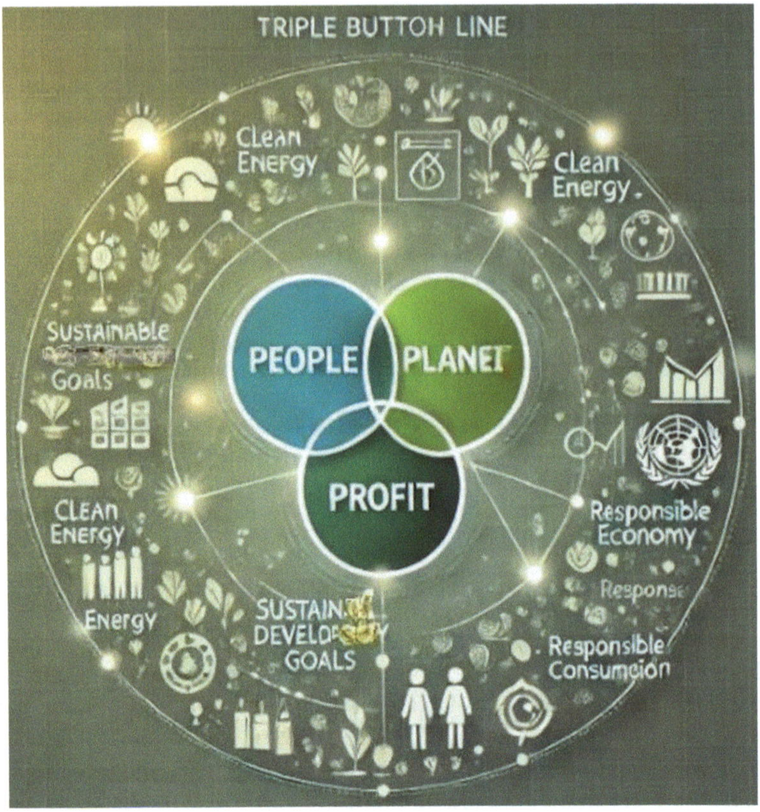

Fig. 15.1 Triple bottom line

15.4.2 External Criteria

1. Alignment with UAE National Sustainability Strategy (Weight: 0.1): Assesses the framework's alignment with countrywide sustainability desires and priorities.
2. Local Benchmarking (Weight: 0.1): This degree measures the framework's functionality to residences near contextual factors and benchmarks.
3. International Benchmarking (Weight: 0.05): This evaluation evaluates how the framework contains sustainability necessities and allows global evaluation and alignment framework incorporates global sustainability standards and enables international comparison and alignment.

Each framework was evaluated against the desired standards, and scores were calculated based on performance. Here is a detailed breakdown of the rankings and analysis for each criterion. Based on the weighted criteria and rankings, each framework's last rankings are as follows:

- THE Impact Rankings: 7.15

Table 15.3 Calculated scores for each framework

	Criteria	Weights	THE Impact Rankings	UI Green Metrics	STARS	QS Stars	QS Sustainability
Internal criteria	Alignment with Liwa Strategy		14 out of 18	9 out of 18	12 out of 18	16 out of 18	13 out of 18
		0.3	8	2	6	10	7
	Best for our Operations			Environmental sustainability		Focused more on Operational excellence	ESGs
		0.25	4	6	8	8	4
	Focus on Net Zero Carbon Emissions		No focus on Zero carbon Emissions	Towards zero Carbon emission through Transportation +Electricity (39%)	No focus on Zero carbon Emissions	Carbon neutral certification (4%)	Net Zero Commitment (1%)
		0.2	7	6	4	7	8
External Criteria	Alignment with UAE National Sustainability Strategy		11 out of 11	9 out of 11	11 out of 11	7 out of 11	8 out of 11
		0.1	10	8	10	6	8
	Local Benchmarking		6 out of 6	3 out of 6	1 out of 6	4 out of 6	6 out of 6
		0.1	10	5	3	8	10
	International Benchmarking		28 out of 30	1 out of 30	11 out of 30	0 out of 30	28 out of 30
		0.05	7	3	5	2	7
	Score		7.15	4.75	6	7.6	7

- UI Green Metrics: 4.75
- STARS: 6.00
- QS Stars: 7.60
- QS Sustainability: 7.00

After evaluating the various sustainability ranking frameworks, the following observations were made:

15.4.3 *The Impact Rankings*

The framework aligns with the countrywide sustainability method and performs excellently in local benchmarking. However, it should be more aware of operational excellence and attaining zero carbon emissions. While the Impact Rankings provide a properly rounded framework, there may be better matches for Liwa College's particular operational excellence and carbon neutrality needs. This framework excels in operational excellence metrics. However, it must showcase higher performance in global benchmarking and align more with institutional strategy. While well-suited for measuring operational metrics, it wishes a broader alignment with strategic and global dreams.

15.4.4 UI Green Metrics

This framework competes with environmental, social, and governance (ESG) standards. However, it has a confined focus on reaching net zero carbon emissions and neighborhood benchmarking. While STARS presents a stable framework for ESG metrics, it desires to enhance it in precise regions, which can be essential for Liwa College.

15.4.5 STARS

This framework demonstrates high alignment with the institutional approach, makes a specialty of operational excellence, and emphasizes carbon neutrality. However, it shows only slight overall performance in countrywide alignment and neighborhood benchmarking. The complete method of QS Stars, balanced with strategic alignment, operational excellence, and carbon neutrality, makes it the first-class overall match for Liwa College.

15.4.6 QS Stars

The quantitative analysis shows that the QS Stars framework acquired the highest average score of 7.60, making it the most appropriate alternative for Liwa College. This framework aligns nicely with the organization's approach, focuses on ESGs, and is committed to carbon neutrality.

15.4.7 QS Sustainability

This framework is devoted to achieving internet zero emissions and performs exceedingly well in neighborhood benchmarking. However, its alignment with institutional strategy and worldwide benchmarking might be milder. While mainly vital for carbon neutrality goals, it could align more with institutional strategy than

QS Stars. The quantitative analysis shows that the QS Stars framework acquired the highest average score of 7.60, making it the most appropriate alternative for Liwa College.

This framework aligns nicely with the organization's approach, focuses on ESGs, and is committed to carbon neutrality. Additionally, it performs nicely in the neighborhood and worldwide benchmarking, ensuring relevance and applicability across distinctive contexts. By adopting the QS Stars framework, Liwa College can optimize its sustainability projects, advantage visibility, and foster stakeholder partnerships. This choice aligns with the group's particular context, targets, and strategic priorities, positioning it to reap its long-term sustainability goals effectively.

Calculating the weighted scores is as follows:

The Impact Rankings:

$$(8*0.3)+(6*0.25)+(4*0.2)+(10*0.1)+(6*0.1)+(8*0.05)=7.15$$

UI Green Metrics:

$$(6*0.3)+(8*0.25)+(6*0.2)+(8*0.1)+(4*0.1)+(6*0.05)=6.50$$

The methodical and systematic approach guarantees an intensive and obvious assessment, yielding valuable insights to aid in choosing the most suitable sustainability framework for the company. This procedure additionally identifies precise strengths and regions for development within every framework, guiding the organization to make well-informed decisions that align with its sustainability targets.

15.5 Recommendations

- Adopt QS Stars: Given its highest overall score and alignment with Liwa College's strategic goals.
- Monitor and Adjust: Continuously monitor the performance under the QS Stars framework and make necessary adjustments to improve alignment with emerging sustainability goals.
- Stakeholder Engagement: Enhance partnerships with industry collaborators, governmental agencies, and non-profit organizations to strengthen sustainability initiatives.
- Continuous Improvement: Utilize feedback and performance metrics to drive continuous improvement in sustainability practices.

By following those pointers, Liwa College could be well-located to gain its sustainability dreams and preserve a leading position in sustainable practices inside the education quarter. These scores provide a quantitative foundation for comparing the performance of each framework in meeting Liwa College's unique priorities and requirements. The framework with the best score may be considered the most suitable choice for adoption, considering the group's particular context, objectives, and strategic priorities. Based on the assessment results, the QS stars are an excellent framework for our institution.

15.6 Conclusion

In the end, this research work has comprehensively examined distinguished sustainability metrics frameworks and proposed a singular assessment framework to aid groups in selecting the most suitable framework for their needs. By reading criteria consisting of visibility, cognizance of internet-zero carbons emissions, international and nearby benchmarking, prices, and compliance with national and institutional techniques, we have provided an established approach to comparing the alignment and effectiveness of sustainability frameworks. The utility of this evaluation framework in Liwa College's case observation has confirmed its sensible application in guiding organizational selection-making. Through a weighted scoring gadget, we have quantitatively assessed each framework's overall performance in meeting Liwa College's priorities, offering valuable insights for strategic making plans and implementation. Amidst the escalating complexities faced using constructed surroundings professionals, compounded via demographic shifts, climate trade, and intensifying aid usage, there is a developing want for extra robust frameworks to guide decision-making procedures in the direction of a sustainable destiny.

This work proposes mixing a sustainability decision-making framework within an interpretative approach to systems thinking. The conceptual framework is then applied to assess planning and improvement decision-making methods for sustainability within the Liwa College area. Our studies emphasize the numerous and complicated nature of sustainability frameworks available to companies today. While every framework offers unique strengths and advantages, no single framework emerges as universally advanced. Instead, the most efficient choice relies upon the company's specific context, targets, and values. Ultimately, this study adds to the increasing repository of sustainability size and reporting knowledge by providing a realistic framework for selection-making. While supplying valuable insights, the analysis sheds light on the efficacy and constraints of current sustainability metrics frameworks. By empowering organizations to make knowledgeable alternatives, we aim to foster extra responsibility, transparency, and development toward a more sustainable future for all.

References

Bandura A (1977) Self-efficacy: toward a unifying theory of behavioral change. Psychol Rev 84(2):191

Barth M, Godemann J, Rieckmann M, Stoltenberg U (2007) Developing key competencies for sustainable development in higher education. Int J Sustain High Educ 8(4):416–430. https://doi.org/10.1108/14676370710823582

Besong F, Holland C (2015) The dispositions, abilities and behaviours (DAB) framework for profiling learners' sustainability competencies in higher education. J Teach Educ Sustain 17(1):5–22

Bocken NMP, Short SW, Rana P, Evans S (2014) A literature and practice review to develop sustainable business model archetypes. J Clean Prod 65:42–56

Elkington J (1994) Towards the sustainable corporation: Win-win-win business strategies for sustainable development. Calif Manag Rev 36(2):90–100

Evans T (2019) Competencies and pedagogies for sustainability education: a roadmap for sustainability studies program development in colleges and universities. Sustainability 11(19):5526. https://doi.org/10.3390/su11195526

Focus on Net Zero Carbon Emissions: United Nations (2015) Paris agreement. https://unfccc.int/process-and-meetings/the-paris-agreement/the-paris-agreement

Geels FW (2002) Understanding the dynamics of technological transitions. A co-evolutionary and socio-technical analysis

James R, Ballinger (2023) Quality and practices for sustainability in higher education—an impact ranking approach. In: Perspectives and trends in education and technology, Smart innovation, systems and technologies. Springer, pp 297–310. https://doi.org/10.1007/978-981-19-6585-2_27

Kaplan RS, Norton DP (2008) The execution premium: linking strategy to operations for competitive advantage. Harvard Business Press

Kosta K (2019) Institutional sustainability assessment. In: Encyclopedia of sustainability in higher education. Springer International Publishing, Cham, pp 974–979

Lozano R, Lukman R, Lozano FJ, Huisingh D, Lambrechts W (2013) Declarations for sustainability in higher education: becoming better leaders, through addressing the university system. J Clean Prod 48:10–19. https://doi.org/10.1016/j.jclepro.2011.10.006

Mol APJ, Spaargaren G (2000) Ecological modernisation theory in debate: a review. Environ Polit 9(1):17–49

National Sustainability Strategy Document (n.d.) Alignment with the national sustainability strategy: Government of [Country]

Porter ME (1990) The competitive advantage of nations. Free Press

Rich N, Bateman N (2003) Beyond the lean paradigm: a survey of lean production frameworks. J Manuf Technol Manage 14(4):4–10

Sayed A, Asmuss M (2013) Benchmarking tools for assessing and tracking sustainability in higher educational institutions: identifying an effective tool for the University of Saskatchewan. Int J Sustain High Educ 14(4):449–465

Sepasi S, Rahdari A, Rexhepi G (2018) Developing a sustainability reporting assessment tool for higher education institutions: the University of California. Sustain Dev 26(6):672–682

Sheila Mae C, Lourdes L, Dylan Josh Domingo L, Rosheilla May M (2023) A proposed framework for sustainable and impactful higher education institutions. ACE. https://doi.org/10.22492/issn.2186-5892.2023.22

Simon M, Angela D (2023) Assessing the suitability of sustainability frameworks for embedding sustainability in higher education curricula: pragmatism versus transformation. Int J Sustain High Educ 24(9):318–334. https://doi.org/10.1108/ijshe-08-2020-0315

Thomas I (2004) Sustainability in tertiary curricula: what is stopping it happening? Int J Sustain High Educ 5(1):33–47. https://doi.org/10.1108/14676370410517387

Wiek A, Withycombe L, Redman CL (2011) Key competencies in sustainability: a reference framework for academic program development. Sustain Sci 6(2):203–218. https://doi.org/10.1007/s11625-011-0132-6

Yturzaeta JEG (2023) Assessing sustainability initiatives in higher education institutions. J Manage Global Sustain 8(1):8

Chapter 16
Dissecting Job Roles for a Circular Economy

Siddharth Varma and Ruchika Gupta

Abstract While development has been a key concern for every country in the world we are paying a price for this development, largely in the shape of climate change and environmental issues. The current model of development based on linear supply chains needs to be replaced with circular supply chains which aim at zero waste. A circular economy (CE) shifts the focus from only economic prosperity to prosperity along with sustainability. This paradigm shift will throw up new requirements from organizations to achieve zero waste through activities like reuse, recycling, repair and refurbishing. This will, in turn, require new skills from the workforce also. Performance evaluation metrics of employees would need to be modified. This chapter tries to understand the requirements from job roles in a circular economy. Based on these requirements certain performance evaluation metrics for employees have also been suggested.

Keywords Circular economy · Sustainability · Zero waste · Workforce skills · Performance metrics

16.1 Introduction

Development has been a key concern for every country in this world. This development, however, comes at a cost. The cost manifests as climate change and environmental issues including pollution and waste generation. The current model of development is based on linear supply chains (take-make-use-dispose model). This model produces waste at every step in the supply chain. Approximately 11.2 billion tonnes worth of solid waste is collected throughout the globe. About 5% of all

S. Varma
IMI Delhi, New Delhi, Delhi, India

R. Gupta (✉)
Mangalmay Institute of Management & Technology, Greater Noida, Uttar Pradesh, India

K. Mathiyazhagan et al. (eds.), *Humanizing Circular Supply Chain Management*, Circular Economy and Sustainability,
https://doi.org/10.1007/978-3-031-86562-6_16

global gases that are harmful to our environment result from the decay and breakdown of organic materials found within this waste (UNEP 2019). These greenhouse gases trap heat and lead to increase in the temperature of earth. The increase in temperature further causes changes in precipitation and increase in sea levels. We need to cut down on greenhouse gas emissions if we must save our planet. The world needs to cut down on 42% of greenhouse gas emissions by 2030 to limit the increase in temperature to 1.5 °C (UNEP 2023).

As per World bank Report, each individual produces an average daily waste that weighs 0.74 kg. As for the total waste generated globally, 34% comes from well off nations. For these countries, daily waste generation per individual is expected to grow by 19% by 2050 while in case of less developed and average-income nations it would jump by over 40%. Global waste may grow to 3.40 billion tonnes during the same period. Region wise waste generation has been depicted in Fig. 16.1.

As a natural process climate may change over decades or more. But the rapid warming which the earth is experiencing today is unprecedented. Humans are largely responsible for this warming which is mainly due to the green house gases generated as a result of burning fossil fuels. According to the United Nations Report of May 2024, the earth is about 1.1 °C warmer than it was in the 1800s. If carbon dioxide emissions continue at the current rate global temperature could increase by 4.4 °C by the end of the century. This is beyond the safe limit for temperature of this earth to increase. Floods, droughts and increase in sea levels are all consequences of climate change.

The obsession for development has also led to pollution of land, water and air. As stated by European Environment Agency, air pollution is the greatest threat to environmental health in Europe, which causes early death and disease (EEA 2023). According to a publication released by Fuller et al. (2022), nearly nine million people die every year due to pollution.

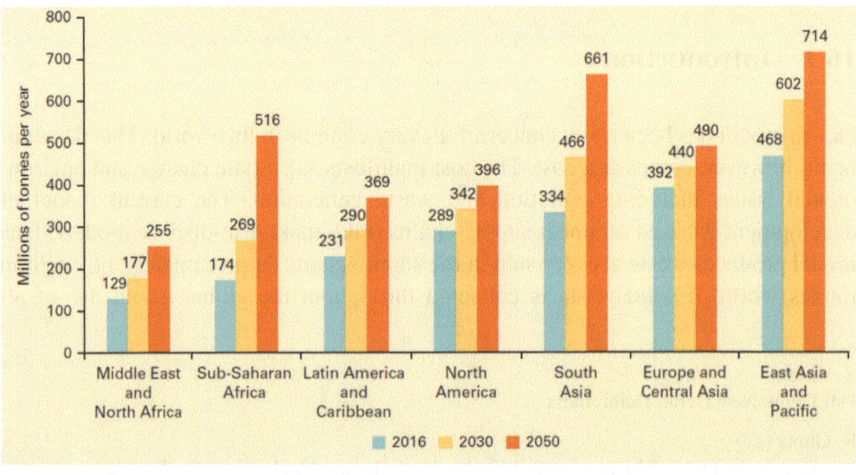

Fig. 16.1 Projected waste generation, by region (millions of tonnes/year). (Source: World Bank n.d.)

Pollutants can travel for long distances and eventually end up in the soil where they are diluted and remain for some time. The European Environment Agency (EEA 2023) states that soil is termed contaminated when presence of toxic elements affect negatively on human well-being or ecological system. When chemicals or microorganisms contaminate any water body, water quality is degraded and becomes toxic to humans or the environment.

16.2 Circular Economy: The Way Out

The European Parliament (2023) provides a definition of circular economy (CE) as a system that prioritizes the exchange, rental, reuse, repair, refurbish and recycle of materials and things for extended time frames. A CE results in lower requirement of resources. Since life cycle of products increases, there is lesser requirement of raw material which means less requirement of natural resources. Moreover, energy requirement is also reduced. Lesser energy usage finally results in lower emissions of green house gases.

The circular economy is based on Circular Supply Chains (CSCs) which are an alternate to the conventional Linear Supply Chains. Circular supply chains try to aim for zero waste throughout the supply chain.

The Ellen McArthur Foundation (EMF) defines the Circular Economy (CE) as "*looking beyond the current take, make and dispose extractive industrial model; the circular economy is restorative and regenerative by design. The biological materials or nutrients become part of the biosphere and can be reused as production inputs. Technical materials are designed for material recovery through repair, refurbishing, remanufacturing, and recycling*".

The Fig. 16.2 depicts three different types of supply chains: Circular, Linear and Closed Loop. Linear supply chains are conventional supply chains which are accompanied by waste generation at every step. Close loop supply chains utilize some of the waste generated within the same supply chain. The rest may still go as waste. It may not be possible to utilize all waste within the same supply chain. Circular supply chains try to utilize the waste generated in the supply chain either within the same or a different supply chain, thus attempting to have zero waste.

While several countries are making efforts in moving towards a CE, Finland is the first country to have developed a blueprint for making its economy circular. The strategy prepared in 2016 aims that by 2025, Finland would be one of the world's leaders in the circular economy (SITRA 2016). In India, the Circular Economy Cell (CE Cell) was launched within NITI Aayog in September 2022. Furthermore, ten sector-specific action plans for promoting Circular Economy have been prepared by NITI Aayog for implementation.

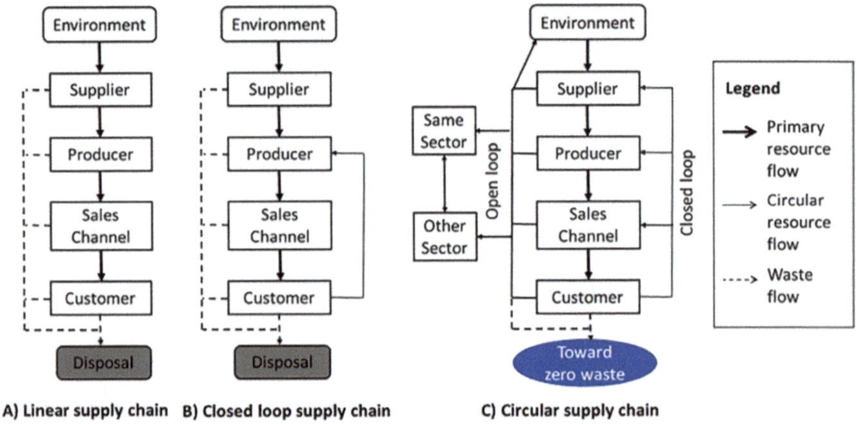

Fig. 16.2 Linear (**A**), closed loop (**B**) and circular supply chains (**C**). (Source: Farooque et al. 2019)

16.3 Need for the Study

The CE presents a different paradigm of the economy. The focus shifts from economic prosperity only to prosperity along with sustainability for the planet. This can cause major changes in the existing economic set up. The CE throws up new requirements from organizations with its focus on zero waste which requires extensive reuse, recycling, repair and refurbishing in order to increase the life of products. As a result, new skills are expected from the workforce who have also to be familiar with new concepts related to CE. New job roles will come into existence while old ones may also need to be modified in line with the principles of CE.

A job role is a position in any firm accompanied with responsibilities and expectations. Defining these roles helps communicate the job expectations with the employee and also evaluate their performance. An understanding of new job roles would help firms in hiring the right people and implement circularity initiatives. This would facilitate transitioning to a CE. By understanding the new job roles, this study can be helpful to firms who are trying to transform to a circular supply chain from a linear one.

16.4 Economic and Social Dimensions of CE

According to the Ellen McArthur Foundation, a circular economy is a systemic approach to economic development designed to benefit businesses, society, and the environment. While CE aims at zero waste with a focus on resource efficiencies prolonging use of products as long as possible through reuse, recycle, refurbish and remanufacture; there may be other consequences of pursuing a circular economy. Transitioning to a circular economy may have social as well as economic implications. The economic dimensions can easily be understood. A CE requires efficient utilization of resources which leads to economic benefits. Also, it aims at minimizing waste through reuse, recycle etc., which again leads to reduce requirement of resources and brings in monetary gains.

Implementing a CE may bring in other benefits related to social sustainability. These may be in the form of social equity, improved working conditions and most importantly job creation. The CE contribution to the social dimensions has been studied mostly in terms of job creation opportunities (Aguilar-Hernandez et al. 2021). Implementing a CE may also result in improved quality of life through dignifying work conditions like formalizing commoning for waste pickers (Carenzo et al. 2022). Liveability in cities can be improved using digitally enabled CE initiatives (Lekan and Rogers 2020). A CE may also help in diversity. For example, cooperatives or associations may result in 'necessity-driven circularities' (Flamand 2022). Formulation of a CE policy and governmental discourse in countries like Norway have resulted from a quest for equity (Hermann et al. 2022). According to the European Environment Agency (2024) circular economy can reduce the impacts of production and consumption on biodiversity hence aiding to conservation of biodiversity. According to Deloitte, producing locally reduces dependence on foreign supplies and reduces risks due to disruption from natural disasters. Local production also results in shorter supply chains which can help in climate adaptation.

16.5 Circular Economy and the Job Scenario

A prominent social dimension of a circular economy is creation of new jobs. A circular economy can result in a number of new job opportunities due to activities like recycling and waste to energy conversion. According to a 2019 report by Circle Economy and Erasmus Happiness Economics Research Organization (EHERO 2019), circular employment includes those sectors which are either connected directly to a circular economy or those that help it indirectly. Jobs that contribute indirectly to circular jobs are called indirectly circular jobs. According to Nichols (2015), a British study commissioned by an environmental management firm showed that closed-loop society could boost British economy and its employment levels by 2025. Circular economy could create jobs and boost GDP by £29bn.

Torres (2017) posits that in a data-driven circular economy, the recycling and reuse of accumulated waste can lead to resource conservation and job creation. The Netherlands has already attained a significant milestone, with 8.1% of all occupations being circular as revealed by a research conducted by Circle Economy in 2015. A new resolution acknowledging the potential of the social economy which spearheads circular job creation has just been passed by the European Union (Barna et al. 2023). The transition to a circular economy model in India has the capacity to generate a significant number of employment opportunities. Niti Aayog CEO Amitabh Kant states that the circular economy can create 1.4 crore employment opportunities during the next 5–7 years (Livemint 2019). The World Economic

Forum predicts that India has the potential to create up to 50 million jobs in the green economy (Chakravorti and Dalmia 2023). ILO (International Labour Organization) also estimates that 24 million jobs could result by 2030 due to shift towards green and sustainable practices. Out of these, 18 million jobs would be specifically associated with resource efficient sectors like waste management, energy efficiency and renewable energy. Additionally, the International Labour Organization (ILO) reported that around 1.2 billion employees may necessitate upskilling in order to facilitate transitioning towards a circular economy (ILO 2018).

The impact of circular economy on work and associated competence will be significant. The circular economy cannot be built without experts who understand the topic well (Degerman et al. 2023).

Burke and Grodach (2024) have discussed the challenges of ensuring access to core circular economy jobs, such as repair and recycling while building circular cities. Small and medium-sized companies can provide opportunities for employment by participating in transformational activities like repair and reuse, which are also low tech (D'Urzo and Campagnaro 2023).

In their study, Aguilar-Hernandez et al. (2021) created many scenarios and determined that by 2030, the adoption of circular economy practices might lead to a mutually beneficial outcome, resulting in minimal or gradual changes in gross domestic product and employment, while significantly lowering CO_2 emissions. A study by Singh and Singh (2019) examined the relationship between organizational citizenship behavior, organizational justice, psychological empowerment, and job satisfaction with respect to the circular economy. They concluded that organizations should provide employees with fair and empowering environment and recognize the importance of organizational citizenship behaviour to support circular economy.

Napathorn (2022) states that companies in various sectors in Thailand actively facilitate the acquisition of environmentally conscious knowledge and skills among their personnel, enabling them to effectively carry out environmentally friendly tasks. According to Romero de Brito et al. (2022), reverse logistics of WEEE (Waste from Electrical and Electronic Equipment) as well as EoL EEE (Electronic and Electrical Equipment) presents an opportunity which has not been exploited before in Brazil but can create new jobs and generate income. Circular economy can be fostered through improvement of outsourcing strategies as well as training and motivating employees to overcome issues of knowledge, capacity as well as monetary and financial constraints (Zaidi and Chandra 2024).

16.6 Methodology

This study tries to understand the new job requirements which would result due to transitioning to a circular economy. Once the job requirements are understood, an attempt has also been made to define the metrics for performance evaluation of the new job requirements. A detailed literature review has been carried out to understand how job roles in supply chains may get affected due to adoption of a circular economy.

The authors try to define job requirements for each job role in an organization based on this literature review. Specific KRAs have been derived from the inputs gained from the literature review. Based on the KRAs, the performance metrics have been defined for achieving the new job requirements. As far as possible, the metrics for performance evaluation have been kept quantitative to facilitate easy measurement.

16.7 Literature Review

A circular economy would make new demands from existing job roles, especially in a manufacturing organization. The following literature review tries to understand how job roles across the supply chain may get affected due to a company transitioning towards a circular economy.

16.7.1 Purchase Officer/Purchase Manager

Circular procurement means purchase of items and services considering the principles of the circular economy. Apart from cost and quality purchasers have also to take sustainability into account (Schulze et al. 2019). According to the Ellen McArthur Foundation procurement for circular economy covers the following.

a) Strategy: This process assists in determining the decision rationale that should be examined before to initiating the sourcing activity, once a business need has been identified. b) Sourcing: The process of incorporating circular economy principles into sourcing activities aims to produce ideas for sourcing items and materials that promote and incentivize the most effective circular practices. c) Management: By establishing a circular economy partnership with suppliers, this method facilitates the management of ongoing performance and ensures the creation of mutual value.

Green procurement, as per GEP, an esteemed organization in procurement and supply chain, indicates acquisition of products with small negative effects on the ecosystem. Recycling items, energy effective systems and clean technologies are some examples of what this entails. This implies that buying choices are made primarily based on environmental factors rather than monetary aspects concerning products and services. For any procurement office to carry out proper purchases they need to know how materials affect circularity (Deloitte 2021). Oscar et al. (2024) considered the impact of Green Construction Procurement Practices (GCPPs) on Circular Economy (CE) efficiency in the construction industry. They particularly focused on determining sustainable procurement practices which may promote a CE. They concluded that green purchasing implementation has a significant influence over the circular economy in building sector. Jurij et al. (2023) researched infrastructure purchasing from Circular Economy (CE) viewpoint. It was revealed that awareness creation among primary stakeholders about CE principles is very important because education alignment is necessary for achieving CE objectives.

16.7.2 Quality Manager

Quality is important for any organization whether into manufacturing or services. According to Nagarali (2023), Quality Assurance ensures products are designed and manufactured to be robust, reliable, and able to withstand extended use and multiple cycles of reuse, refurbishment, and remanufacturing. An illustrative case study presented by Dezi et al. (2022) showed that the common factor among TQM, CE and social enterprises are economic, social and environmental sustainability. Marica et al. (2019) studied the role of certifications for implementing environmentally sustainable practices in the Italian luxury furniture industry.

16.7.3 Design Officer

Product design can have a profound impact on any circularity initiative. Circular product design requires designing out waste from the product. Also, the product needs to be designed for durability and for repair and reuse.

As per the Ellen MacArthur Foundation, conventional design methods prioritize the requirements of the end user, whereas a design for the circular economy considers not just the user but also the overall system in which the design will operate. Maintenance, reuse, remanufacturing and recycling are the essential features that should be prioritized by designs so as to allow goods, components and materials to keep circulating within an economy. Marina et al. (2021) carried out a research that aimed at exploring and examining main design strategies for developing circular products and how these were related to new product development. The research findings showed that taking up a circular product design practice brought forth fresh ideas such as having multiple usage cycles; infusing emotional durability; using biomimicry among others. Den Hollander et al. (2017) stated that eco-design and circular product design have important distinctions. Hapuwatte and Jawahir (2021) propose a new methodology which combines elements of design for sustainability and circularity. Dokter et al. (2020) have listed the major considerations for circular design of products. They stated that products must be designed for the following.

- Dis-assembly and re-assembly
- User trust
- Attachments
- Customization
- Long life
- Upgrades
- Align lifespan with maintenance frequency

16.7.4 Production Manager

As per the Ellen McArthur Foundation, continuous circulation of goods and materials via maintenance, reuse, refurbishing, remanufacturing, recycling, and composting, is what constitutes a circular economy. Jensen (2024) states that manufacturers are obliged to make circularity part of their mode of operation so that they can consider product end-of-life stage and enable reintegration of resources back into supply chain. Turner et al. (2019) examined the potential adoption of new manufacturing technologies such as additive manufacturing or 3D printing to establish sustainable production-construction systems. With 3D printing repairs, spare and replacements, and custom modifications become possible.

Antonioli et al. (2022) studied the ability of sustainable production choices to affect the economy and concluded that within a short period it can be hard to benefit economically only from circular innovations. The fundamental principles underpinning the European Waste Legislation are established in waste hierarchy (Van Ewijk and Stegemann 2016) (Fig. 16.3).

The foundation of 'Lean' philosophy lies in curtailing inefficiencies. The main aim is to optimize the utilization of resources and remove the processes which do not add value (Netland 2013). Meanwhile, Green Lean aims for achieving both ecological and fiscal efficiency in manufacturing operations (Kurdve et al. 2014).

According to Lindahl (2024), 'circular production strategy', 'management systems' and 'external standards', continuously and independently evolve as a network.

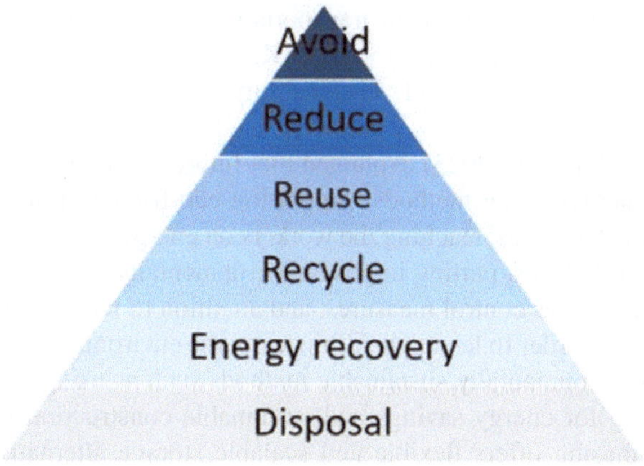

Fig. 16.3 Waste hierarchy. (Adapted from EU 2023)

16.7.5 Waste Management Officer

This is a relatively novel job profile and several manufacturing firms may not possess it at present in their organization structure. This position coordinates and supervises the collection, disposal and recycling of the waste. According to Workbred (2023), a UK company a Waste Management Officer supervises and organizes garbage disposal, treatment as well as recycling within an organization or community. In addition to that, he/she can formulate policies concerning waste management, ensure compliance with environmental laws and advocate garbage disposal plus recycling among the populace. He/she should also ensure that legal requirements related to waste management are upheld. Evinemi and Afon (2016) made a study on waste collection, storage, and disposal in Nigeria. They suggested that public awareness and environmental education be given to all managers and waste management officials working in health institutions. Helman (2016) researched the economic aspects of recycling and the effects of incorrectly recycled materials. In their work, Yuan et al. (2018) developed a model to investigate factors that influence project managers' intentions of reducing wastage. It was found that attitude was the most significant predictor of these intentions, followed by perceived behavioral control and subjective norms.

16.7.6 Logistics Manager

According to van Luu (2023) there are nine variables for Green Logistics and Sustainable Supply Chain practices: compliance with environmental regulations, reduction in energy consumption, reduction in hazardous materials, green IT and communication, green transportation, green manufacturing, green storage, green procurement, reverse logistics. Karaman et al. (2020) have shown that efficiency of green logistics and sustainability reporting practices have a strong positive correlation.

Ali et al. (2023) explained that future green warehousing will include: adopting energy-saving methods, supporting eco-friendly transportation, building compatible buildings, teaching the workers on energy saving habits of life, promoting green workplaces, putting in place environmentally conscious guidelines and procedures as well as control measures, and attention to waste management for disposal.

In order to lessen their impact on the environment, most warehouses are utilizing environmentally sustainable methods such as using solar power, appropriate lighting for energy savings and sustainable construction materials. On-demand warehousing offers flexible and scalable storage alternatives that do away with long term leases.

The SAP definition of a warehouse management system (WMS) is software that helps businesses manage all aspects of a warehouse, from the moment goods come in to the moment they're shipped out. AI algorithms can help in maximizing uptime by optimizing warehouse layouts, streamlining picking routes, and predicting maintenance needs.

16.7.7 Marketing Manager

Marketing can help a company build resilient brands. It can help shape the experiences, products, and business models that will drive the circular economy forward.

Green marketing is an appropriate approach that one can use in order to promote a circular economy. It entails promoting or selling a product made out of eco-friendly materials or with less harmful effects on the environment (Ingavale and Gaikwad 2011). Mostaghel et al. (2023) focused on marketing theory and practice within the CE in their study so as to establish if there are any specific barriers to conduct marketing in the CE. Lu et al. (2023) formed a framework that offers insights on how green advertising can help a linear economy transform into a circular economy in hospitality industry. Rybanský and Jánoš (2019) have evaluated the use of communication tools in changing consumer behaviour for Slovakia to have a circular economy.

16.7.8 Human Resources Manager

A report by the European Commission (2019) states that it is possible for one to become a Circular Human Resources Manager if he understands well its fundamental principles. This gives them the ability to change HR policies from linearity into circularity. Such managers can also make use of staff competencies and skills therein while promoting circular approaches.

In their study, Subramanian and Suresh (2022) discovered that all facets of organizational performance were positively influenced by green human resource management. According to the Green HR Management Institute, Green HRM is the execution of HRM practices aimed at fostering sustainability in resource utilization within organizations and more broadly advocating for environmental sustainability. This means that every aspect related to recruitment, selection, training, remuneration, growth and promotion of workforce should integrate the green goals. For instance, Chiappetta Jabbour et al. (2019) considered a broader GHRM framework that could help companies make a transition towards circular economy by incorporating human inputs in circularity. A research conducted by Dibia et al. (2020) looked into six case studies so as to examine some of the larger societal implications associated with Circular Economy (CE) such as creation of jobs possibilities, better quality of life and changes in organizational culture.

16.7.9 *Refurbishing Manager*

Refurbishment refers to the extension of the life of a product through repair or replacement of components. The refurbishment of high value products can be of considerable economic importance for related firms. Hence, a new job profile of a Refurbishing Manager/Repair Manager is required for any product firm trying to transition to a circular supply chain.

Refurbishing of products may be advantageous for the surroundings and economy as well. It reduces raw materials and energy requirements while also providing customers with a more economical option. Repair of products is a related concept providing similar benefits to the environment and economy.

Refurbishing enterprises are hesitant to embrace the economic potential of their industry due to concerns about the cannibalization effect it may have on sales of new products. An empirical investigation by Kurdhi et al. (2023), showed that there is a linear relation between cannibalization/switching effects and price. On the other hand, Richter et al. (2023) carried out case studies to offer an all-round analysis of factors that promote or hinder spare part harvesting while identifying main stakeholders involved in the process. For instance, wrong disposal methods for computer waste pose serious dangers to our environment and human health. The right disposal techniques which include reusing, reuse and recycling need be employed so as to minimize these threats. Accordingly, Yeow and Wee (2018) made an investigation into what motivates Malaysian customers to reuse/recycle/refurbish old computers/electronics. They also proposed ways through which these activities could be encouraged more widely. It was noted there are many vendors who do not yet provide this option; hence consumers think feel that this principle of reduce, reuse and recycle (3Rs) lacks facilities to put these into practice.

According to Egbu (1997), it is essential for refurbishing firms to be innovative with respect to their processes, products, technology, service delivery and even within the market they operate. Also, managing refurbishing effectively, requires a team with requisite skills, knowledge and competence.

16.8 Research Gaps and Objectives

From the literature review it seems that no study has been carried out which tries to understand job roles emanating from a circular economy. Explicit KRAs for such job roles are missing. The requirements from new job roles or modified job roles needs to be understood clearly in order to recruit people with the right skills and knowledge. The modified job roles would also require a different way for performance evaluation of employees.

This study answers the following research questions to fill the gap of understanding job roles in a circular economy.

- What would be the new requirements from job roles in a circular economy?
- How would performance evaluation of these job roles be carried out?

16.9 New Job Requirements and Performance Evaluation Criteria

On the basis of the inputs gained from extant literature, the new job requirements in an organization can be identified. The following section provides the current requirements and the additional requirements due to a company going circular. The performance evaluation criteria have also been listed.

16.9.1 Purchase Officer

16.9.1.1 Current Requirements

- Purchase of raw materials and components
- Conforming to quality requirements of the organization
- Timely deliveries
- Reduction in procurement costs
- Reduction in inventory holding costs.

16.9.1.2 Additional Requirements for CE

- Reduced requirements of material
- Procuring material and components which can be reused/recycled
- Using recycled material as long as it is conforming to quality requirements of the company
- Reducing transportation to reduce use of fossil fuels
- Using clean fuels for transportation.
- Use of renewable energy

16.9.1.3 Performance Evaluation Metrics

- Reduction in material requirements (kg)
- Percentage of material and components procured which can be reused/recycled (Percentage by number of materials and components/percentage by weight/percentage by value)
- Percentage of material used which is made up of recycled material
- Reduction in transportation cost of material/components transported from supplier to warehouse (Rs per MT)
- Clean fuel used as percentage of total fuel requirement
- Percentage of energy coming from renewable sources

16.9.2 Quality Manager

16.9.2.1 Current Requirements

- Ensuring quality of procured material/components
- Developing quality standards for purchased material/components

16.9.2.2 Additional Requirements for CE

- Obtaining certifications for circular economy (ISO 59020/ISO 59004/ISO 59010)
- Ensuring quality of repaired/refurbished products
- Ensuring products are reliable and durable
- Ensuring products can be repaired/refurbished/recycled/remanufactured

16.9.2.3 Performance Evaluation Metrics

- Value of repaired/refurbished/remanufactured/recycled products (Rs)
- Percentage of rejected products repaired or refurbished (percentage by value in Rs)
- Certifications obtained for circular economy
- Reliability and durability of the product

16.9.3 Design Officer

16.9.3.1 Current Requirements

- Designing of products
- Ensuring Design for Manufacturability
- Understanding customer requirements and attention to detail
- Creativity

16.9.3.2 Additional Requirements for CE

- Reduced energy requirement
- Use of recyclable material
- Modular design which can enable repair and remanufacturing
- Durability to ensure enhanced product life
- Easy dis-assembly and re-assembly of products
- User trust must be maintained
- Easy use of attachments

- Facilitate customization
- Facilitate upgrades
- Align lifespan and maintenance frequency

16.9.3.3 Performance Evaluation Metrics

- Time taken for dis-assembly and re-assembly
- Percentage reduction in energy requirements for producing a product (percentage of energy in kWH/percentage of energy by value in Rs)
- Percentage of products which use recyclable material
- Life of products in years
- Time required to carry out repairs (hours)
- Percentage of products which have modular design for customization, upgrades and attachments

16.9.4 Production Manager

16.9.4.1 Current Requirements

- Ensuring production as per plan provided
- Capacity utilization
- Ensuring quality of product
- Minimizing downtime

16.9.4.2 Additional Requirements for CE

- Minimize wastage
- Reuse good parts of rejected products
- Recycle/repair rejected products
- Reduce energy consumption
- Use renewable energy
- Use 3D printing wherever feasible
- Employ lean manufacturing

16.9.4.3 Performance Evaluation Metrics

- Percentage reduction in material (percentage by weight/percentage by value in Rs)
- Percentage of parts reused out of rejected products/value of reused parts
- Value of products sent for repair/refurbishing
- Value of products recycled

- Reduction in energy consumption per unit produced (kWH)
- Percentage of energy for production coming from renewable sources

16.9.5 *Waste Management Officer*

16.9.5.1 Current Requirements

Nil (new job role).

16.9.5.2 Additional Requirements for CE

- Coordinating with various departments: purchase, production, quality, distribution, for managing waste through reuse, recycling, repair and refurbishing.
- Responsible for collection of EoL product from customers for reuse, repair or recycling
- Finding new uses for rejected products as inputs for other industries
- Compliance with government regulations regarding waste management/disposal
- Dissemination of information to the general public regarding disposal of waste and recycling
- Sensitizing employees regarding waste management

16.9.5.3 Performance Evaluation Metrics

- Value of waste across the supply chain recycled/reused/repaired/refurbished
- Percentage of EoL product reused/recycled/repaired/refurbished
- Value of EoL product reused/recycled/repaired/refurbished
- Public awareness campaigns carried out
- Employee sensitization activities carried out

16.9.6 *Logistics Manager*

16.9.6.1 Current Requirements

- Receipt and despatch of material
- Loading/unloading at warehouse
- Proper maintenance of inventory in warehouse
- Record keeping of inventory
- Physical audits of inventory

16.9.6.2 Additional Requirements for CE

- Reduction in inventory waste
- Maximize use of renewable energy
- Energy efficient operations including efficient building design
- Recycling/repair/refurbishment of goods spoilt in inventory
- Route optimization of trucks to reduce use of fuel
- Efficient utilization of space
- Reverse logistics for collecting EoL products for recycling/repair/refurbishment from customer/dealers/distributors
- Employee training for sustainable work culture

16.9.6.3 Performance Evaluation Metrics

- Percentage reduction of waste in inventory (value in Rs)
- Percentage of energy coming from renewable resources
- Percentage reduction in energy requirements
- Percentage of goods spoilt in inventory which are recycled/repaired/refurbished
- Transportation cost per km per kg
- Utilization of space (Value of goods in Rs per square foot)
- Value of EoL goods taken back from customers
- Employee trainings done for promoting a sustainable work culture

16.9.7 Marketing Manager

16.9.7.1 Current Requirements

- Developing and executing marketing strategies
- Advertising
- Competitor analysis
- Preparing budget and monitoring it
- Managing employees in the department

16.9.7.2 Additional Requirements for CE

- Use marketing communication to promote sustainable products and services and initiatives taken by the company
- Educate customers and enhance awareness for circular economy among customers/dealers/distributors
- Educate and train marketing staff for circular economy

- Educate customers for returning EoL products so that company can reuse/repair/recycle/refurbish these products
- Selling refurbished/repaired products

16.9.7.3 Performance Evaluation Metrics

- Value of refurbished/repaired products sold
- Campaigns organized to educate customers/dealers/distributors regarding circular economy/return of EoL products
- Trainings done for marketing staff

16.9.8 Human Resources Manager

16.9.8.1 Current Requirements

- Manage human resources within the organization
- Recruiting
- Training
- Motivating employees
- Compensation and bonuses

16.9.8.2 Additional Requirements for CE

- Incorporate sustainability criteria as part of HR processes: recruitment, training, performance evaluation.
- Organize training programs on sustainability/circular economy for its employees, especially for the HR department.
- Develop a long-term strategy to recruit and develop talent w.r.t circular economy
- Promote circular values as part of organizational culture
- Redeploy/repurpose work force through training. Develop workers to minimize job losses/attrition

16.9.8.3 Performance Evaluation Metrics

- No. of employees recruited considering sustainability as one of the criteria
- Training programs organized to educate employees on circularity/sustainability
- Integration of sustainability in performance evaluation
- No. of employees redeployed/repurposed through training
- Reduction in attrition rates
- Strategy formulation to develop and recruit talent w.r.t. circular economy

16.9.9 Refurbishing Manager

16.9.9.1 Current Requirements

Nil (new job role).

16.9.9.2 Additional Requirements for CE

- Receive goods from various departments for refurbishing/repair
- Refurbish/repair products to satisfaction of customer
- Ensure employees have requisite knowledge and necessary skills and knowledge

16.9.9.3 Performance Evaluation Metrics

- Quality of refurbished goods
- Value of goods reused/refurbished/recyled (in Rs)
- Profits earned through reused/refurbished goods
- Establishing facilities for reduce/reuse/recycle
- Training done for workforce for requisite skills and knowledge for refurbishment, health and safety and quality.

16.10 Conclusion

The study has attempted to understand the changing requirements of job roles while transitioning to a circular economy, specifically for a product based organization. A detailed description of the modified/new job roles has been done which includes the new responsibilities (KRAs) and performance evaluation metrics against these KRAs (KPIs). Performance evaluation metrics suggested are largely quantitative which can make it easy to implement a performance evaluation system.

16.11 Implications to Practitioners

This study tries to delve into the new requirements from job roles in an organization which is trying to transition from a linear supply chain to a circular supply chain. Such a clear description of job roles will be helpful to the top management and HR Managers for promoting circular values in a company which would contribute to a circular economy. These would be especially helpful while recruiting employees for various job roles by clearly defining the new responsibilities of employees. By communicating the expectations clearly to prospective candidates, organizations can minimize mismatch between expectations of employees and the organization. The performance evaluation metrics are quantitative which will lead to objective assessment of the performance of the employees in terms of his or her contribution to circularity. The new job roles also provide top management who are desiring to transition to a circular supply chain, food for thought related to organization redesign which may be required due to the change in job roles.

References

Aguilar-Hernandez GA, Dias Rodrigues JF, Tukker A (2021) Macroeconomic, social and environmental impacts of a circular economy up to 2050: a meta-analysis of prospective studies. J Clean Prod 278:123421. https://doi.org/10.1016/J.JCLEPRO.2020.123421

Ali SS, Kaur R, Khan S (2023) Evaluating sustainability initiatives in warehouse for measuring sustainability performance: an emerging economy perspective. Ann Oper Res 324(1/2):461–500

Antonioli D, Ghisetti C, Mazzanti M, Nicolli F (2022) Sustainable production: the economic returns of circular economy practices. Bus Strategy Environ 31(5):2603–2617. https://doi.org/10.1002/bse.3046

Barna C, Zbuchea A, Stănescu S (2023) Social economy enterprises contributing to the circular economy and the green transition in Romania. CIRIEC J Public Soc Cooperat Econ 107:47–69

Burke L, Grodach C (2024) Planning inclusive circular cities. Plan News 51(2):25–27

Carenzo S, Juarez P, Becerra L (2022) Is there room for a circular economy "from below"? Reflections on privatisation and commoning of circular waste loops in Argentina. Local Environ 27(3):1338–1354. https://doi.org/10.1080/13549839.2022.2048258

Chakravorti B, Dalmia G (2023) Is India the world's next great economic power? Harvard Business Review

Chiappetta Jabbour CJ, Sarkis J, Lopes de Sousa Jabbour AB, Scott Renwick DW, Singh SK, Grebinevych O, Kruglianskas I, Filho MG (2019) Who is in charge? A review and a research agenda on the 'human side' of the circular economy. J Clean Prod 222:793–801

Circular Economy (2015) Circular jobs: understanding employment in the circular economy in the Netherlands. https://circulareconomy.europa.eu/platform/sites/default/files/circular_jobs.pdf

D'Urzo M, Campagnaro C (2023) Design-led repair & reuse: an approach for an equitable, bottom-up, innovation-driven circular economy. J Clean Prod 387:135724

de Brito JLR, Ruiz MS, Kniess CT, dos Santos MR (2022) Reverse remanufacturing of electrical and electronic equipment and the circular economy. Revista de Gestão 29(4):380–394

Degerman R, Värre U, Roschier S, Sepponen S, Nurmi J (2023) The impact of the circular economy on jobs and skills. https://www.sitra.fi/en/publications/the-impact-of-the-circular-economy-on-jobs-and-skills/

Deloitte (2021) Procurement in a circular economy benefits beyond sustainability. https://www.deloitte.com/content/dam/Deloitte/ch/Documents/strategy-operations/deloitte-ch-en-procurement-in-a-circular-economy.pdf

Deloitte (n.d.) Circular economy and producing locally. https://www.deloitte.com/global/en/Industries/government-public/perspectives/urban-future-with-a-purpose/circular-economy-and-producing-locally.html

Den Hollander M, Bakker CA, Hultink E (2017) Product design in a circular economy: development of a typology of key concepts and terms: key concepts and terms for circular product design. J Ind Ecol 21:517–525. https://doi.org/10.1111/jiec.12610

Dezi L, Hysa X, Calabrese M, Mercuri F (2022) Open total quality management in the circular economy age: a social enterprise perspective through the case of Patagonia. Total Qual Manag Bus Excell 36(11):1–15. https://doi.org/10.1080/14783363.2022.2051698

Dibia C, Anderson M, Dirpal G, Oruh E (2020) Human resource management and circular economy: a critical perspective. British Academy of Management Conference: Innovation for a Sustainable Future

Dokter G et al (2020) Cards for circularity: towards circular design in practice. IOP Conf Ser Earth Environ Sci 588:042043

Egbu CO (1997) Refurbishment management: challenges and opportunities. Build Res Inf 25(6):338–347

EHERO (2019) Circular HRM: aligning circular economy with human resources management. https://ec.europa.eu/programmes/erasmus-plus/project-result-content/d8a2faab-4b90-4e10-ba11-42f3dafbb20f/IO2_A1_Circular-HRM_EU_Qualification_Framework.pdf

Ellen MacArthur Foundation (n.d.) What is a circular economy? https://www.ellenmacarthurfoundation.org/topics/circular-economy-introduction/overview

Ellen McArthur Foundation (n.d.) Circular economy procurement framework. https://emf.gitbook.io/circular-procurement/-MB3yM1RMC1i8iNc-VYj

EU (2023) EU waste framework directive. https://environment.ec.europa.eu/topics/waste-and-recycling/waste-framework-directive_en

European Commission (2019) Communication from the Commission to the European Parliament, The European Council, The Council, The European Economic and Social Committee and The Committee of the Regions. The European Green Deal COM/2019/640 final

European Environment Agency (2023) How air pollution affects our health. https://www.eea.europa.eu/en/topics/in-depth/air-pollution/eow-it-affects-our-health#:~:text=Air%20pollution%20is%20the%20single,the%20most%20substantial%20health%20impacts

European Environment Agency (2024) The benefits to biodiversity of a strong circular economy. https://www.eea.europa.eu/publications/the-benefits-to-biodiversity

European Parliament (2023) Circular economy: definition, importance and benefits. https://www.europarl.europa.eu/topics/en/article/20151201STO05603/circular-economy-definition-importance-and-benefits

Evinemi OT, Afon AO (2016) Solid waste storage, collection and disposal practices in private health institutions in Ibadan, Nigeria. Manage Res Pract 8(4):53–62

Farooque M, Zhang A, Thurer M, Qu T, Huisingh D (2019) Circular supply chain management: a definition and structured literature review. J Clean Prod 228:882–900. https://doi.org/10.1016/j.jclepro.2019.04.303

Flamand K (2022) The search for opportunity: co-operatives and circular economy. Can J Nonprofit Soc Econ Res 13:6. https://doi.org/10.29173/CJNSER593

Fuller R et al (2022) Pollution and health: a progress update. Lancet Planet Health 6:E535–E547. https://doi.org/10.1016/S2542-5196(22)00090-0

Hapuwatte B, Jawahir I (2021) Closed-loop sustainable product design for circular economy. J Ind Ecol 25:1430–1446. https://doi.org/10.1111/jiec.13154

Helman C (2016) Rethinking recycling. Forbes 198(4):52–56

Hermann RR, Pansera M, Nogueira LA, Monteiro M (2022) Socio-technical imaginaries of a circular economy in governmental discourse and among science, technology, and innovation actors: a Norwegian case study. Technol Forecast Soc Change 183:121903. https://doi.org/10.1016/J.TECHFORE.2022.121903

ILO (2018) World employment and social outlook. https://www.ilo.org/resource/news/24-million-jobs-open-green-economy-0#:~:text=ILO%20flagship%20report%20estimates%20job,moves%20to%20a%20greener%20economy.&text=GENEVA%20(ILO%20News)%20%E2%80%93%20Twenty,a%20new%20ILO%20report%20says

Deepa Ingavale, Anuradha Gaikwad (2011) A study of awareness for green marketing. In Conference: national conference on 'Business strategies for emerging markets' at SIBACA, Lonavala, Pune

Jensen HH (2024) For manufacturers, the circular economy strengthens supply chains. Here's how. https://www.weforum.org/agenda/2024/02/how-manufacturers-could-lead-the-way-in-building-the-circular-economy/

Jurij K, Meath C, Mwangangi LM, MacDonald C, Brockmann A (2023) Implementing circular economy principles in infrastructure procurement to support circular supply chains. Eng Constr Archit Manage 31(11):4645–4658

Karaman AS, Kilic M, Uyar A (2020) Green logistics performance and sustainability reporting practices of the logistics sector: the moderating effect of corporate governance. J Clean Prod 258:120718

Kurdhi NA, Dabadghao SS, Fransoo JC (2023) Revenue management in a refurbishing duopoly with cannibalization. J Oper Manag 69(2):246–260

Kurdve M, Zackrisson M, Wiktorsson M, Harlin M (2014). Lean and green integration into production system models-experiences from a Swedish country, Journal of Cleaner Production. 85:180–189

Lekan M, Rogers HA (2020) Digitally enabled diverse economies: exploring socially inclusive access to the circular economy in the city. Urban Geogr 41:898–901. https://doi.org/10.1080/02723638.2020.1796097

Lindahl E (2024) Circular production management strategies for accelerating the transition from linear to circular, licentiate thesis in industrial engineering and management. KTH Royal Institute of Technology, Stockholm

Livemint (2019) Circular economy likely to generate 1.4 crore jobs in 5–7 years: Amitabh Kant. https://www.livemint.com/news/india/circular-economy-likely-to-generate-1-4-crore-jobs-in-5-7-years-amitabh-kant-1560781052184.html

Lu S-H, Ngo T-S, Tran HTT (2023) Exploring a hierarchical model for circular economy transition through launching green marketing. Sustain Dev 32(3):2119–2138

Marica B, Bravi L, Savelli E (2019) Sustainability and quality management in the italian luxury furniture sector: a circular economy perspective, 2019. Sustainability 11(11):3089

Marina FA, Mesa JA, Jugend D, Pinheiro MAP, Fiorini PDC (2021) Circular product design: strategies, challenges and relationships with new product development. Manage Environ Qual Int J 2021(2):300–329

Mostaghel R, Oghazi P, Lisboa A (2023) The transformative impact of the circular economy on marketing theory. Technol Forecast Soc Change 195:122780. https://doi.org/10.1016/j.techfore.2023.122780

Sanjeev Nagarali (2023) Circular economy and role of QA. https://www.linkedin.com/pulse/circular-economy-role-qa-sanjeev-nagarali/

Napathorn C (2022) The development of green skills across firms in the institutional context of Thailand. Asia Pac J Bus Adm 14(4):539–572

Netland T (2013) Exploring the phenomenon of company-specific production systems: one-best-way or own-best-way? Int J Prod Res 51(4):1084–1097. https://doi.org/10.1080/00207543.2012.676686

Nichols W (2015) Circular economy could create jobs and boost GDP by £29bn. Environ Data Serv 486:43–44

Oscar KK, Adinyira E, Agyekum K (2024) Impact of green construction procurement practices on the success of circular economy in Ghana. Built Environ Project Asset Manage 14(3). https://doi.org/10.1108/BEPAM-06-2023-0109

Richter JL, Svensson-Hoglund S, Dalhammar C, Russell JD, Thidell Å (2023) Taking stock for repair and refurbishing: a review of harvesting of spare parts from electrical and electronic products. J Ind Ecol 27(3):868–888

Rybanský R, Jánoš D (2019) Specific aspect of marketing communication and consumer behaviour in a circular economy. Marketing Identity 2019(1):568–585

Schulze H, Bals L, Johnsen T (2019) Individual competences for sustainable purchasing and supply management (SPSM). Int J Phys Distrib Logist Manag 49(3):287–304

Singh SK, Singh AP (2019) Interplay of organizational justice, psychological empowerment, organizational citizenship behavior, and job satisfaction in the context of circular economy. Manag Decis 57(4):937–952

SITRA (2016) The critical move, Finland's road map to the circular economy 2025. https://www.sitra.fi/en/projects/critical-move-finnish-road-map-circular-economy-2-0/#the-challenge

Subramanian N, Suresh M (2022) The contribution of organizational learning and green human resource management practices to the circular economy: a relational analysis—Part I (2022). Learn Organ 29(5):428–442

Torres O (2017) Harnessing a data-driven circular economy: repurposing billions of tons of waste can create jobs, save resources. Int Trade Forum 3:26–27

Turner C, Moreno M, Mondini L, Salonitis K, Charnley F, Tiwari A, Hutabarat W (2019) Sustainable production in a circular economy: a business model for re-distributed manufacturing. Sustainability 11:4291. https://doi.org/10.3390/su11164291

UNEP (2019) Annual report 2019. https://www.unep.org/resources/unep-annual-report-2019

UNEP (2023) Annual report 2023. https://www.unep.org/resources/annual-report-2023

United Nations (2024). Climate action fast facts. https://www.un.org/en/climatechange/science/key-findings

Van Ewijk S, Stegemann JA (2016) Limitations of the waste hierarchy for achieving absolute reductions in material throughput. J Clean Prod 132:122–128. https://doi.org/10.1016/j.jclepro.2014.11.051

Van Luu T, Chromjaková F, Nguyen HQ (2023) A model of industry 4.0 and a circular economy for green logistics and a sustainable supply chain. Bus Strategy Dev 6:897–920

Workbred (2023) Waste management officer: key skills, qualifications and career path. https://workbred.co.uk/career/waste-management-officer

World Bank (n.d.) What a waste 2.0, a global snapshot of solid waste management to 2050. https://datatopics.worldbank.org/what-a-waste/trends_in_solid_waste_management.html

Yeow PHP, Wee HL (2018) Determinants of consumer behavior regarding reusing, refurbishing, and recycling computer waste: an exploratory study in Malaysia. Int J Bus Inf 13(4):457–488

Yuan H, Wu H, Zuo J (2018) Understanding factors influencing project managers behavioral intentions to reduce waste in construction projects. J Manag Eng 34(6):1–12

Zaidi AA, Chandra R (2024) The challenges to circular economy in the Indian apparel industry: a qualitative study. Res J Text Apparel. https://doi.org/10.1108/RJTA-09-2023-0105

Chapter 17
Discovering the Intangible Dimension of Friendshoring: The Role of Human Capital in Developing Sustainable Supply Chains

Rosa Palladino, Pasquale Sasso, Marta Mazzanti, Valentina Cillo, and Manlio Del Giudice

Abstract In recent years, the concept of friendshoring has gained prominence as a strategic approach to mitigate risks associated with global supply chains. Friendshoring involves relocating supply chain operations to countries that share similar political and economic systems, ensuring stability and reliability.

This chapter delves into the human dimension of friendshoring, highlighting the critical role of human capital in fostering sustainable and resilient global supply chains. Through an exploratory analysis of the literature field, the manuscript explores the role of skills, training, and policies needed to optimize human capital for effective friendshoring.

The chapter aims to enhance the intangible dimension of friendshoring and its role in developing sustainable and human-led global supply chains. Focusing on the critical role of human capital, it provides insights into the strategies, policies, and practices needed to effectively support friendshoring initiatives towards creating holistic models that can support the geopolitical challenges that are compromising markets around the world.

Keywords Friendshoring · Supply chain · Sustainability · Human capital · Geopolitics

R. Palladino (✉)
Department of Human Sciences, IUL Digital University, Florence, Italy
e-mail: r.palladino@iuline.it

P. Sasso · V. Cillo · M. Del Giudice
Department of Management and Economics, Pegaso Digital University, Naples, Italy
e-mail: pasquale.sasso@unipegaso.it; valentina.cillo@unipegaso.it; manlio.delgiudice@unipegaso.it

M. Mazzanti
Link Campus University, Rome, Italy
e-mail: marta.mazzanti@linkstudents.it

K. Mathiyazhagan et al. (eds.), *Humanizing Circular Supply Chain Management*, Circular Economy and Sustainability,
https://doi.org/10.1007/978-3-031-86562-6_17

17.1 Introduction

In recent years, the global economy has faced multiple shocks, such as the COVID-19 pandemic and the resulting disruption of demand, geopolitical tensions, and trade disputes between the United States and China, in addition to the ongoing war between Russia and Ukraine. These events have revealed vulnerabilities in traditional supply chains, prompting companies to explore alternative strategies such as friendshoring.

Friendshoring involves relocating supply chain operations to politically and economically aligned countries to mitigate risks associated with geopolitical tensions and economic instability, reducing the risk of disruptions caused by geopolitical conflicts and trade barriers by moving parts of the supply chain to countries with similar political and economic systems (Vivoda and Matthews 2023).

This trend evokes the restructuring of the perimeter of supply chains, to limit trade with neighboring countries, both geographically and culturally. Friendshoring promotes the sustainability of global supply chains, because it favors greater proximity between senders and recipients of intermediate components or finished products.

Recent research highlights the importance of ensuring greater economic resilience of the supply chain in the face of global disruptions, fostering the rapprochement of partners who share common ideas, values, and goals.

Some authors have highlighted that while supply chain agility expresses the ability to ensure continuity in business operations, resilience indicates the adaptive capacity of a company to return to normal operations after a disruption (Mubarik et al. 2021).

Sheffi (2020) highlights that resilience involves adapting and thriving amidst constant change, which is crucial as supply chains face risks ranging from pandemics to geopolitical upheavals. Indeed, precisely because of the high exposure to vulnerabilities, global supply chains have adopted a series of risk mitigation strategies, linked to the diversification of supply sources, for the maintenance of safety stocks. Indeed, the adoption of digital technologies such as real-time monitoring, predictive analytics and artificial intelligence help to anticipate and respond to supply chain disruptions (Zamani et al. 2023), shaping collaboration strategies and participatory dialogue with suppliers to reduce bottlenecks, developing real partnerships (Koh et al. 2019). In particular, AI is recognized as a major technology enabler that allows supply chain managers to manage operations in a timely and appropriate manner, improving the agility of normal operations, generating positive effects on supply chain resilience (Zamani et al. 2023).

In this scenario, the literature has demonstrated a positive association between human capital management (intended as the collective sum of attributes, including experience, knowledge, enthusiasm, energy and creativity, that employees invest in their workplace) and supply chain resilience, where digital technologies can optimize human performance, through the reduction of routine operations and improving skills (Yamin et al. 2024).

The integration of digital technologies, services and algorithms for the supply chain have become the main competitive factors in the new era generating a large amount of data and information that, once analyzed, improve the efficiency of specific business processes. Therefore, such technologies require a deep connection with the human fabric of the company, permeating every functional, procedural, organizational and strategic aspect, positively impacting performance. According to Zhou et al. (2021) digitalization can be internal or external, where internal digitalization aims to reduce costs and improve the efficiency of internal operational processes. While external digitalization emphasizes the reliance on digital technology to improve interactions with stakeholders, facilitate partnerships with suppliers and other partners, but can also accurately anticipate customer needs and increase customer loyalty.

In fact, employee skills are defined as their ability to learn new concepts, their familiarity with the latest technologies and their readiness to face unprecedented disruptions.

The goal of improving supply chain resilience through friend-shoring will have not only impacts on efficiency and robustness, but also systemic implications, linked to the consideration of non-economic factors in the process of allocating supply chain levels, requalifying commercial relationships in the light of political-institutional, national security and global sustainability assessments.

On the other hand, shortening value chains will be costly in financial terms and will probably require a long time to adapt, especially for value chains with a high level of operational complexity, as the world still remains dependent on raw materials from distant countries.

In this way, friendshoring could constitute an unexpected accelerator of the emergence of more sustainable value chains only by promoting a human-centric approach, which enhances the empathic component of human capital to integrate sustainable values in the supply chain link reallocation process, favoring the development of adaptable knowledge flows to promote the alignment of socio-environmental values among partners.

Therefore, following the paradigm of "digital humanism" that leverages the relationship between human and humanoids (Wagner 2020), friend shoring should foster a collaborative supply chain management approach that enhances the role of human capital to improve the outcomes of technologies in the supply chain, through employee involvement, through training and education activities that integrate the culture of sustainability and integration.

Subramony et al. (2018) highlight that human capital development through targeted education and training programs ensures that employees are equipped with the necessary skills to manage restructured and globally aware supply chains. Furthermore, integrating sustainability into supply chain management is increasingly important to cross production and ecosystem protection objectives. Jabbour et al. (2019) highlight that green supply chain management practices, supported by well-trained human capital, are crucial for long-term sustainability goals, because they foster the integration of a cultural substrate aware of global challenges that require the abandonment of fossil fuel combustion, in favor of new renewable energy sources. This process will generate significant changes in the relationship between the powers, so much so that we can speak of a new human-led "geopolitics of energy transition", to favor new practices that include the efficient use of resources, the reduction of waste and the adoption of renewable energy sources, which require a competent and qualified workforce.

In addition to technical skills, the development of soft skills such as adaptability, cultural awareness and problem solving is essential to navigate the complexities of friendshoring supply chains. Robles (2012) notes that these soft skills are critical to overall job performance and organizational success, especially in diverse and dynamic environments.

Considering these premises, this chapter aims to analyze the intangible dimension of friendshoring, exploring the role of human capital in creating sustainable and resilient supply chains, enhancing the supply network with reliable and "allied" suppliers, and proposing frameworks and strategies through which both economic and sustainable value-based relationships are shared. Focusing on the integration of soft and technical skills, continuous learning, and sustainability practices, this chapter provides insights on how companies can build resilient and sustainable supply chains through a human-centric approach.

17.2 Understanding Friendshoring

Friendshoring is a strategic move where companies relocate their supply chain operations to politically stable countries that share similar economic values. This strategy aims to insulate supply chains from geopolitical disruptions and economic volatility. The COVID-19 pandemic exposed the fragility of global supply chains, prompting a reevaluation of dependence on single-source suppliers. Boschma (2022) explores the impact of the COVID-19 pandemic on global supply chains, emphasizing the need for diversification and the benefits of establishing operations in politically aligned countries. Additionally, increasing geopolitical tensions between major economic powers have underscored the risks of concentrated supply chains. By relocating operations to friendly nations, businesses aim to create a more predictable and secure supply chain environment (Rojas et al. 2022; Attinasi et al. 2023).

Unlike offshoring, which primarily seeks cost advantages by shifting production to low-cost countries, friendshoring prioritizes stability, reliability, and the reduction of geopolitical risks. This approach has gained traction in response to disruptions caused by events such as the US-China trade war and the Russia-Ukraine conflict. These geopolitical events have highlighted the vulnerabilities inherent in having critical supply chains concentrated in regions that may be subject to political instability or economic sanctions (Vivoda and Matthews 2023; Javorcik et al. 2024).

Furthermore, the benefits of friendshoring go beyond mere risk mitigation, as this trend may be a driver for the integration of sustainable values into supply chains. By collaborating with countries that share similar regulatory frameworks, labor standards, and environmental policies, companies can ensure smoother integration and cooperation toward achieving socio-environmental goals. This alignment can lead to smoother operations, better compliance with international standards, and potentially faster dispute resolution mechanisms. Additionally, friendshoring can foster stronger diplomatic and economic ties between countries, contributing to a more collaborative international environment. Moreover, friendshoring can contribute to regional economic development by encouraging investment in infrastructure and skills development in partner countries. This can lead to more balanced global economic growth and reduce economic disparities between developed and developing nations (Rojas et al. 2022).

In essence, friendshoring represents a shift towards a more resilient and cooperative global trade framework. It acknowledges the importance of political and economic alliances in ensuring the continuity and security of global supply chains. As businesses and governments navigate an increasingly complex and interconnected world, friendshoring offers a strategic pathway to achieving long-term stability and prosperity (Rojas et al. 2022; Attinasi et al. 2023).

However, there is no shortage of concerns related to this trend, which risks creating closures in world trade, generating a new protectionism through the increase of higher prices and uneven economic growth among different countries (Vivoda 2023).

These reasons justify the need to explore mediation solutions that guarantee a balance between geopolitical vulnerabilities, market stabilization needs and inclusive and sustainable economic growth.

17.3 Geopolitical and Economic Drivers

The drive towards friendshoring is fueled by geopolitical uncertainties, trade wars, and the need for economic security. Geopolitical tensions and trade conflicts have exposed the vulnerabilities of global supply chains heavily reliant on specific regions. The US-China trade tensions and the Russia-Ukraine conflict have particularly highlighted the risks of over-reliance on certain regions for critical supplies (European Council 2023). These geopolitical disruptions have significantly impacted global supply chains, prompting businesses to seek more stable and politically aligned alternatives (Boschma 2022). Geopolitical risks can lead to supply chain disruptions, increased costs, and regulatory challenges. For example, trade policy uncertainty can deter investment and trade flows, leading companies to reassess their supply chain strategies (Handley and Limão 2017). The US-China trade war, with its tariffs and trade barriers, forced companies to relocate operations to mitigate increased costs and disruptions (Evenett 2019). The Russia-Ukraine conflict has further underscored the need for supply chain diversification. This conflict has caused significant disruptions in the supply of critical commodities such as energy, metals, and agricultural products, impacting global supply chains and driving companies to consider friendshoring as a strategy to reduce dependency on unstable regions (Toygar and Yildirim 2023). Economic security is another critical driver of friendshoring. Companies recognize the need to protect their supply chains from economic shocks and ensure continuity of operations. Research highlights the importance of supply chain resilience in the face of economic volatility. By relocating operations to countries with stable economic and political environments, companies can reduce the risk of supply chain disruptions and enhance resilience (Miroudot and Cadestin 2017). Moreover, friendshoring can foster stronger economic and diplomatic ties between countries. Economic integration through friendshoring can lead to more collaborative international relations, enhancing the stability and predictability of global supply chains.

These goals require strengthening human skills, enhancing the role of sensitivity and empathy in the face of the pervasive development of technologies that aim to maximize the role of the supply chain while neglecting the integration of sustainable and resilient values.

17.4 The Role of Human Capital in Friendshoring

17.4.1 Human Capital as a Key Enabler

At a time when geopolitical tensions threaten to undermine the traditional functioning of markets, friendshoring offers an alternative to relocate some stages of the value chain to countries that share the same value system—or the same interests—in reference to a geopolitical alignment with the country that relocates.

Human capital, defined by the skills, knowledge, and experience possessed by individuals, plays a pivotal role in the success of friendshoring initiatives.

Human capital refers to the ability and efficiency of people to transform raw materials and capital into goods and services (Mincer 1984). The development of HC is important for its intrinsic value as well as its instrumental value, as individuals are considered to be repositories of knowledge, expertise, and experience. Although the conceptual definition of human capital is clear, its measurement is difficult because it is virtually impossible to observe individual skills and even more difficult to design a metric that is comparable across individuals and countries.

Highlighting the social and economic importance of human capital theory, Becker (1993) observed that the most valuable of all capitals is investment in human beings, through education and training programs, to compete in globalized knowledge sectors.

Beyond the technical aspects, some authors have highlighted the role of human capital in managing supply chain disruptions and the consequences on long-term resilience, demonstrating how the lack of skills of employees during periods of stress poses major challenges to the recovery of the disrupted SC, such as changes in work routines, anxieties and fears among employees and changes in work skills in changed working conditions (Gu et al. 2023).

According to the literature, SCR is a multidimensional capability, which includes the development of different types of resources and organizational capabilities, combining and adapting their resources, capabilities and processes (El Baz and Ruel 2021). In this sense, companies must reconfigure their production systems, operational processes and supply chain activities to withstand the impact of risk and restore business as usual, through a high degree of involvement of human capital integrated into business processes.

The effectiveness of relocating supply chains depends significantly on the workforce's capability to adapt and thrive in new environments. According to Becker (1993), investments in human capital, such as education and training, are crucial for enhancing workforce productivity and organizational performance. This is particularly relevant in the context of friendshoring, where the successful integration of new supply chains hinges on the ability of the local workforce to meet the operational demands of relocated businesses. Investing in human capital is essential for the long-term success of friendshoring. As noted by Wright and McMahan (1992), effective human resource management practices can lead to sustained competitive advantage. Companies that invest in training and development for their employees in new locations can better leverage local talents and ensure a smoother transition and integration of operations. This investment not only improves productivity but also fosters loyalty and reduces turnover, which is essential for maintaining stable supply chains. Moreover, a well-trained and competent workforce can significantly enhance the effectiveness of relocated operations, reducing operational risks and improving the quality of products and services offered. Huselid (1995) highlights how high-performance human resource management practices are associated with greater productivity and better financial performance. This reinforces the importance of investing in human capital as an integral part of friendshoring strategies. Further studies support this view. Pfeffer (1998) emphasizes that organizations adopting a strategic approach to human resource management, by investing in their human capital, tend to achieve superior performance compared to their competitors. Practices that value human capital, such as continuous training and professional development, help create a more innovative and responsive work environment, capable of quickly adapting to market changes and new geopolitical challenges. More recent studies continue to confirm the importance of human capital. For example, a work by Delery and Roumpi (2017) discusses how strategic human resource management practices can positively influence organizational performance, highlighting the importance of alignment between business strategies and personnel management policies. Additionally, Jiang et al. (2012) demonstrate that well-designed human resource management practices can lead to significant improvements in business performance through increased employee motivation and skills.

17.4.2 Skills and Competencies for Sustainable Supply Chains

To develop a competitive advantage, it is important to leverage HC as a competitive weapon for long-term survival and sustainability, ensuring that employees have the knowledge, skills, and abilities they need to work effectively in a rapidly changing and complex environment.

More in details, the shift towards friendshoring necessitates a diverse set of skills and competencies that are crucial for the success of supply chain relocation initiatives, according to a logic of value integration oriented towards sustainability. Technical skills in manufacturing and logistics are essential, but equally important are soft skills such as adaptability, cultural awareness, and problem-solving abilities. Additionally, knowledge of sustainable practices is fundamental for integrating sustainability into supply chain operations. Recent studies highlight the importance of these varied skill sets in creating resilient and efficient supply chains.

Successful friendshoring requires a workforce proficient in various domains:

- Technical skills: expertise in manufacturing, logistics, and supply chain management. According to a study by Mangan and Lalwani (2016), technical proficiency in these areas is fundamental for maintaining operational efficiency and ensuring the smooth transition of supply chains to new locations. The ability to leverage advanced technologies and manage complex logistics networks is particularly vital for addressing the logistical challenges that arise during relocation. Adequate mastery of digital technologies and innovations such as the Internet of Things (IoT) and artificial intelligence can significantly enhance operational efficiency and goods traceability;
- Soft skills: adaptability, cultural awareness, and problem-solving abilities. These skills are essential for navigating the challenges associated with relocating supply chains. As noted by Sheffi (2015), adaptability and cultural awareness help employees effectively manage cross-cultural interactions and adjust to new operational environments. Problem-solving abilities are crucial for addressing unforeseen issues and ensuring continuous improvement in supply chain processes. The ability to work in diverse teams and communicate effectively across language and cultural barriers is equally important for ensuring productive and harmonious collaboration.
- Sustainability knowledge: Understanding of sustainable practices and the ability to implement them in supply chain operations. The growing emphasis on sustainability in global trade requires a workforce that is knowledgeable about environmental impact and resource efficiency. A study by Gold et al. (2013) underscores the importance of sustainability competencies in enhancing the environmental performance of supply chains. Employees must be equipped to integrate sustainable practices into daily operations and long-term strategies. This includes waste management, efficient resource use, and the adoption of renewable energy sources. The ability to implement circular economy practices and reduce the carbon footprint of the supply chain is crucial for meeting regulatory and consumer expectations.

Moreover, the integration of sustainability into supply chains is increasingly seen as a competitive advantage. According to Carter and Rogers (2008), sustainable supply chain practices not only contribute to environmental and social goals but also enhance operational efficiency and corporate reputation. This alignment of sustainability with business objectives requires employees to possess a deep understanding of sustainable supply chain management principles. The ability to measure and communicate the impact of sustainability initiatives is essential for ensuring transparency and credibility. Recent research by Dubey et al. (2017) further elaborates on the critical role of green human resource management (GHRM) in fostering a sustainability-oriented culture within organizations. GHRM practices, such as training programs focused on environmental management and sustainability, are essential for developing the necessary competencies among employees. These practices also include incentivizing sustainable behaviors through reward and recognition systems.

To achieve sustainable friendshoring, organizations must invest in comprehensive training and development programs that encompass these skill areas. By fostering a workforce proficient in technical, soft, and sustainability skills, companies can build more resilient and adaptable supply chains capable of thriving in diverse geopolitical landscapes. The ability to develop a versatile and well-prepared workforce is fundamental to addressing the complexities of friendshoring and ensuring operational continuity in a rapidly evolving world.

17.5 Enhancing Human Capital for Sustainable Friendshoring

17.5.1 Training and Development

Effective training and development programs are essential for equipping the workforce with the necessary skills for friendshoring. Continuous education and upskilling are crucial to meet evolving industry demands, focusing on technical skills, sustainability practices, and soft skills to create a well-rounded and capable workforce. Technical training should cover areas such as manufacturing, logistics, and supply chain management, with an emphasis on advanced technologies like automation, artificial intelligence, and data analytics, which are vital for enhancing operational efficiency (Subramony et al. 2018). Incorporating sustainability into training programs is increasingly important as organizations aim to reduce their environmental footprint. Training in practices such as energy efficiency, waste reduction, and resource management helps employees integrate these principles into their daily operations. Green training programs foster a culture of sustainability and improve environmental performance (Jabbour and Jabbour 2016). Employees trained in sustainability can contribute to reducing carbon emissions and promoting eco-friendly practices. Soft skills, including adaptability, cultural awareness, and problem-solving, are critical for the successful transition to friendshoring. Training programs should enhance these skills to help employees navigate cultural differences and adapt to new environments. Soft skills are as important as technical skills for overall job performance and organizational success, enabling effective collaboration, communication, and problem-solving in dynamic and multicultural work settings (Robles 2012). To maintain a competitive edge, organizations must promote a culture of lifelong learning and continuous improvement. Providing opportunities for employees to upgrade their skills and knowledge continuously is key to employee engagement, retention, and organizational adaptability to changing market conditions (Noe et al. 2017). Lifelong learning initiatives ensure the workforce remains agile and capable of meeting industry challenges. Effective training and development should be comprehensive and integrated into the organizational strategy. Systematic training programs lead to significant improvements in employee performance, job satisfaction, and overall organizational outcomes (Aguinis and Kraiger 2009). Training programs should be tailored to the specific needs of the organization and aligned with strategic goals, including successful friendshoring implementation.

17.5.2 Educational and Training Programs

Developing tailored educational and training programs is essential for enhancing human capital in the context of friendshoring. These programs should be created in collaboration with industry experts, educational institutions, and policymakers to address specific needs. Programs should focus on core skills required for supply chain operations, including advanced manufacturing techniques, logistics management, and the use of technology. Mastering technologies like automation, artificial intelligence, and data analytics is critical. Subramony et al. (2018) emphasize the importance of developing technical skills to drive productivity and innovation, while Frank et al. (2019) discuss incorporating Industry 4.0 technologies to enhance supply chain efficiency and responsiveness. Training should also emphasize sustainability practices, such as energy efficiency, waste reduction, and resource management. Jabbour and Jabbour (2016) highlight the importance of green training programs for fostering a culture of sustainability within organizations. Employees equipped with these skills can help companies achieve environmental goals and enhance corporate social responsibility. Developing soft skills like adaptability, cultural awareness, and problem-solving is crucial for navigating the complexities of friendshored supply chains. Robles (2012) notes that these skills are essential for overall job performance and organizational success, enabling effective collaboration and communication in diverse and dynamic environments. Promoting a culture of lifelong learning is vital for maintaining a competitive and adaptable workforce. Companies should encourage continuous education and provide opportunities for upskilling and reskilling. Noe et al. (2017) highlight that continuous learning and development are key to employee engagement, retention, and organizational adaptability to changing market conditions. Training and development should be integrated into the organizational strategy. Aguinis and Kraiger (2009) state that systematic training programs lead to significant improvements in performance, job satisfaction, and organizational outcomes. These programs should be tailored to organizational needs and aligned with strategic goals, including the successful implementation of friendshoring initiatives. Collaboration with industry experts, educational institutions, and policymakers is essential for developing relevant and effective training programs as work evolves, it is critical to continuously assess and address emerging skills gaps. The World Economic Forum (2020) reports that rapid technological advancements and market shifts necessitate regular updates to training curricula to reflect the latest industry requirements. Leveraging digital learning platforms can enhance training accessibility and effectiveness. Deng and Yu (2023) demonstrate that digital learning tools, such as e-learning modules and virtual simulations, significantly improve knowledge retention and practical application of skills. These platforms offer the flexibility needed to accommodate diverse learning styles and schedules.

17.6 Conclusions

Friendshoring is having a significant development as a response to the instability of markets, disrupted by the geopolitical turbulence of recent years. In addition to mitigating risks related to the vulnerability of economic relations, the distribution of supply chain links in friendly countries has the potential to strengthen a culture of values based on sustainability, through the integration of socio-environmental values along supply chains.

However, friend-shoring also involves several risks, because trading only with countries with similar levels of development risks eliminating the comparative advantage of production in countries with different levels of development. In fact, friend-shoring would tend to exclude poor countries that need world trade the most to become richer and more democratic. Furthermore, another danger is related to the construction of barriers between countries and between people, which fuel divisions and inequalities.

This chapter has highlighted the fundamental role of the human dimension of friendshoring, underlining the need for a holistic approach to the development of human capital for the integration of socio-environmental values in supply chain restructuring initiatives in friendly countries. As is known, digital transformation requires an iterative approach, which creates a synergic combination between technologies and the human factor. Organizations in every sector should adopt digital approaches that work with people, not independently of people, promoting integration mechanisms aimed at building skills and change management policies, capable of achieving sustainability goals.

As the global economy continues to evolve, the importance of human capital in driving sustainable supply chain practices will only grow. By prioritizing the development and empowerment of human capital, companies can build resilient, sustainable, and friendshoring-aligned supply chains. Sustainability is a critical component of friendshore supply chains. Integrating sustainable practices not only improves supply chain resilience, but also aligns with global efforts to address environmental and social challenges. In an effort to achieve a synergistic balance of economic growth, technological development, and human resource optimization, human capital plays a critical role in integrating sustainability into supply chains, so employees must be equipped with the knowledge and tools to implement sustainable practices, such as reducing carbon footprints, managing waste, and ensuring ethical sourcing. Sustainability training programs should be a critical component of employee development, to foster a culture of collaboration and social belonging that can drive responsible and inclusive development. These programs should cover topics such as environmental stewardship, resource efficiency, and ethical supply chain practices. Empowering employees to take the lead in sustainability efforts, while valuing sensitivity and empathy, can lead to significant improvements in their areas of responsibility, ensuring that friendshoring operations are not only efficient but also sustainable and oriented towards the integration of socio-environmental values. Therefore, companies must adopt practices that prioritize the development of human capital, ensuring that their workforce is equipped to support sustainably oriented friendshoring initiatives. Involving employees in decision-making processes and recognizing their contributions promotes a sense of belonging and commitment to the integration of noble values. Creating channels for employee feedback and involving them in the implementation of friendshoring strategies is essential to nurture the flow of knowledge, skills and relationships that ensure the right orientation of friendshoring in the vulnerable context of the global market.

17.7 Policy Recommendations

Policy makers have a crucial role to play in supporting the orientation of friendshoring practices, to promote the development of sustainable values in supply chains in the light of the evolving global market. Effective policies should focus on several key areas, starting from funding technical and vocational education to promoting a human-centric approach to supply chain management. Investment in these areas ensures a steady supply of skilled workers and should align with industry needs, particularly in sustainable supply chain management. McGuinness et al. (2018) highlight that aligning educational programs with labor market demands enhances employability and productivity, aiding the transition to friendshoring by providing the necessary skills for workforce adaptation to new technologies.

Creating frameworks to facilitate the movement of skilled labor across borders is also essential. Policies should support the recognition of qualifications and provide incentives for skilled workers to relocate to friendshored supply chain hubs. Boucher and Cerna (2014) emphasize that international labor mobility addresses skill shortages and enhances innovation. Governments can attract and retain talent by establishing mutual recognition agreements and streamlining immigration processes for skilled workers. Encouraging companies to invest in employee training and development through tax incentives and grants is another key strategy. Financial incentives, as noted by De Grip and Sauermann (2012), can significantly increase employers' willingness to invest in training programs. Policymakers should design tax credits and grants that offset training costs, making it financially feasible for companies to offer ongoing education and development opportunities, thereby enhancing workforce capabilities and fostering global market innovation and competitiveness.

Collaboration between the public and private sectors is crucial for the successful implementation of human capital development policies. Public-private partnerships can align educational curricula with industry needs, ensuring training programs are relevant and effective. Casady et al. (2020) indicate that such partnerships improve skill development and economic growth outcomes. Governments should encourage collaboration between educational institutions, businesses, and industry associations to create a cohesive workforce development strategy.

Policies should also promote lifelong learning by providing opportunities for adult education and professional development. The rapid pace of technological change necessitates continuous skill upgrading. Field (2015) asserts that lifelong learning policies are essential for maintaining workforce adaptability and productivity. Governments can support lifelong learning through initiatives such as subsidized training programs, flexible learning opportunities, and recognition of prior learning and work experience. To support remote learning and digital training programs, investing in robust digital infrastructure is essential. Policymakers should focus on improving internet access and digital literacy, especially in underserved areas. Van Deursen and van Dijk (2019) highlight the importance of digital skills for participating in the modern workforce. Enhancing digital infrastructure ensures all individuals have access to online educational resources and training opportunities.

Policies should also aim to close gender and diversity gaps in technical and vocational education. Encouraging the participation of underrepresented groups in STEM (science, technology, engineering, and mathematics) fields can diversify the talent pool and drive innovation. UNESCO (2017) reports that gender-responsive education policies significantly impact economic growth and social development. Governments should implement initiatives that promote inclusivity and equal opportunities in education and training programs.

17.8 Challenges and Future Research Directions

Despite the benefits, several challenges hinder the development of human capital in friendshoring: first, limited funding for education and training programs is a significant obstacle, so governments and companies need to allocate sufficient resources to develop and implement effective training programs.

Second, resistance to change may hinder the adoption of new socially and environmentally oriented friendshoring practices, aimed at creating a holistic and humane model for building sustainable supply chains. Finally, there is a need to mitigate cultural barriers inherent in managing a culturally diverse workforce, which presents challenges in communication and collaboration. Companies need to promote cultural awareness and inclusiveness to overcome these barriers.

Future research should focus on exploring innovative approaches to humanizing supply chains, enhancing the role of human capital to integrate sustainability in friendshoring through the use of technology in training. Policy makers should continue to refine and adapt policies to support the changing needs of the global workforce, given geopolitical dynamics and the imperatives associated with the UN 2030 Agenda.

Future research should investigate the use of technology, such as virtual reality and online learning platforms, to enhance training programs to provide scalable and flexible solutions.

The rise of remote working has implications for friendshoring, so research should explore how remote working can be integrated into friendshoring supply chains and its impact on human capital development. Policy makers should continuously evaluate and update policies to address emerging challenges and opportunities in human capital development. This includes adapting to changes in the global economic landscape and technological advances.

Finally, friendshoring combines geopolitical and technical dimensions in decision-making, so future research should explore the role of human capital in managing complex decision-making processes, which affect national security, political balances, and global development.

References

Aguinis H, Kraiger K (2009) Benefits of training and development for individuals and teams, organizations, and society. Annu Rev Psychol 60(1):451–474

Attinasi, M. G., Boeckelmann, L., & Meunier, B. (2023). Friend-shoring global value chains: a model-based assessment. Economic Bulletin Boxes, European Central Bank, vol. 2.l. https://www.ecb.europa.eu/press/economic-bulletin/focus/2023/html/ecb.ebbox202302_03~d4063f8791.en.html

Becker GS (1993) Human capital: a theoretical and empirical analysis with special reference to education. University of Chicago Press

Boschma R (2022) Global value chains from an evolutionary economic geography perspective: a research agenda. Area Dev Policy 7(2):123–146

Boucher A, Cerna L (2014) Current policy trends in skilled migration policy. Int Migr 52(3):21–25

Carter CR, Rogers DS (2008) A framework of sustainable supply chain management: moving toward new theory. Int J Phys Distribut Logist Manage 38(5):360–387

Casady CB, Eriksson K, Levitt RE, Scott WR (2020) (Re)defining public-private partnerships (PPPs) in the new public governance (NPG) paradigm: an institutional maturity perspective. Public Manag Rev 22(2):161–183

De Grip A, Sauermann J (2012) The effects of training on own and co-worker productivity: evidence from a field experiment. Econ J 122(560):376–399

Delery JE, Roumpi D (2017) Strategic human resource management, human capital, and competitive advantage: is the field going in circles? Hum Resour Manag J 27(1):1–21

Deng X, Yu Z (2023) A meta-analysis and systematic review of the effect of chatbot technology use in sustainable education. Sustain For 15(4):2940

Dubey R, Gunasekaran A, Ali SS (2017) Exploring the relationship between leadership, operational practices, institutional pressures, and environmental performance: a framework for green supply chain. Int J Prod Econ 187:252–263

El Baz J, Ruel S (2021) Can supply chain risk management practices mitigate the disruption impacts on supply chains resilience and robustness? Evidence from an empirical survey in a COVID-19 outbreak era. Int J Prod Econ 233:107972

European Council (2023) The effect of sanctions on Russia and Ukrainian export routes. EC

Evenett SJ (2019) Protectionism, state discrimination, and international business since the onset of the Global Financial Crisis. J Int Bus Policy 2:9–36

Field J (2015) Lifelong learning and the new educational order. Trentham Books

Frank AG, Dalenogare LS, Ayala NF (2019) Industry 4.0 technologies: implementation patterns in manufacturing companies. Int J Prod Econ 210:15–26

Gold S, Seuring S, Beske P (2013) Sustainable supply chain management and inter-organizational resources: a literature review. Corp Soc Responsib Environ Manag 20(4):205–221

Gu M, Zhang Y, Li D, Huo B (2023) The effect of high-involvement human resource management practices on supply chain resilience and operational performance. J Manage Sci Eng 8(2):176–190

Handley K, Limão N (2017) Policy uncertainty, trade, and welfare: theory and evidence for China and the United States. Am Econ Rev 107(9):2731–2783

Huselid MA (1995) The impact of human resource management practices on turnover, productivity, and corporate financial performance. Acad Manag J 38(3):635–672

Jabbour CJC, Jabbour AB (2016) Green human resource management and green supply chain management: linking two emerging agendas. J Clean Prod 112:1824–1833

Jabbour CJC, de Sousa Jabbour ABL, & Sarkis J (2019) Unlocking effective multi-tier supply chain management for sustainability through quantitative modeling: Lessons learned and discoveries to be made. International Journal of Production Economics, 217, 11–30.

Javorcik B, Kitzmüller L, Schweiger H, Yıldırım MA (2024) Economic costs of friendshoring. World Econ 47:2871–2908

Jiang K, Lepak DP, Hu J, Baer JC (2012) How does human resource management influence organizational outcomes? A meta-analytic investigation of mediating mechanisms. Acad Manag J 55(6):1264–1294

Koh L, Orzes G, Jia FJ (2019) The fourth industrial revolution (Industry 4.0): technologies disruption on operations and supply chain management. Int J Oper Prod Manag 39(6/7/8):817–828

Mangan J, Lalwani C (2016) Global logistics and supply chain management. Wiley

McGuinness S, Pouliakas K, Redmond P (2018) Skills mismatch: concepts, measurement, and policy approaches. J Econ Surv 32(3):985–1015

Mincer J (1984) Human capital and economic growth. Econ Educ Rev 3(3):195–205

Miroudot S, Cadestin C (2017) Services in Global value chains: from inputs to value-creating activities. In: OECD trade policy papers, No. 197. OECD Publishing

Mubarik MS, Naghavi N, Mubarik M, Kusi-Sarpong S, Khan SA, Zaman SI, Kazmi SHA (2021) Resilience and cleaner production in industry 4.0: role of supply chain mapping and visibility. J Clean Prod 292:126058

Noe RA, Clarke ADM, Klein HJ (2017) Learning in the twenty-first-century workplace. Annu Rev Organ Psych Organ Behav 4(1):245–275

Pfeffer J (1998) The human equation: building profits by putting people first. Harvard Business School Press

Robles MM (2012) Executive perceptions of the top 10 soft skills needed in today's workplace. Bus Commun Q 75(4):453–465

Rojas M, Routh A, Sherwood J, Buckley J, Keyal A (2022) Reshoring and "friendshoring" supply chains. Deloitte insights: government trends. https://www2.deloitte.com/us/en/insights/industry/public-sector/government-trends/2022/reshoring-global-supply-chains.html. Accessed 10 Oct 2022

Sheffi Y (2015) The power of resilience: how the best companies manage the unexpected. MIT Press

Sheffi Y (2020) The new (ab)normal: reshaping business and supply chain strategy beyond Covid-19. MIT Press

Subramony M, Segers J, Chadwick C, Shyamsunder A (2018) The practice of skill development: bridging the gap between skill demands and workforce competency. Hum Resour Manag 57(2):557–575

Toygar A, Yildirim U (2023) Examining the effects of the Russia-Ukraine conflict on global supply chains. In: Handbook of research on war policies, strategies, and cyber wars. IGI Global, pp 184–199

UNESCO (2017) Cracking the code: girl's and women's education in science, technology, engineering, and mathematics (STEM). UNESCO Publishing

van Deursen AJAM, van Dijk JAGM (2019) The first-level digital divide shifts from inequalities in physical access to inequalities in material access. New Media Soc 21(2):354–375

Vivoda V (2023) Friend-shoring and critical minerals: exploring the role of the minerals security partnership. Energy Res Soc Sci 100:103085

Vivoda V, Matthews R (2023) "Friend-shoring" as a panacea to Western critical mineral supply chain vulnerabilities. Miner Econ 37:463–476

Wagner DN (2020) Augmented human-centered management. Human resource development for highly automated business environments. J Hum Res Manage 23(1):13–27

Wright PM, & McMahan GC (1992) Theoretical perspectives for strategic human resource management. Journal of management, 18(2), 295–320.

World Economic Forum (2020) The future of jobs. WEF

Yamin MA, Almuteri SD, Bogari KJ, Ashi AK (2024) The influence of strategic human resource management and artificial intelligence in determining supply chain agility and supply chain resilience. Sustainability 16(7):2688

Zamani ED, Smyth C, Gupta S, Dennehy D (2023) Artificial intelligence and big data analytics for supply chain resilience: a systematic literature review. Ann Oper Res 327(2):605–632

Zhou D, Yan T, Dai W, Feng J (2021) Disentangling the interactions within and between servitization and digitalization strategies: a service-dominant logic. Int J Prod Econ 238:108175

Chapter 18
Data-Driven Cohesion: Leveraging BI for Cross-Functional Alignment in Supply Chain Decision-Making

Bhajneet Kaur, Gunjan Chhabra, and Kamini Rai

Abstract Business Intelligence (BI) tools are becoming increasingly prevalent in Supply Chain Management (SCM), resulting in the transformation of many organizational practices. This study explores the acceptance of BI tools in SCM, focusing on how not merely strategic planning but also strategic levels decisions are heavily influenced by these systems. In addition to this, the transformation that Business Intelligence tools have brought is providers supply chain departments are able to break down large data sets and gain insights into designing of inventory management systems forecasting demand effectively, optimizing logistics which ultimately results in informed decisions. Technological Readiness, Data Governance and Organizational Support were significant facilitators in the implementation of BI tools for SCM as uncovered by this study. It also highlights the advantages as well disadvantages of BI integrating supply chains with an example which tells how it will help in predicting, allocating resources and making data-based strategies for a better supply chain. The results give some idea that organizations will find it beneficial to their competitive advantage when aligning the broader business objectives with supply chain practices by fostering a culture of continuous improvement through efficiency BI tools in SCM. Within the growing field of supply chain management research relating to BI, this study proposes a conceptual framework, helps advance what is known about how these tools can transform strategic decision-making in a way that provides valuable takeaways for practicing internists and business leaders whose aim it is to leverage data more effectively within their organizations.

B. Kaur (✉)
Rukmini Devi Institute of Advanced Studies (RDIAS), New Delhi, India

Sunstone Education Pvt Ltd, Gurugram, India

G. Chhabra
Sunstone Education Pvt Ltd, Gurugram, India

K. Rai
Rukmini Devi Institute of Advanced Studies, New Delhi, Delhi, India

K. Mathiyazhagan et al. (eds.), *Humanizing Circular Supply Chain Management*, Circular Economy and Sustainability,
https://doi.org/10.1007/978-3-031-86562-6_18

Keywords Business intelligence · Decision making · Supply chain professionals · Data-driven insights · Supply chain analytics

18.1 Introduction

In today's fast moving business environment, incorporating Business Analytics (BA) are a basic piece of organizations need to improve vital decisions in Supply Chain Analytics (SCA). BA is further used to represent a wide spectrum of methods and technology, which are implemented with the goal of evaluation company data in order that understanding can be gleaned for future planning as well as investments (Kristoffersen et al. 2021). The application of this approach extends as far as being industry agnostic (applicable in wide sectors such healthcare/finance/logistics) since the methodology allows to perform an analysis aided by statistical instruments through a data-driven approach. On the other hand, SCA focusses only on making supply chain operations more flexible and efficient through analysis of supplier performance, inventory management and logistics efficiency.

BA and SCA complement each other to provide organizations with data-driven insights for day-to-day operations, strategic level too. Timely and informative Business Intelligence (BI) system is a key part of ensuring that the data made available to businesses be used effectively. They facilitate gathering, storing and analysing of data both internally (from within the firm) as well external to it which help businesses improve their supply chain performances (Choi et al. 2022; Bawack and Ahmad 2021). Supply chain managers can use real-time analytics, for example, to pinpoint logistical bottlenecks and predict changes in demand based on past sales data.

Also, advanced analytical methods, including predictive modeling and data mining, allow organizations to take a look at the future and anticipate emerging market patterns and customer requirements. This is crucial in supply chain management, where swift actions are able to boost business performance and yield a competitive edge (Sousa et al. 2024). With proper alignment of BA with SCA, firms will be able to fine-tune the supply chain processes and in the long run improve the overall performance of the business through efficient allocation of resources.

As if there is any doubt that we live in an era of Big Data's dominance, BI systems' time has never been more pronounced than now. Organizations struggle with the volume of data coming from various places like social platforms or responses from clients. The information sourced from BI matters to managers who have to find a way of achieving the business targets. Therefore spending on BI systems makes business sense if great returns like quality decisions and increased productivity are realized (Sousa et al. 2024). An organization can make use of these tools in its supply chain in order to work through large amounts of data concerning suppliers and their respective logistics and this can assist them in making more rational choices on the distribution of resources. In addition, the incorporation of analytics into supply chain management implies the new way of doing things. Effective future demand planning, risk asses going further helped by detailed data interpretation. Data analysis entails the examination of a data set in order to evaluate new trends that may emerge and formulate strategies as a result. To illustrate, anticipatory models can be used to gauge the degrees of optimal stock and supplier who can be trusted to be there for the business based on projections of demand variability.

The challenges posed by globalization and technological changes make analytics very important for SCA. Firms that manage to combine such frameworks can expect to grow consistently as their decisions are strategically relevant. In this research, we define the boundaries of business intelligence, within which we aim to address the most important boundaries in arch supply chain analytics through business analytics.

18.2 Literature Review

The introduction of Supply Chain Analytics (SCA) is reshaping management practice especially with regard to Supply Chain Management (SCM). As SCM involves basic business processes like sourcing, manufacturing and selling, analytics needs to be harnessed in order to reduce costs and improve efficiency. Considering that a large percentage of the costs incurred by an organization stems from supply chain activities, the implementation of such measures becomes a necessity in management. There is a relationship between Business Intelligence systems and SCA in the evolution of SCA. They have the processes necessary to organize and distribute large volume of information and provide decision-making support and forecasting. The use of business intelligence systems has led to increased new information since these systems integrate data from various sources. This enhances knowledge on various decisions and strategies as it aligns the operational focus with the strategic objectives of the organization. Ülkü and Mansouri (2023) stress that SCA enhances the speed of response, intelligence and efficiency in decision making within the diverse industries such as transportation and assisting in procurement activities.

18.2.1 Key Considerations for Effective Supply Chain Analytics

To maximize the benefits of Supply Chain Analytics (SCA), several conditions have to be fulfilled. Organizational Change should take place beforehand—otherwise, it may take much longer and may be costly to implement change and is likely to have driven better performance and higher sustainability anyways. **Data Governance** is again, important; in order for analytics to be useful, high-quality data must be present and employed efficiently within the company. It is about defining roles: Data Owners, who are responsible for the appropriate usage of data; Data Users, who hold out access for usability; and Data Stewards, who deal with data management on a daily basis and are in between business and IT sides. Additionally, while change needs to take place, **Organizational Support** from senior management is of the utmost importance in the integration of BI systems into the business. There is also an essential contribution of the leaders by distributing the resources, determining the targets to be achieved in processes, and being involved in the activities that are required for the implementation. Finally, **Technology Readiness** is essential in this regard; Companies must spend on IT and resource development to keep up with the constant changes in technology. With high technology readiness, organizations are able to fully exploit BI, which gives them a huge edge over their competitors. Together, these factors form the backbone of a robust and successful SCA strategy.

18.2.2 Aligning Supply Chain Strategies with Business Objectives

Aligning supply chain strategies with broader business goals is crucial for staying ahead in a competitive market. By using supply chain analytics (SCA), managers can pinpoint areas for improvement and craft strategies that outshine their competitors. In recent years, there has been significant growth in the use of analytics within supply chains, especially between 2020 and 2023. Supply chain managers now rely heavily on metrics to set clear performance goals and leverage predictive analytics to forecast demand more accurately, helping them make smarter, data-driven decisions (Seyedan and Mafakheri 2020).

18.2.3 Addressing Inefficiencies Through Advanced BI Systems

Supply chain processes can often run into inefficiencies when data isn't properly integrated. Advanced Business Intelligence (BI) systems help solve this problem by providing real-time data, enabling better decision-making in logistics. Companies that track key metrics like supplier performance and inventory turnover can identify and fix inefficiencies, ultimately leading to smoother operations and better overall results.

18.2.4 Benefits of Supply Chain Analytics in Strategic Decision Making

Supply chain analytics plays a crucial role in enhancing strategic decision-making processes within organizations. By leveraging data-driven insights, businesses can optimize their operations, mitigate risks, and improve overall efficiency. Here are the key benefits mentioned in Table 18.1.

Table 18.1 Benefits of supply chain analytics in strategic decision making

Improved decision-making speed and accuracy	Supply chain analytics enables organizations to make faster and more informed decisions by providing real-time data and predictive insights. This capability is essential in today's fast-paced business environment, where timely responses to market changes can significantly impact competitiveness
Enhanced demand forecasting	Utilizing advanced analytics allows companies to better predict customer demand by analyzing historical data and market trends. This leads to optimized inventory levels, minimizing stockouts and excess inventory, which are critical for maintaining operational efficiency and reducing costs (Team Digital Defynd 2023)
Risk management and mitigation	Analytics tools can identify potential risks within the supply chain by recognizing patterns and trends. By forecasting disruptions, organizations can develop contingency plans to mitigate these risks, ensuring smoother operations despite uncertainties
Cost reduction	Businesses can identify inefficiencies and areas for improvement through detailed analysis of supply chain operations. This insight allows for more strategic planning and resource allocation, ultimately leading to significant cost savings across various supply chain functions (IBM 2023)
Better collaboration across stakeholders	Analytics fosters improved communication and collaboration among supply chain partners by providing a single source of truth. This transparency enables all stakeholders—suppliers, vendors, and logistics partners—to make coordinated decisions that enhance overall supply chain performance (IBM 2023)

18.3 Challenges in Supply Chain Analytics for Decision Making

Supply chain analytics is essential for improving decision-making processes, yet several challenges impede its effective implementation and utilization. Below are the key challenges along with relevant citations mentioned in Table 18.2.

18.4 Strategies for Effective Implementation of Supply Chain Analytics

Successfully implementing supply chain analytics requires a strategic approach that aligns with organizational goals, fosters a data-driven culture, and utilizes advanced tools and techniques. A key strategy is **demand forecasting and inventory management**, where accurate demand predictions reduce costs by 10% to 15% and improve service levels by 5% to 10%. For example, Walmart leverages precise demand forecasts to optimize inventory turnover, enhancing overall efficiency. **Real-time visibility and tracking** powered by the Internet of Things (IoT) is another transformative approach. IoT devices and sensors provide real-time insights into supply chain operations, enabling proactive issue resolution. Despite its

Table 18.2 Challenges in supply chain analytics for decision making

Data quality and integration	Inaccurate Data: Poor data quality can lead to misleading insights, affecting decision-making. Inconsistent and incomplete data can result in flawed analytics, making it crucial for organizations to ensure high-quality data collection and integration from various sources (GraceBlood LLC 2023)
	Integration Issues: Integrating data from disparate systems (e.g., ERPs, CRMs) poses significant challenges. Companies often struggle with data silos, complicating the analysis process (ProVenta International 2023)
Complexity of supply chain networks	Dynamic Environments: The complexity of managing logistics data from multiple sources can be overwhelming. Organizations need robust tools for data merging, representation, and simulation to gain valuable insights (SCM Dojo 2023)
Limited predictive capabilities	Outdated Methods: Inadequate predictive analytics capabilities hinder accurate forecasting, leading to inefficiencies in supply chain operations. Continuous refinement of models and investment in advanced analytics technologies are necessary to enhance predictive accuracy (RateLinx 2023)
Integration with legacy systems	Obsolete Technologies: Many organizations still rely on outdated systems that complicate the integration of modern analytics solutions. Transitioning to more sophisticated systems is vital for leveraging advanced analytics effectively (River Logic 2023)

potential, only 6% of companies achieve full visibility, highlighting a major area for growth.

Supplier performance optimization is also crucial, as data-driven insights enable companies to evaluate suppliers based on key metrics like reliability and quality, leading to cost savings. Amazon, for example, uses data analytics to optimize its supplier network, ensuring that products are delivered efficiently and meet high standards. Route optimization and logistics efficiency are additional advantages of supply chain analytics. By utilizing advanced route optimization, Amazon reduces logistics costs, improves resource allocation, and speeds up delivery times. As a result, the route optimization software market is projected to grow significantly, thanks to its positive impact on logistics operations.

Finally, **risk management and mitigation** through predictive analytics helps identify potential disruptions. By analyzing historical and real-time data, businesses can anticipate and address issues proactively. Coca-Cola, for example, uses predictive analytics to manage supply chain bottlenecks, ensuring smoother operations (EBS 2023). Together, these strategies illustrate the transformative potential of supply chain analytics in enhancing decision-making and operational efficiency (Fig. 18.1).

The Supply Chain Analytics conceptual framework presented here is designed to illustrate how various foundational elements interact with analytical tools to drive strategic decision-making in supply chain management (SCM).

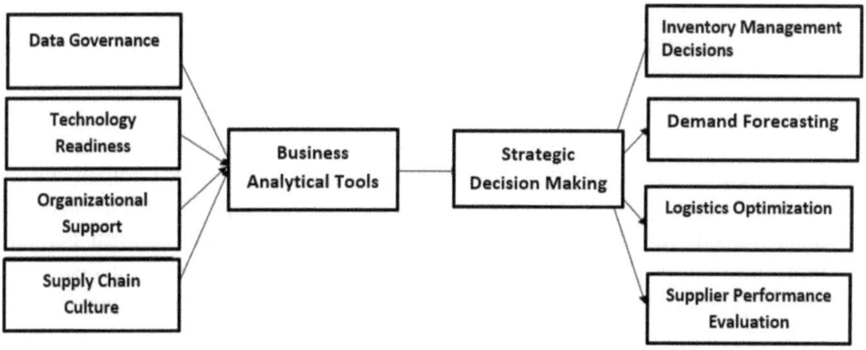

Fig. 18.1 Conceptual framework: supply chain analytics tools in decision making

18.5 Key Components of Supply Chain Analytics

To effectively implement supply chain analytics (SCA), several foundational elements must be in place. Data Governance is critical—it ensures that the data collected across the supply chain is accurate, reliable, and up to date. With strong data governance, companies can trust their data to make important decisions, like accurately tracking product quantities across warehouses to avoid costly discrepancies. Technology Readiness refers to an organization's ability to support advanced analytics, which involves having the right software, hardware, and infrastructure in place. For example, companies that use Internet of Things (IoT) devices to track shipments in real-time show a high level of technological preparedness. Operational Support is necessary for ensuring that analytics are smoothly integrated into daily processes like procurement, production, and distribution. For instance, aligning warehouse teams with analytics teams helps optimize delivery schedules. A Data-Driven Supply Chain Culture encourages decision-makers to rely on data analysis rather than intuition. For example, a company may use data-driven forecasting to plan inventory during peak seasons, instead of just relying on past experience. Business Analytical Tools such as predictive analytics, machine learning, and simulation models are essential for processing and analyzing data. These tools help optimize operations, such as predicting demand fluctuations and preventing overstocking or stockouts.

18.6 How Analytics Enhance Strategic Decision-Making

Once these foundational elements are in place, supply chain decision-making improves in several key areas:

Inventory Management: Analytics help align stock levels with real-time and forecasted demand, reducing holding costs and ensuring products are available when needed. For example, a retailer may adjust inventory at each store based on historical sales data and market trends. Supplier Performance Evaluation: Data-driven metrics allow companies to assess suppliers based on delivery times, defect rates, and other key performance indicators (KPIs). This helps companies make decisions like renegotiating contracts or switching suppliers if necessary. For example, if a supplier is consistently late with deliveries, the company might choose a more reliable alternative. Demand Forecasting: By using historical data, market trends, and real-time analytics, companies can forecast demand more accurately. This leads to better production and procurement decisions, reducing the risks of overproduction or stockouts. A beverage company, for instance, might use past sales data combined with weather forecasts to adjust production levels for the summer season. Logistics Optimization: Analytics help determine the most efficient delivery routes, reducing shipping costs and delivery times, which enhances customer satisfaction. A logistics company might use route optimization software to find the fastest and most cost-effective delivery routes for its fleet.

18.7 Types of Supply Chain Analytics

Supply chain analytics employs three main types of analysis to drive decision-making:

Descriptive Analysis uses historical data to understand current performance metrics like inventory turnover, supplier lead times, and logistics efficiency. This helps companies see trends and patterns in their supply chain operations. **Predictive Analysis** uses statistical methods like correlation and regression to forecast future outcomes, such as the effects of improving supplier relationships or increasing buffer stock on supply chain efficiency. **Prescriptive Analysis** offers actionable strategies to improve supply chain performance, such as enhancing supplier relationships, adjusting inventory policies, or improving transportation routes based on predictive insights.

18.8 Conceptual Framework for Effective Supply Chain Analytics

The framework in Fig. 18.1 that supports the successful use of SCA tools involves four main elements: Data Governance, Organizational Culture, Organizational Support, and Technology Readiness. Data Governance ensures that supply chain data is accurate, secure, and accessible, allowing for reliable insights. It involves roles such as data owners who ensure data integrity, data users who manage access, and data stewards who oversee daily data quality. Proper governance provides a solid foundation for effective decision-making. Organizational Culture plays a crucial role. A culture that values data-driven decisions and embraces technological innovation creates an environment that supports the integration of SCA tools. When employees across departments are encouraged to use data insights, SCA tools can be more effectively used to optimize logistics, manage risks, and enhance efficiency. Senior Management Support is necessary for successful SCA implementation. Leadership must allocate resources, set clear objectives, and actively engage in decision-making to ensure SCA initiatives get the funding and attention they need. Support from top management also helps minimize resistance to change by providing training and managing transitions.

Technology Readiness ensures that the organization has the necessary IT infrastructure and real-time data systems and that employees are trained to use advanced analytics tools. A high level of technology readiness allows for the seamless integration of SCA tools, leading to improved demand forecasting, optimized transportation and better inventory management.

18.9 Conclusion

In conclusion, the effective use of Supply Chain Analytics tools is key to enhancing strategic decision-making within supply chain management. By leveraging descriptive, predictive, and prescriptive analytics, organizations can gain a deeper understanding of demand forecasting, inventory management, supplier performance, and logistics optimization. The integration of data governance, a data-driven culture, senior management support, and technology readiness is crucial for maximizing the benefits of SCA tools. These factors ensure that supply chain data is reliable, decisions are data-driven, and the necessary resources and infrastructure are in place for successful implementation. As supply chains become increasingly complex, the use of analytics will be critical for making informed, strategic decisions that drive efficiency, resilience, and long-term success. Organizations that invest in strong data governance, foster a culture of innovation, and maintain technological readiness will be well-positioned to harness the full potential of supply chain analytics.

References

Bawack RE & Ahmad MO (2021) Understanding Business Analytics continuance continuance in agile information system development projects: An expectation-confirmation perspective. Information Technology & People, 34(2), 550–575

Choi TM, Kumar S, Yue X, & Chan HL (2022) Disruptive technologies and operations management in the Industry 4.0 era and beyond. Production and operations management, 31(1), 9–31.

GraceBlood LLC (2023) Challenges in supply chain analytics and management. https://graceblood.com/blog/what-are-the-top-supply-chain-analytics-problems/

IBM (2023) What is supply chain analytics? https://www.ibm.com/topics/supply-chain-analytics

Kristoffersen E, Mikalef P, Blomsma F & Li J (2021) The effects of business analytics capability on circular economy implementation, resource orchestration capability, and firm performance. International Journal of Production Economics, 239, 108205.

Oracle Corporation. (2023, November 17). Announcing Oracle E-Business Suite 12.2.13, EBS Enterprise Command Centers – November 2023 Update, and Other 2023 Innovations [PDF]. Oracle. https://www.oracle.com/a/ocom/docs/applications/ebusiness/ebs-12-2-13-and-ecc-november-2023.pdf Oracle+1Oracle+1

ProVenta International (2023) The impact of data analytics on supply chain decision making. https://proventainternational.com/the-impact-of-data-analytics-on-supply-chain-decision-making/

RateLinx (2023) Problems in supply chain big data analytics: systems or data? https://www.ratelinx.com/resources/problems-in-big-data-supply-chain-analytics/

River Logic (2023) The biggest problem with supply chain analytics, and how you can fix it. https://riverlogic.com/?blog=the-biggest-problem-with-supply-chain-analytics-and-how-you-can-fix-it

SCM Dojo (2023) Supply chain analytics: benefits, challenges & architecture. https://www.scmdojo.com/supply-chain-analytics/

Seyedan M, Mafakheri F (2020) Predictive big data analytics for supply chain demand forecasting: methods, applications, and research opportunities. J Big Data 7(1):53. https://doi.org/10.1186/s40537-020-00329-2

Sousa R et al (2024) The impact of business intelligence on organizational performance: a review. J Strateg Inf Syst 14(12):199–214

TeamDigitalDefynd(2023)Roleofsupplychainanalyticsindecisionmaking[2024].DigitalDefynd. https://digitaldefynd.com/IQ/role-of-supply-chain-analytics-in-decision-making/

Ülkü MA, Mansouri B (2023) Supply chain analytics: overview, emerging issues, and research outlook. In: Sarkis J (ed) The Palgrave handbook of supply chain management. Palgrave Macmillan. https://doi.org/10.1007/978-3-030-89822-9_80-1

Chapter 19
Investigating the Hindrances Linked with Procurement of Used Products: A Circular Approach

Shefali Srivastava, Lakshmi Shetty, Ashish Dwivedi, and Suchi Dubey

Abstract The circular economy is regarded as a pivotal advancement in achieving sustainability and mitigating the environmental impact associated with waste generation within the electrical and electronic equipment (EEE) sector, which is known for generating substantial volumes of e-waste. This study seeks to investigate the hindrances to the adoption of circular economy principles within the procurement processes for used products in the EEE sector. A comprehensive literature review has delineated potential hindrances to the acquisition of second-hand products as well as challenges in the application of circular economy principles. The research methodology employed a synthesis of expert interviews and the fuzzy TOPSIS method to systematically prioritize the identified hindrances to the procurement of second-hand products. The findings of the study indicate that consumer perception constitutes the most significant hindrance to the purchase of used products, succeeded by technological constraints and the insufficiency of support and incentives within the EEE sector. In the fuzzy TOPSIS analysis, the identified hindrances are systematically organized in a coherent manner, taking into account the inherent uncertainties and ambiguities present in expert evaluations. Consequently, this has profound implications for businesses, policymakers, and consumers, providing actionable recommendations for the formulation of targeted strategies and interventions aimed at addressing these critical hindrances. This could expedite the transi-

S. Srivastava
Operations and Decision Sciences, Christ University, Bengaluru, Karnataka, India

L. Shetty
Symbiosis Centre for Management Studies, Bengaluru, Symbiosis International (Deemed University), Pune, Maharashtra, India

A. Dwivedi (✉)
Jindal Global Business School, O.P. Jindal Global University, Sonipat, Haryana, India

S. Dubey
Symbiosis International University, Dubai, UAE

Symbiosis International (Deemed University), Pune, India
e-mail: sdubey@siu-dubai.ac.ae

K. Mathiyazhagan et al. (eds.), *Humanizing Circular Supply Chain Management*, Circular Economy and Sustainability,
https://doi.org/10.1007/978-3-031-86562-6_19

tion towards a circular economy in the EEE sector. While this study contributes to the academic discourse surrounding circular economy principles, it also acknowledges inherent limitations, which may arise from potential biases in expert perspectives and underscores the necessity for further research to investigate the roles of various stakeholders throughout the different stages of the product life cycle.

Keywords Circular economy · Hindrances · Electrical and electronic equipment · Fuzzy TOPSIS and transition

19.1 Introduction

In the current global context, the paradigm of Circular Economy (CE) has gained considerable importance, particularly within electrical and electronic equipment (EEE) sector. CE has put foremost concern on sustainability through the advocacy of practices aimed at waste reduction, resource conservation and value creation via. perpetual use of resources (Qazi and Appolloni 2022). This shift in paradigm is considered not only essential to reduce the negative environmental impacts like resource depletion and climate change but also to make the economy more resilient and to establish social justice lie at the very core (Sharma et al. 2021a, b). The transition towards CE within the EEE sector is increasingly urgent, given the rapid escalation of electronic waste (e-waste), which poses significant risks to both environmental integrity and public health (Sarkhoshkalat et al. 2024). Industries are encouraged to adopt strategies that prioritize design for reuse, refurbishment, and recycling, thereby facilitating procurement processes that promote the prolongation of product life cycles through minimized waste generation (Kouloumpis et al. 2023). Innovative business models underpin this transformational process, providing the economic impetus to collect useful materials from e-waste and finally evolve into a more sustainable and efficient framework for resource management (Goyal et al. 2023). It was in the midst of strong challenges, like climate change and resource insufficiency, that have gripped the global scene, whereby integrating principles of CE into the EEE industry is not beneficial but also a necessity (Farooque et al. 2023). This scenario presents a timely opportunity to reconsider the methodologies surrounding product design, usage, and disposal in order to enhance systemic efficiency in minimal generation of waste, optimal use of resources and substantially reduced environmental footprints (Bag et al. 2021). This integrated approach not only supports the EEE industry in improving its sustainability credentials but also makes a strong contribution to attaining broader socio-economic and environmental stewardship objectives (Migliore et al. 2020).

The need for the study thus focuses on the hindrances to the implementation of circularity principles within the EEE industry, considering that this industry greatly contributes to global waste production and resource consumption (Parajuly et al. 2020). In this regard, an increase in e-waste volume with growing EEE industry is projected. Such conditions present formidable challenges to environmental sustainability and public health (Shaharudin et al. 2023). Hindrances to circularity have to be understood and such understanding must make sense out of the uncertainties in material recovery, the regulatory complexities and consumer attitudes so that effective strategies may be developed to get over such hindrances (Kim et al. 2022). In this direction, the use of techniques such as Fuzzy Technique for Order Preference by Similarities to Ideal Solution (TOPSIS) could help significantly in prioritizing such hindrances and thus empowering stakeholders with the execution of very informed decisions supporting sustainability goals. As businesses increasingly acknowledge the financial benefit of circular approaches, this study will be able to lead into a 'shift' in business models toward innovation that lengthens the life span and enhances resource-use efficiency of products (Xu et al. 2022). In brief, this is an essential study for the development of a low-carbon EEE industry in resonance with the global goals of sustainability while meeting the current challenges associated with waste management and resource conservation (Chakrabarty and Nandi 2021). The following Research Questions (RQs) are aimed at understanding the main obstacles and potential advantages of circularity in the EEE industry and, in doing so, provides an overall research framework in this domain.

RQ1. What are the potential hindrances in procuring used products and implementing CE principles in EEE industry and how can these hindrances be prioritized using the Fuzzy TOPSIS method?

RQ2. How can CE practices adoption in the EEE industry contribute to reducing electronic waste and enhancing resource efficiency and what role do stakeholder perceptions play in this transition?

The subsequent sections of this study are delineated as follows: literature review focuses on the dimensions of CE and the hindrances associated with sourcing used products and their underlying frameworks. Section 19.3 elaborates on the criteria pertinent to circularity, accompanied by a detailed exposition of the research methodology employed. Sections 19.4 and 19.5 illustrate the research findings derived from the application of the fuzzy TOPSIS method, followed by an in-depth discussion of the implications. Last section reflects the conclusions and outlines potential avenues for future research.

19.2 Literature Review

The challenges in field of CE integration in building ventures consist of a lack of awareness, incentive and engagement from stakeholders, alongside the absence of incentives and prevailing procurement tactics (Corsini et al. 2024). Furthermore, the utilization of e-procurement in small and medium enterprises are impeded by limited utilization of electronic commerce by suppliers and a reluctance to allocate substantial resources to e-procurement (Nikolaou et al. 2022). Moreover, in China, a single unified governmental procurement system does not exist, which leads to problems in macroeconomic functions and market structures that further enhance issues of corruption (Richter et al. 2023). Despite the fact that public procurement functions as a critical policy instrument in fostering innovation and sustainability in the Dutch economic context, it is still feared that existing sustainability criteria may hamper innovation (Andhov et al. 2020). Therefore, there is a need to adjust procurement approaches in a manner that encourages and promotes the procurement of sustainable innovative goods (Neessen et al. 2021a, b).

Since 2015, the Circular Procurement Practices (CPP) and models have been core enablers of sustainability and circularity across industries. Based on the pivotal role that green construction procurement practices play in driving CE success within the construction industry (Sönnichsen and Clement 2020). This is where the impacts of practices like on-site waste management come into play and commitment by stakeholders. Furthermore, it becomes imperative to mention that circular principles have to be integrated into supplier selection processes. It has become very critical in terms of supporting environmentally friendly practices across supply chains (Polonsky et al. 2022). Again, it goes on implicitly bringing forth the case for consistency with the CE principle at the time of choosing a supplier (Neessen et al. 2021a, b). Moreover, circular public procurement has been recognized as one of the most pivotal strategies for advancing and facilitating sustainability within public spending practices (Wijayasundara et al. 2022). More specifically, focusing on furniture flows shows that strategic decision-making, centralized governance structures and follow-up systems are required. These have been core elements of how to effectively facilitate circular flows and resource efficiency (Hunka et al. 2021). Such insights jointly help in building up comprehensive procurement models that, by their very essence of trend, seek to integrate green principles, lifecycle analysis and CE concepts. The overarching objectives are to mitigate environmental impacts and to encourage the adoption of sustainable industrial methodologies (Ibitz 2020).

Research on CPP and models points to the deficiencies in available literature, thus underlining a systemic approach that shows how these issues related to Circular Business Models (CBM) are interlinked (Rizos and Bryhn 2022). While enhancing economic, environmental and social dimension in manufacturing sector through CPP Management which is well-documented, there is an air of uncertainty over the effectiveness in their implementation and integration to the supply chains (Pollard et al. 2023). Moreover, no consensus has been reached on which would be the optimum benchmarking approach for Circular Supply Chains (CSC), therefore, measuring progress and efficiently implementing circular business practices is difficult. The identified gaps underline the need for further research that could enable the development of holistic models and frameworks in the future. These can be tools that will make an organization effectively adopt and implement CPP, considering innovation, collaboration and sustainability principles.

The hindrances to CE practices in procuring used products are many, as pinpointed by the research findings. Those hindrances are related to technological constraints, regulatory ambiguity, organizational, supply chain complexities, lack of regulations, large initial investment cost, fragmented supply chains and lack of appropriate CSC formation (Menon and Ravi 2022). In contrast, drivers are multi-stakeholder coordination, regulatory alignment, organizational redesign, supply chain integration, economic incentives, technological advancement, innovation in circular tool, market and customer demand, environmental awareness and product design for circularity (Bressanelli et al. 2021). These will have to be overcome, or in some cases utilized, in order for the CE practices in procurement of used products to be successful. This literature review scrutiny underlines eight such hindrances in procuring used products as summarized in Table 19.1.

The scholarly void pertaining to the hindrances encountered in the procurement of pre-owned products within CE paradigm is primarily ascribed to the insufficient awareness, knowledge, and incentivizing mechanisms among stakeholders in the construction industry, which consequently impedes the application of CE principles (Kumar et al. 2021). CPP play an immensely pivotal role in enhancing operational efficiency and sustainability, especially when integrated with the advanced technologies characteristic of Industry 4.0, including artificial intelligence and big data analytics (Khan et al. 2023). Thus, it is capable of significantly contributing to the advancement of economic, environmental, and social aspects within industrial organizations. The elements of digital traceability and collaborative efforts serve as essential facilitators in endeavour to bolster sustainability outcomes pertinent to CPP (Guldmann and Huulgaard 2020). Moreover, the knowledge obtained from the European Union (EU) framework initiatives points to the fact that circularity considerations have to be introduced at the design stage of products, in production processes and at end-of-life phases. This presents valuable insights for policymakers who intend to push for progress in circular material policy, if there is a likelihood of successfully mitigating climate change (Mhatre et al. 2021).

19.3 Research Methodology

The existing literature, in combination with expert interviews and the Fuzzy TOPSIS method, investigates hindrances in procuring used products and opportunities in the application of principles concerning CE within the EEE industry (Toker and Görener 2023). Literature research is conducted in the beginning of the study to identify the possible hindrances of circularity within the EEE industry pertaining to technological, regulatory and economic feasibility (Chen et al. 2023). Semi-structured interviews were systematically administered with industry experts, policy makers and academicians to validate the identified hindrances and to gain further insights. Fuzzy TOPSIS method is used to rank hindrances with respect to the criteria reduce, re-use, recycle and recover (Sharma et al. 2020). In this, fuzzy logic will be used, which further helps in incorporating linguistic variables and uncertainty in expert judgments. The last aspect considered in the study is the potential advantages of practices in CE within EEE sector with respect to reduced generation of waste, resource efficiency and new business opportunities (Sagnak et al. 2021). The results are analysed in the context of the perceptions of stakeholders to reach an overall understanding of the transition towards circularity in the EEE industry (Table 19.2).

Table 19.1 Identifying hindrances in procuring used products

Hindrances	Description	References
Legal barriers	Second-hand products are governed by intricate regulations on transport, storage and processing. Inconsistencies in regulations, ambiguous frameworks of liability and concerns related to intellectual property create major legal barriers to circular procurement	Kazancoglu et al. (2022), Wrålsen et al. (2021), Svensson-Hoglund et al. (2021)
Quality assurance and testing	Quality verification and used products functionality requires specialized skills and equipment and is thus costly and complex, an additional burden on the procurement process. Ensuring that robust quality assurance mechanisms are in place could be very important in building consumer trust and acceptance	Abbasi et al. (2022), Kumar et al. (2021), Althaf and Babbitt (2021), MahmoumGonbadi et al. (2021)
Inconsistent demand and supply	Second-hand products are irregularly available used products, therefore, make it difficult to plan the supply and meet demand. An irregular supply chain generates uncertainty in material flow and planning that retards the development of viable CBMs	Pan et al. (2022), Nag et al. (2021), Farooque et al. (2019), Marrucci et al. (2019)
Reverse logistics	Collecting storage and transportation of used products could be complicated and costly, logistically speaking. This is true, more so for large or dangerous goods. For this reason, reverse logistics systems should be developed in order to make possible flow of materials back into supply chain	Ding et al. (2023), Mallick et al. (2023), Mishra et al. (2023), Azadnia et al. (2021), Dutta et al. (2021), Julianelli et al. (2020)
Consumer perception	Some consumers will be hesitant to buy used produce due to its quality, safety and hygiene concerns. Counteracting the negative perceptions and rebuilding trust in CE initiatives hold the key to build a market for reused product materials	Nag et al. (2021), Perez-Castillo and Vera-Martinez (2021), Islam et al. (2021), Dutta et al. (2021), Pisitsankkhakarn and Vassanadumrongdee (2020)
Technological limitation	Low-quality technologies for refurbishing, remanufacturing and recycling used products will strand the options available for procurement professionals. The necessity here is investment in frontiers of innovation in circular technologies	Hina et al. (2022), Mendoza et al. (2022), Shojaei et al. (2021), Sharma et al. (2021a, b)

(continued)

Table 19.1 (continued)

Hindrances	Description	References
Resistance to change	Greater challenges in shifting to a CE approach include how to overcome entrenched mindsets and established procurement practices in organizations. The cultivation of a culture emphasizing sustainability and innovation is imperative for promoting the adoption of CE practices	Geissdoerfer et al. (2023), Menon and Ravi (2022), Khan et al. (2022), Milios (2021), Agrawal et al. (2021)
Lack of support and incentives	Inadequate financial incentives, tax breaks, or governmental support for procurement that involves used products can undermine the growth of a robust CE	Brändström et al. (2024), Rizos and Bryhn (2022), Qazi and Appolloni (2022), Stumpf et al. (2021)

Table 19.2 Profile of the respondents

Age Group	30–40–9
	40–50–16
	Above 50–17
Gender	Male—32
	Female—10
Geographic location	North region—13
	South region—11
	East region—8
	West region—10
Job responsibility	Circular supply chain consultants—9
	Consumer—8
	Non-Governmental Organizations (NGOs)—7
	Academicians/Professors—4
	Legal experts—6
	Recycling company mangers—8

19.3.1 TOPSIS Approach

TOPSIS constitutes a widely recognized methodology within the realm of multi-criteria decision-making (MCDM), enabling decision-makers to evaluate and prioritize alternatives grounded in their nearness to an optimal solution. TOPSIS approach was conceived in the 1980s, premised on the foundation that the superior alternative to the prevailing issue must exhibit the minimal geometrical distance from the positive ideal solution (which is deemed most favourable outcome) and the maximal distance from the negative ideal solution, which typically exemplifies most unfavourable scenario (Akcan and Taş 2019). This methodological framework entails the formulation of a decision matrix, the execution of data normalization, the computation of the weighted normalized decision matrix, and the evaluation of both an ideal and a negative ideal solution. It is step-by-step procedure allowing for an effective evaluation of alternatives against different criteria; hence, it gets very useful in quite complex decision problems with trade-offs (Ogunnusi et al. 2023).

19.3.2 Fuzzy TOPSIS

Unlike traditional TOPSIS, incorporation of fuzzy logic serves to adeptly address uncertainty and vagueness within the decision-making process. Husain et al. (2021) studied that in practical contexts there is usually imprecision or vagueness regarding the data used in making a decision. Frequently, expert assessments are articulated not in precise numerical terms. Fuzzy TOPSIS inherently possesses the capability to manage such uncertainties through the utilization of fuzzy numbers and linguistic variables, thus rendering it more adaptable in complex decision-making scenarios (Rani et al. 2020). This becomes very relevant considering the principles of CE within the EEE industry, where different stakeholders may have a variety of different perceptions and priorities about hindrances to implementation (Kumar et al. 2022). In this respect, the application of fuzzy TOPSIS will facilitate a deeper comprehension of the nuances embedded in expert perspectives and stakeholder positioning, coming out with a more robust assessment of the challenges and opportunities that an infringement-based, leading towards transition-oriented path to a CE (Rizos and Bryhn 2022).

19.4 Results

Hwang and Yoon in 1981 proposed fuzzy TOPSIS which is a popular and widely used method for MCDM used to rank the alternative in a fuzzy environment.

19.4.1 Phase 1: Creating Decision Matrix

In the current investigation, drawing upon the extant literature, we have delineated four criteria (Leong et al. 2023) and eight alternatives, which are ranked utilizing the Fuzzy TOPSIS methodology. Table 19.3 delineates types of criteria and the weights assigned to each criterion.

Table 19.4 shows the fuzzy scale used in this study.

Considering the presence of multiple expert participants in the evaluation process, the criteria are assessed for each alternative, and the decision matrix reflects the arithmetic mean of all experts' evaluations, as depicted in Table 19.5.

Table 19.3 Determining CE criteria characteristics

CE criterion	Types	Weights
Reduce	+	(0.250, 0.250, 0.250)
Reuse	+	(0.250, 0.250, 0.250)
Recycle	+	(0.250, 0.250, 0.250)
Recovery	+	(0.250, 0.250, 0.250)

Table 19.4 Assigning fuzzy scale

Code	Linguistic terms	L	M	U
1	Very Low	1	1	3
2	Low	1	3	5
3	Medium	3	5	7
4	High	5	7	9
5	Very High	7	9	9

Table 19.5 Calculating decision matrix based on arithmetic mean

Hindrances	Reduce	Reuse	Recycle	Recover
Legal barriers	(4.400, 6.000, 7.200)	(2.600, 4.200, 5.800)	(2.800, 4.000, 5.800)	(3.000, 5.000, 6.600)
Quality assurance and testing	(1.600, 3.400, 5.400)	(3.600, 5.400, 7.200)	(2.600, 4.200, 6.000)	(3.000, 4.200, 5.800)
Inconsistent demand and supply	(3.800, 5.400, 7.200)	(1.800, 3.000, 5.000)	(3.000, 4.800, 6.600)	(3.200, 4.800, 6.600)
Reverse logistics	(2.600, 3.800, 5.600)	(3.800, 5.400, 6.800)	(2.800, 4.200, 6.000)	(3.000, 4.800, 6.600)
Consumer perception	(3.600, 5.600, 7.400)	(3.800, 5.400, 7.000)	(3.800, 5.000, 6.400)	(5.600, 7.600, 8.200)
Technological limitation	(4.600, 6.400, 7.600)	(2.800, 4.600, 6.600)	(4.400, 6.000, 7.000)	(4.000, 5.600, 7.000)
Resistance to change	(1.800, 3.600, 5.600)	(3.000, 4.600, 6.200)	(2.600, 4.600, 6.600)	(2.800, 4.400, 6.200)
Lack of support and incentives	(4.000, 6.000, 8.000)	(3.200, 5.000, 6.600)	(3.400, 5.000, 7.000)	(2.200, 4.200, 6.200)

19.4.2 Phase 2: Creating the Normalized Decision Matrix

The normalized decision matrix is formulated through the application of both the positive and negative ideal solutions, utilizing the subsequent equation for its derivation:

$$\tilde{r}_{ij} = \left(\frac{a_{ij}}{c_j^*}, \frac{b_{ij}}{c_j^*}, \frac{c_{ij}}{c_j^*} \right); \quad c_j^* = \max_i c_{ij}; \quad \text{Positive ideal solution} \tag{19.1}$$

$$\tilde{r}_{ij} = \left(\frac{a_j^-}{c_{ij}}, \frac{a_j^-}{b_{ij}}, \frac{a_j^-}{a_{ij}} \right); \quad a_j^- = \min_i a_{ij}; \quad \text{Negative ideal solution} \tag{19.2}$$

Table 19.6 displays the normalized decision matrix.

19.4.3 Phase 3: Development of the Weighted Normalized Decision Matrix

Based on the varying weights assigned to the criteria, the weighted normalized decision matrix is derived by multiplying the weight of each criterion within the normalized fuzzy decision matrix, in accordance with the specified formula.

$$\tilde{v}_{ij} = \tilde{r}_{ij}.\tilde{w}_{ij} \tag{19.3}$$

"where $\tilde{w}_{ij}$ denotes weight of criterion c_j".

Table 19.7 presents weighted normalized decision matrix calculation.

Table 19.6 Calculating normalized decision matrix

Hindrances	Reduce	Reuse	Recycle	Recover
Legal barriers	(0.595, 0.811, 0.973)	(0.361, 0.583, 0.806)	(0.424, 0.606, 0.879)	(0.366, 0.610, 0.805)
Quality assurance and testing	(0.216, 0.459, 0.730)	(0.500, 0.750, 1.000)	(0.394, 0.636, 0.909)	(0.366, 0.512, 0.707)
Inconsistent demand and supply	(0.514, 0.730, 0.973)	(0.250, 0.417, 0.694)	(0.455, 0.727, 1.000)	(0.390, 0.585, 0.805)
Reverse logistics	(0.351, 0.514, 0.757)	(0.528, 0.750, 0.944)	(0.424, 0.636, 0.909)	(0.366, 0.585, 0.805)
Consumer perception	(0.486, 0.757, 1.000)	(0.528, 0.750, 0.972)	(0.576, 0.758, 0.970)	(0.683, 0.927, 1.000)
Technological limitation	(0.575, 0.800, 0.950)	(0.424, 0.697, 1.000)	(0.629, 0.857, 1.000)	(0.571, 0.800, 1.000)
Resistance to change	(0.225, 0.450, 0.700)	(0.455, 0.697, 0.939)	(0.371, 0.657, 0.943)	(0.400, 0.629, 0.886)
Lack of support and incentives	(0.500, 0.750, 1.000)	(0.485, 0.758, 1.000)	(0.486, 0.714, 1.000)	(0.314, 0.600, 0.886)

Table 19.7 Calculating weighted normalized decision matrix

Hindrances	Reduce	Reuse	Recycle	Recover
Legal barriers	(0.149, 0.203, 0.243)	(0.090, 0.146, 0.201)	(0.106, 0.152, 0.220)	(0.091, 0.152, 0.201)
Quality assurance and testing	(0.054, 0.115, 0.182)	(0.125, 0.188, 0.250)	(0.098, 0.159, 0.227)	(0.091, 0.128, 0.177)
Inconsistent demand and supply	(0.128, 0.182, 0.243)	(0.063, 0.104, 0.174)	(0.114, 0.182, 0.250)	(0.098, 0.146, 0.201)
Reverse logistics	(0.088, 0.128, 0.189)	(0.132, 0.188, 0.236)	(0.106, 0.159, 0.227)	(0.091, 0.146, 0.201)
Consumer perception	(0.122, 0.189, 0.250)	(0.132, 0.188, 0.243)	(0.144, 0.189, 0.242)	(0.171, 0.232, 0.250)
Technological limitation	(0.144, 0.200, 0.238)	(0.106, 0.174, 0.250)	(0.157, 0.214, 0.250)	(0.143, 0.200, 0.250)
Resistance to change	(0.056, 0.113, 0.175)	(0.114, 0.174, 0.235)	(0.093, 0.164, 0.236)	(0.100, 0.157, 0.221)
Lack of support and incentives	(0.125, 0.188, 0.250)	(0.121, 0.189, 0.250)	(0.121, 0.179, 0.250)	(0.079, 0.150, 0.221)

Table 19.8 Calculating positive and negative ideal solutions

CE criterion	Positive	Negative
Reduce	(0.149, 0.203, 0.250)	(0.054, 0.115, 0.182)
Reuse	(0.132, 0.188, 0.250)	(0.063, 0.104, 0.174)
Recycle	(0.144, 0.189, 0.250)	(0.098, 0.152, 0.220)
Recovery	(0.171, 0.232, 0.250)	(0.091, 0.128, 0.177)

19.4.4 Phase 4: Determine the Fuzzy Positive Ideal Solution (FPIS, A) and the Fuzzy Negative Ideal Solution* (FNIS, A⁻)

The FPIS and FNIS pertaining to the alternatives can be represented as follows:

$$A^* = \left\{\tilde{v}_1^*, \tilde{v}_2^*, \ldots, \tilde{v}_n^*\right\} = \left\{\left(|i \in B\right), \left(|i \in C\right)\right\} \quad (19.4)$$

$$A^- = \left\{\tilde{v}_1^-, \tilde{v}_2^-, \ldots, \tilde{v}_n^-\right\} = \left\{\left(|i \in B\right), \left(|i \in C\right)\right\} \quad (19.5)$$

"where $\tilde{v}_i^*$ is the max value of i for all the alternatives and $\tilde{v}_1^-$ is the min value of i for all the alternatives. *B* and *C* represent the positive and negative ideal solutions, respectively".

The positive and negative ideal solutions are detailed in Table 19.8.

19.4.5 Phase 5: Calculation of Distance Between Each Alternative and the Fuzzy Positive Ideal Solution A*and the Distance Between Each Alternative and the Fuzzy Negative Ideal Solution A⁻

The distances measured between each alternative and the FPIS, as well as between each alternative and the FNIS, are computed respectively as follows:

$$S_i^* = \sum_{j=1}^{n} d\left(\tilde{v}_{ij}, \tilde{v}_j^*\right) \quad i = 1, 2, \ldots, m \tag{19.6}$$

$$S_i^- = \sum_{j=1}^{n} d\left(\tilde{v}_{ij}, \tilde{v}_j^-\right) \quad i = 1, 2, \ldots, m \tag{19.7}$$

"d is the distance between two fuzzy numbers, when given two triangular fuzzy numbers (a_1, b_1, c_1) and (a_2, b_2, c_2), e distance between the two can be calculated as follows":

$$d_v\left(\tilde{M}_1, \tilde{M}_2\right) = \sqrt{\frac{1}{3}\left[\left(a_1 - a_2\right)^2 + \left(b_1 - b_2\right)^2 + \left(c_1 - c_2\right)^2\right]} \tag{19.8}$$

"Note that $d\left(\tilde{v}_{ij}, \tilde{v}_j^*\right)$ and $d\left(\tilde{v}_{ij}, \tilde{v}_j^-\right)$ are crisp numbers".

Table 19.9 illustrates the distances from both the positive and negative ideal solutions.

19.4.6 Phase 6: Calculating the Closeness Coefficient and Ranking the Alternatives

The closeness coefficient associated with each alternative is determined through the application of the ensuing equations:

Table 19.9 Calculating distance from positive and negative ideal solutions

Hindrances	Distance from positive ideal	Distance from negative ideal
Legal barriers	0.154	0.14
Quality assurance and testing	0.209	0.081
Inconsistent demand and supply	0.182	0.112
Reverse logistics	0.177	0.119
Consumer perception	0.026	0.267
Technological limitation	0.02	0.186
Resistance to change	0.183	0.017
Lack of support and incentives	0.092	0.108

$$CC_i = \frac{S_i^-}{S_i^+ + S_i^-} \tag{19.9}$$

The derived closeness coefficient for each alternative, accompanied by the relevant ranking sequence, is delineated in Table 19.10, wherein the optimal alternative is recognized as being closest to the FPIS and the furthest from the FNIS.

19.5 Discussion

Table 19.10 and Fig. 19.1 illustrate the findings derived from the fuzzy TOPSIS analysis, which are instrumental in comprehending prioritization of hindrances to procuring used products and implementing CE principles in EEE. According to the ranking order, the hindrance that comes first is consumer perception with a Ci of 0.912, Thus, this analysis indicates the proximity to the positive ideal solution while

Table 19.10 Calculating closeness coefficient

Hindrances	Ci	rank
Legal barriers	0.476	4
Quality assurance and testing	0.279	7
Inconsistent demand and supply	0.380	6
Reverse logistics	0.401	5
Consumer perception	0.912	1
Technological limitation	0.905	2
Resistance to change	0.087	8
Lack of support and incentives	0.541	3

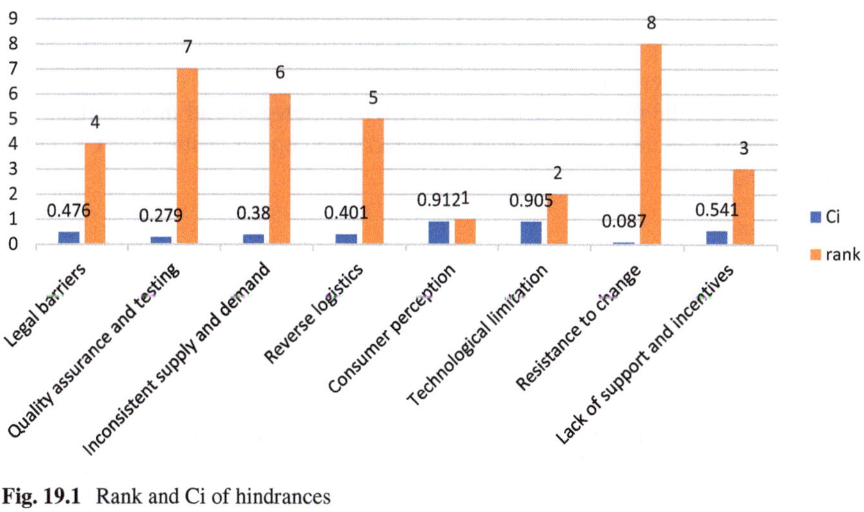

Fig. 19.1 Rank and Ci of hindrances

simultaneously demonstrating the distance from the negative ideal solution. The result demonstrates that success of CE initiatives within the EEE industry is highly pegged to consumer attitude and trust. Willingness by consumers to buy used or refurbished products is an integral element in creating a ready market for the materials after reuse, hence increasing product lifetimes (Dumitrica et al. 2023). This high ranking would underscore the need for targeted marketing campaigns, educational programs and the development of quality assurance mechanisms to make practices easier to understand by consumers and engender public confidence (Islam et al. 2021).

The second and third most important hindrances are technological limitations and lack of support and incentives with a value of Ci = 0.905 and Ci = 0.541, respectively. Technological hindrance alone, like the lack of recycling infrastructure and the inability to recover valuable materials from complex electronic products, acts as a hindrance in applying CE principles (Shojaei et al. 2021). Such hindrances will only be surmounted by proper investments in research and development, through industrial-academic-government collaborations that propel innovation in remanufacturing and recycling technologies (Hina et al. 2022). One of the important hindrances against practices of CE has been lack of support and incentives in the form of tax breaks, subsidies or schemes of extended producer responsibility (Rizos and Bryhn 2022). Policymakers and regulators must develop enabling frameworks and provide financial incentives to businesses so that investment in CE-oriented initiatives is economically feasible (Centobelli et al. 2021).

The remaining hindrances, in order of importance, are legal barriers (Ci = 0.476), reverse logistics (Ci = 0.401), Inconsistent demand and supply (Ci = 0.38), quality assurance and testing (Ci = 0.279) and resistance to change (Ci = 0.087). Such findings suggest that while these hindrances are of less importance compared to consumer perception, technological limitations and lack of support, they still deserve some attention and focused interventions in EEE sector to create an enabling environment for CE (Pollard et al. 2021).

The fuzzy TOPSIS analysis offers a sturdy, orderly way of prioritizing hindrances to procuring used products and CE implementation in EEE sector, considering that some uncertainty and ambiguity usually characterize expert judgments. Such results can be used in the design of tailored strategies and interventions that would mitigate the most serious hindrances and speed up a transition to a CE within the EEE industry.

19.5.1 Implications

Implication of this study regarding the hindrances to implementation of CE principles within EEE sector are significantly impactful for commercial enterprises, policymakers and consumers. These results provide a clear pathway of targeted interventions through the identification and prioritization of key hindrances—such as consumer perception, technological limitations and lack of support and incentives—to facilitate a shift towards a CE. Knowing how to overcome such hindrances can help businesses develop strategies for improving product design, quality assurance and consumer engagement, thereby raising market acceptance for reused and refurbished products (Pajunen and Holuszko 2022). Such insights will be useful for policymakers to formulate enabling policy frameworks containing financial incentives for such circular practices in a bid to spur innovation and attract investment in clean technologies. The consumer education and awareness regarding their role in creating positive perceptions of CE initiatives is emphasized in this study through shaping drivers of demand for the success of sustainable products (Rizos and Bryhn 2022). Overall, the study findings will enhance the academic discourse around the concept of CE and practically guide the stakeholders toward sustainability and resource efficiency in EEE industries.

19.6 Conclusion

The current investigation underscores the hindrances associated with the procurement of second-hand products as well as the integration of CE practices within EEE sector, thereby necessitating the adoption of strategic initiatives aimed at facilitating a transition. Using fuzzy TOPSIS analysis, consumer perception, technological constraints and lack of support and incentives were identified as paramount, thus posing major possible hindrances if the uptake of circular approaches is not checked. The results will then be very informative to the stakeholders: enterprises, policymakers and consumers. However, some limitations of this study should be mentioned. The reliance on expert opinion might thus open up potential biases and the narrow geographical scope might eventually limit the generalization of the results. Subsequent research endeavours should seek to elucidate in greater specificity the respective contributions of diverse stakeholders across all stages of the product life cycle and the actual efficacy of various interventions in overcoming the impediments that have been identified. Longitudinal studies provide invaluable insights into the dynamics of transformation within EEE sectors as they progress towards a more sustainable and CE framework. This investigation has contributed to the rapidly expanding knowledge regarding strategies pertinent to CE and further substantiates the discourse with practical recommendations on how to promote sustainability within EEE sector.

References

Abbasi GA, Chee Keong KQ, Kumar KM, Iranmanesh M (2022) Asymmetrical modelling to understand purchase intention towards remanufactured products in the circular economy and a closed-loop supply chain: an empirical study in Malaysia. J Clean Prod 359:132137. https://doi.org/10.1016/j.jclepro.2022.132137

Agrawal R, Wankhede VA, Kumar A, Luthra S (2021) Analysing the roadblocks of circular economy adoption in the automobile sector: reducing waste and environmental perspectives. Bus Strateg Environ 30(2):1051–1066. https://doi.org/10.1002/bse.2669

Akcan S, Taş MA (2019) Green supplier evaluation with SWARA-TOPSIS integrated method to reduce ecological risk factors. Environ Monit Assess 191(12):736. https://doi.org/10.1007/s10661-019-7884-3

Althaf S, Babbitt CW (2021) Disruption risks to material supply chains in the electronics sector. Resour Conserv Recycl 167:105248. https://doi.org/10.1016/j.resconrec.2020.105248

Andhov M, Caranta R, Stoffel T, Grandia J, Janssen WA, Vornicu R et al (2020) Sustainability through public procurement: the way forward–reform proposals. SSRN Electron J. https://doi.org/10.2139/ssrn.3559393

Azadnia AH, Onofrei G, Ghadimi P (2021) Electric vehicles lithium-ion batteries reverse logistics implementation barriers analysis: a TISM-MICMAC approach. Resour Conserv Recycl 174:105751. https://doi.org/10.1016/j.resconrec.2021.105751

Bag S, Dhamija P, Gupta S, Sivarajah U (2021) Examining the role of procurement 4.0 towards remanufacturing operations and circular economy. Prod Plan Control 32(16):1368–1383. https://doi.org/10.1080/09537287.2020.1817602

Brändström J, Jazairy A, Roos Lindgreen E (2024) Barriers to adopting circular business models: a cross-sectoral analysis. Bus Strateg Environ 33(5):4331–4350. https://doi.org/10.1002/bse.3653

Bressanelli G, Pigosso DCA, Saccani N, Perona M (2021) Enablers, levers and benefits of circular economy in the electrical and electronic equipment supply chain: a literature review. J Clean Prod 298:126819. https://doi.org/10.1016/j.jclepro.2021.126819

Centobelli P, Cerchione R, Esposito E, Passaro R, Shashi. (2021) Determinants of the transition towards circular economy in SMEs: a sustainable supply chain management perspective. Int J Prod Econ 242:108297. https://doi.org/10.1016/j.ijpe.2021.108297

Chakrabarty A, Nandi S (2021) Electronic waste vulnerability: circular economy as a strategic solution. Clean Techn Environ Policy 23(2):429–443. https://doi.org/10.1007/s10098-020-01976-y

Chen Z, Yildizbasi A, Wang Y, Sarkis J (2023) Safety in lithium-ion battery circularity activities: a framework and evaluation methodology. Resour Conserv Recycl 193:106962. https://doi.org/10.1016/j.resconrec.2023.106962

Corsini F, de Bernardi C, Gusmerotti NM, Frey M (2024) Introducing the circular assessment of suppliers (CAoS) tool: a Kraljic matrix-based tool to facilitate circular procurement in private organizations. J Clean Prod 452:142085. https://doi.org/10.1016/j.jclepro.2024.142085

Ding L, Wang T, Chan PW (2023) Forward and reverse logistics for circular economy in construction: a systematic literature review. J Clean Prod 388:135981. https://doi.org/10.1016/j.jclepro.2023.135981

Dumitrica C-D, Grigorescu A, Davidescu AA (2023) Circular economy a push and pull mechanism: recycling starts before purchasing. J Clean Prod 430:139363. https://doi.org/10.1016/j.jclepro.2023.139363

Dutta P, Talaulikar S, Xavier V, Kapoor S (2021) Fostering reverse logistics in India by prominent barrier identification and strategy implementation to promote circular economy. J Clean Prod 294:126241. https://doi.org/10.1016/j.jclepro.2021.126241

Farooque M, Zhang A, Thürer M, Qu T, Huisingh D (2019) Circular supply chain management: a definition and structured literature review. J Clean Prod 228:882–900. https://doi.org/10.1016/j.jclepro.2019.04.303

Farooque M, Zhang A, Choi T-M, Hartley JL, Liu Y (2023) Antecedents and performance outcomes of circular procurement: an empirical study in China. J Purch Supply Manag 30:100882. https://doi.org/10.1016/j.pursup.2023.100882

Geissdoerfer M, Santa-Maria T, Kirchherr J, Pelzeter C (2023) Drivers and barriers for circular business model innovation. Bus Strateg Environ 32(6):3814–3832. https://doi.org/10.1002/bse.3339

Goyal M, Singh K, Bhatnagar N (2023) Circular economy conceptualization for lithium-ion batteries- material procurement and disposal process. Chem Eng Sci 281:119080. https://doi.org/10.1016/j.ces.2023.119080

Guldmann E, Huulgaard RD (2020) Barriers to circular business model innovation: a multiple-case study. J Clean Prod 243:118160. https://doi.org/10.1016/j.jclepro.2019.118160

Hina M, Chauhan C, Kaur P, Kraus S, Dhir A (2022) Drivers and barriers of circular economy business models: where we are now, and where we are heading. J Clean Prod 333:130049. https://doi.org/10.1016/j.jclepro.2021.130049

Hunka AD, Linder M, Habibi S (2021) Determinants of consumer demand for circular economy products. A case for reuse and remanufacturing for sustainable development. Bus Strateg Environ 30(1):535–550. https://doi.org/10.1002/bse.2636

Husain Z, Maqbool A, Haleem A, Pathak RD, Samson D (2021) Analyzing the business models for circular economy implementation: a fuzzy TOPSIS approach. Oper Manag Res 14(3–4):256–271. https://doi.org/10.1007/s12063-021-00197-w

Ibitz A (2020) Assessing Taiwan's endeavors towards a circular economy: the electronics sector. Asia Europe Journal 18(4):493–510. https://doi.org/10.1007/s10308-019-00568-w

Islam MT, Huda N, Baumber A, Shumon R, Zaman A, Ali F, Hossain R, Sahajwalla V (2021) A global review of consumer behavior towards e-waste and implications for the circular economy. J Clean Prod 316:128297. https://doi.org/10.1016/j.jclepro.2021.128297

Julianelli V, Caiado R, Scavarda LF, Cruz SPMF (2020) Interplay between reverse logistics and circular economy: critical success factors-based taxonomy and framework. Resour Conserv Recycl 158:104784. https://doi.org/10.1016/j.resconrec.2020.104784

Kazancoglu I, Kazancoglu Y, Kahraman A, Yarimoglu E, Soni G (2022) Investigating barriers to circular supply chain in the textile industry from Stakeholders perspective. Int J Log Res Appl 25(4–5):521–548. https://doi.org/10.1080/13675567.2020.1846694

Khan SA, Mubarik MS, Paul SK (2022) Analyzing cause and effect relationships among drivers and barriers to circular economy implementation in the context of an emerging economy. J Clean Prod 364:132618. https://doi.org/10.1016/j.jclepro.2022.132618

Khan SA, Laalaoui W, Hokal F, Tareq M, Ahmad L (2023) Connecting reverse logistics with circular economy in the context of industry 4.0. Kybernetes 52(12):6279–6320. https://doi.org/10.1108/K-03-2022-0468

Kim CH, Kuah ATH, Thirumaran K (2022) Morphology for circular economy business models in the electrical and electronic equipment sector of Singapore and South Korea: findings, implications, and future agenda. Sustain Prod Consum 30:829–850. https://doi.org/10.1016/j.spc.2022.01.006

Kouloumpis V, Konstantzos GE, Chroni C, Abeliotis K, Lasaridi K (2023) Does the circularity end justify the means? A life cycle assessment of preparing waste electrical and electronic equipment for reuse. Sustain Prod Consum 41:291–304. https://doi.org/10.1016/j.spc.2023.08.008

Kumar P, Singh RK, Paul J, Sinha O (2021) Analyzing challenges for sustainable supply chain of electric vehicle batteries using a hybrid approach of Delphi and best-worst method. Resour Conserv Recycl 175:105879. https://doi.org/10.1016/j.resconrec.2021.105879

Kumar A, Gaur D, Liu Y, Sharma D (2022) Sustainable waste electrical and electronic equipment management guide in emerging economies context: a structural model approach. J Clean Prod 336:130391. https://doi.org/10.1016/j.jclepro.2022.130391

Leong SY, Lee SY, Koh TY, Ang DTC (2023) 4R of rubber waste management: current and outlook. J Mater Cycles Waste Manage 25(1):37–51

MahmoumGonbadi A, Genovese A, Sgalambro A (2021) Closed-loop supply chain design for the transition towards a circular economy: a systematic literature review of methods, applications and current gaps. J Clean Prod 323:129101. https://doi.org/10.1016/j.jclepro.2021.129101

Mallick PK, Salling KB, Pigosso DCA, McAloone TC (2023) Closing the loop: establishing reverse logistics for a circular economy, a systematic review. J Environ Manag 328:117017. https://doi.org/10.1016/j.jenvman.2022.117017

Marrucci L, Daddi T, Iraldo F (2019) The integration of circular economy with sustainable consumption and production tools: systematic review and future research agenda. J Clean Prod 240:118268. https://doi.org/10.1016/j.jclepro.2019.118268

Mendoza JMF, Gallego-Schmid A, Velenturf APM, Jensen PD, Ibarra D (2022) Circular economy business models and technology management strategies in the wind industry: sustainability potential, industrial challenges and opportunities. Renew Sust Energ Rev 163:112523. https://doi.org/10.1016/j.rser.2022.112523

Menon RR, Ravi V (2022) An analysis of barriers affecting implementation of sustainable supply chain management in electronics industry: a Grey-DEMATEL approach. J Model Manag 17(4):1319–1350. https://doi.org/10.1108/JM2-02-2021-0042

Mhatre P, Panchal R, Singh A, Bibyan S (2021) A systematic literature review on the circular economy initiatives in the European Union. Sustain Prod Consum 26:187–202. https://doi.org/10.1016/j.spc.2020.09.008

Migliore M, Talamo C, Paganin G (2020) Circular economy and sustainable procurement: the role of the attestation of conformity. In: Strategies for circular economy and cross-sectoral exchanges for sustainable building products. Springer, pp 159–173. https://doi.org/10.1007/978-3-030-30318-1_6

Milios L (2021) Overarching policy framework for product life extension in a circular economy: a bottom-up business perspective. Environ Policy Gov 31(4):330–346. https://doi.org/10.1002/eet.1927

Mishra A, Dutta P, Jayasankar S, Jain P, Mathiyazhagan K (2023) A review of reverse logistics and closed-loop supply chains in the perspective of circular economy. BIJ 30(3):975–1020. https://doi.org/10.1108/BIJ-11-2021-0669

Nag U, Sharma SK, Govindan K (2021) Investigating drivers of circular supply chain with product-service system in automotive firms of an emerging economy. J Clean Prod 319:128629. https://doi.org/10.1016/j.jclepro.2021.128629

Neessen PCM, Caniëls MCJ, Vos B, de Jong JP (2021a) How and when do purchasers successfully contribute to the implementation of circular purchasing: a comparative case-study. J Purch Supply Manag 27(3):100669. https://doi.org/10.1016/j.pursup.2020.100669

Neessen PCM, de Jong JP, Caniëls MCJ, Vos B (2021b) Circular purchasing in Dutch and Belgian organizations: the role of intrapreneurship and organizational citizenship behavior towards the environment. J Clean Prod 280:124978. https://doi.org/10.1016/j.jclepro.2020.124978

Nikolaou IE, Tsalis TA, Vatalis KI (2022) A framework to integrate circular economy principles into public procurement. In: Circular economy and sustainability. Elsevier, pp 279–291. https://doi.org/10.1016/B978-0-12-819817-9.00020-X

Ogunnusi M, Salman H, Laing R (2023) TOPSIS analysis for sustainable redevelopment potential of abandoned infrastructure in Nigeria. Built Environ Project Asset Manage 13(1):73–88. https://doi.org/10.1108/BEPAM-11-2021-0140

Pajunen N, Holuszko ME (2022) Circular economy in electronics and the future of e-waste. In: Electronic waste. Wiley, pp 299–314. https://doi.org/10.1002/9783527816392.ch13

Pan X, Wong CWY, Li C (2022) Circular economy practices in the waste electrical and electronic equipment (WEEE) industry: a systematic review and future research agendas. J Clean Prod 365:132671. https://doi.org/10.1016/j.jclepro.2022.132671

Parajuly K, Fitzpatrick C, Muldoon O, Kuehr R (2020) Behavioral change for the circular economy: a review with focus on electronic waste management in the EU. Resour Conserv Recycl X 6:100035. https://doi.org/10.1016/j.rcrx.2020.100035

Perez-Castillo D, Vera-Martinez J (2021) Green behaviour and switching intention towards remanufactured products in sustainable consumers as potential earlier adopters. Asia Pac J Mark Logist 33(8):1776–1797. https://doi.org/10.1108/APJML-10-2019-0611

Pisitsankkhakarn R, Vassanadumrongdee S (2020) Enhancing purchase intention in circular economy: empirical evidence of remanufactured automotive product in Thailand. Resour Conserv Recycl 156:104702. https://doi.org/10.1016/j.resconrec.2020.104702

Pollard J, Osmani M, Cole C, Grubnic S, Colwill J (2021) A circular economy business model innovation process for the electrical and electronic equipment sector. J Clean Prod 305:127211. https://doi.org/10.1016/j.jclepro.2021.127211

Pollard J, Osmani M, Grubnic S, Díaz AI, Grobe K, Kaba A, Ünlüer Ö, Panchal R (2023) Implementing a circular economy business model canvas in the electrical and electronic manufacturing sector: a case study approach. Sustain Prod Consum 36:17–31. https://doi.org/10.1016/j.spc.2022.12.009

Polonsky MJ, Wijayasundara M, Noel W, Vocino A (2022) Identifying the drivers and barriers of the public sector procurement of products with recycled material or recovered content: a systematic review and research propositions. J Clean Prod 358:131780. https://doi.org/10.1016/j.jclepro.2022.131780

Qazi AA, Appolloni A (2022) A systematic review on barriers and enablers toward circular procurement management. Sustain Prod Consum 33:343–359. https://doi.org/10.1016/j.spc.2022.07.013

Rani P, Mishra AR, Rezaei G, Liao H, Mardani A (2020) Extended Pythagorean fuzzy TOPSIS method based on similarity measure for sustainable recycling partner selection. Int J Fuzzy Syst 22(2):735–747. https://doi.org/10.1007/s40815-019-00689-9

Richter JL, Svensson-Hoglund S, Dalhammar C, Russell JD, Thidell Å (2023) Taking stock for repair and refurbishing: a review of harvesting of spare parts from electrical and electronic products. J Ind Ecol 27(3):868–881. https://doi.org/10.1111/jiec.13315

Rizos V, Bryhn J (2022) Implementation of circular economy approaches in the electrical and electronic equipment (EEE) sector: barriers, enablers and policy insights. J Clean Prod 338:130617. https://doi.org/10.1016/j.jclepro.2022.130617

Sagnak M, Berberoglu Y, Memis İ, Yazgan O (2021) Sustainable collection center location selection in emerging economy for electronic waste with fuzzy best-worst and fuzzy TOPSIS. Waste Manag 127:37–47. https://doi.org/10.1016/j.wasman.2021.03.054

Sarkhoshkalat MM, Afkham A, Bonyadi Manesh M, Sarkhosh M (2024) Circular economy and the recycling of e-waste. In: New technologies for energy transition based on sustainable development goals. Springer Nature, pp 319–354. https://doi.org/10.1007/978-981-97-2527-4_16

Shaharudin MR, Zailani S, Tan K-C, Cross J, Hotrawaisaya C (2023) Fostering closed-loop supply chain orientation by leveraging strategic green capabilities for circular economy performance: empirical evidence from Malaysian electrical and electronics manufacturing firms. Environ Dev Sustain. https://doi.org/10.1007/s10668-022-02832-3

Sharma M, Joshi S, Kumar A (2020) Assessing enablers of e-waste management in circular economy using DEMATEL method: an Indian perspective. Environ Sci Pollut Res 27(12):13325–13338. https://doi.org/10.1007/s11356-020-07765-w

Sharma NK, Govindan K, Lai KK, Chen WK, Kumar V (2021a) The transition from linear economy to circular economy for sustainability among SMEs: a study on prospects, impediments, and prerequisites. Bus Strateg Environ 30(4):1803–1822. https://doi.org/10.1002/bse.2717

Sharma M, Joshi S, Govindan K (2021b) Issues and solutions of electronic waste urban mining for circular economy transition: an Indian context. J Environ Manag 290:112373. https://doi.org/10.1016/j.jenvman.2021.112373

Shojaei A, Ketabi R, Razkenari M, Hakim H, Wang J (2021) Enabling a circular economy in the built environment sector through blockchain technology. J Clean Prod 294:126352. https://doi.org/10.1016/j.jclepro.2021.126352

Sönnichsen SD, Clement J (2020) Review of green and sustainable public procurement: towards circular public procurement. J Clean Prod 245:118901. https://doi.org/10.1016/j.jclepro.2019.118901

Stumpf L, Schöggl J-P, Baumgartner RJ (2021) Climbing up the circularity ladder? A mixed-methods analysis of circular economy in business practice. J Clean Prod 316:128158. https://doi.org/10.1016/j.jclepro.2021.128158

Svensson-Hoglund S, Richter JL, Maitre-Ekern E, Russell JD, Pihlajarinne T, Dalhammar C (2021) Barriers, enablers and market governance: a review of the policy landscape for repair of consumer electronics in the EU and the U.S. J Clean Prod 288:125488. https://doi.org/10.1016/j.jclepro.2020.125488

Toker K, Görener A (2023) Evaluation of circular economy business models for SMEs using spherical fuzzy TOPSIS: an application from a developing countries perspective. Environ Dev Sustain 25(2):1700–1741. https://doi.org/10.1007/s10668-022-02119-7

Wijayasundara M, Polonsky M, Noel W, Vocino A (2022) Green procurement for a circular economy: what influences purchasing of products with recycled material and recovered content by public sector organisations? J Clean Prod 377:133917. https://doi.org/10.1016/j.jclepro.2022.133917

Wrålsen B, Prieto-Sandoval V, Mejia-Villa A, O'Born R, Hellström M, Faessler B (2021) Circular business models for lithium-ion batteries: stakeholders, barriers, and drivers. J Clean Prod 317:128393. https://doi.org/10.1016/j.jclepro.2021.128393

Xu L, Jia F, Yan F, Chen L (2022) Circular procurement: a systematic literature review. J Clean Prod 365:132845. https://doi.org/10.1016/j.jclepro.2022.132845

Chapter 20
Understanding the Challenges to Humanising Circular Supply Chains: A Sustainable Perspective

Lakshmi Shetty, Shefali Srivastava, Ashish Dwivedi, Dindayal Agrawal, and Shri Krishna Pandey

Abstract Sustainability attainment process is considered equal to transitioning to Circular Supply Chains (CSCs) as it minimizes wastes, optimizes resource usage and reduces environmental impact. However, humanising CSCs has to be based on a sustainable approach, including social, environmental and economic dimensions. Incidentally, the present study recognizes some challenges which are potentially associated with humanising CSCs. Further, the identified challenges are prioritised and analysed adopting Fuzzy VIKOR method. The most important findings from the study reflect that the major challenges are changing relationships and resistance to change. The organizational culture and mindset is crucial in the successful adoption of circular practices. Moreover, metrics and frameworks lack in measuring progress and guaranteeing constant implementation of circular initiatives. Moreover, the process of transition is further complicated by technology limitations and costs, hence requiring investment in innovative solutions and alternative financing models. Equally important in the creation of effective CSCs is engaging new partners and developing open loops, cross-sector collaboration and adapting to the change in relationship. Such challenges put forward that the common goals of sustainability attained would, at the same time, mean giving due prominence to collaboration and

---

L. Shetty
Symbiosis Centre for Management Studies, Bengaluru, Symbiosis International (Deemed University), Pune, Maharashtra, India

S. Srivastava
Operations and Decision Sciences, Christ University, Bengaluru, Karnataka, India

A. Dwivedi (✉)
Jindal Global Business School, O.P. Jindal Global University, Sonipat, Haryana, India

D. Agrawal
Institute of Management Technology, Ghaziabad, Uttar Pradesh, India

S. K. Pandey
Manipal Academy of Higher Education, Dubai, UAE
e-mail: skpandey@manipaldubai.com

K. Mathiyazhagan et al. (eds.), *Humanizing Circular Supply Chain Management*, Circular Economy and Sustainability,
https://doi.org/10.1007/978-3-031-86562-6_20

relationship-building amongst the stakeholders. In effectively humanising CSCs, organizations should consider a multi-pronged approach that includes strategic planning, engaging stakeholders and technology innovation. Addressing the challenges identified in humanising CSCs and putting the best practices into consideration could help improve efforts on sustainability among organisations and increase the prospects of a more circular economy.

Keywords Humanising · Circular supply chains · Dimensions · Sustainable · Fuzzy VIKOR

20.1 Introduction

Humanising Circular Supply Chains (CSCs) concept signifies a notable advancement in supply chain management (SCM), emphasizing incorporation of sustainability principles into fundamental operation of enterprises (Rašković and Nagar 2024). With industries encountering growing demands to diminish their ecological footprint and embrace more sustainable approaches, the transformation from conventional linear supply chains to CSCs has gained traction (Longoni et al. 2024). The study elucidates the most pertinent dimensions of humanising CSCs within the framework of sustainability, including challenges, advantages, and the necessity for a comprehensive approach that integrates social, economic and environmental considerations (Kopnina et al. 2024).

In the context of a global economy, the adverse effects of resource depletion, environmental degradation, and climate change, CSCs have emerged as a significant paradigm likely to facilitate the attainment of sustainability (Luthra et al. 2018). In contrast to the conventional linear model of material resource flow characterized by the take-make-dispose approach in supply chains, CSCs adopt a "closed-loop cradle-to-cradle model" wherein resources are perpetually cycled through processes of reuse, refurbishment, and recycling (Bals et al. 2024). This paradigm shift not only mitigates waste generation but also enhances resource utilization efficiency, thereby aligning business practices with the overarching goals of sustainable development. As articulated by Ellen MacArthur Foundation (2015), Circular Economy (CE) is conceptualized as a regenerative system "designed to minimize resource input, waste production, emissions, and energy dissipation through innovative design and management". Nevertheless, there are barriers to this migration to CSCs, which are obligatorily necessary to be overcome in order for one to unlock their potentials (Kalverkamp 2018). Such challenges involve a number of implications including the need for new relationships between members of the supply chain which are structurally flexible to interact with a broad range of stakeholders and open major loops spanning organizational boundaries (Amiri et al. 2022). Furthermore, effective inter-sectoral cooperation, performance evaluation, adherence to regulations, consumer behaviour, technological constraints, financial limitations and reluctance to change were further highlighted as blockades to the humanisation of CSCs (Shang et al. 2022).

To navigate these details, it is imperative to establish benchmarks for appraising the efficacy of CSCs. The key criteria will include metrics that can assess things like sustainability outcomes, stakeholder engagement levels, technology adoption rates, regulatory compliance, consumer awareness, financial viability, flexibility and adaptability and collaboration across sectors (Longoni et al. 2024; Khan et al. 2021). Systematic assessment of these benchmarks can enable an organization to identify barriers to circularity and come up with effective strategies for enhancing the sustainability of their CSCs (Tura et al. 2019). This study therefore aims at assessing the challenges of humanising CSCs from the dimension of sustainability, highlighting the benchmarks that would be critical for any evaluation. Any such challenges should be addressed imperatively to ensure a sustainable future and that CSCs make strong contributions to environmental and social welfare.

The objective of the present study is to examine how the challenges associated with humanising CSCs for sustainability necessitate the design and implementation of such supply chains in a manner that minimizes environmental impact and promotes social equity. Among the major objectives are identification of the challenges to transition or change in the circular models, not forgetting the friction to change and failure in regulation. The study deploys fuzzy VIKOR method to rank and analyse such challenges that can help policy-makers prioritize their efforts. Consequently, the current study endeavours to address the following Research Questions (RQs):

RQ1: What are the potential challenges that industries face in humanising CSCs and how these challenges impact the integration of sustainability principles across environmental, social and economic dimensions?

RQ2: How can industries effectively overcome the challenges to humanise their CSCs and what best practices can be implemented to enhance sustainability outcomes?

20.2 Literature Review

Humanising CSCs plays and important role by human actors at various levels, ranging from individual consumers to policymakers, in propelling the CE evolution (Grabowska et al. 2022). This methodology underscores the importance of adopting a person centred framework in scholarly research, highlighting the need of comprehending and influencing individual behaviours towards circularity (Dacre et al 2024). The artificial neural networks as a tool is used for designing circularity accounting networks that bridge the gap between human and non-human actors (Czifra et al. 2023). These networks aim to visually represent the circularity of products by considering factors such as CO_2 emissions and harnessing the potential of Artificial Intelligence to enhance CE accounting practices (Marcon et al. 2023). The incorporation of human centricity within CE framework is deemed essential for promoting sustainable behaviours and realizing the future which is more environmentally conscious (Ren et al. 2023).

This paradigm shift demands a person-centred approach where people cannot be visualised as mere users of circular products; instead they stand at the hub of social contexts (Earley and Hornbuckle 2017). For the circular economy to be successful, peoples' behaviours towards products have to be changed drastically, along with their relationships to material goods. This encompasses the fundamental elements intrinsic to the concept of CE; namely, reduce, reuse, recycle, and recover, when integrated into SCM practices, can effectively address urgent environmental challenges (Vadakkepatt et al. 2021). Furthermore, the promotion of collaboration alongside green supply chain strategies and sustainable practices has been identified as two of the most pivotal actions in realizing CSCs within a business-to-business framework, thereby prioritizing resource rejuvenation and long-term sustainability as central objectives (Grosse et al. 2023).

The existing research gap in humanising CSCs through a sustainability lens is about the need to see a focus on individual behaviours and relationships with material goods as a way forward for advancing CE (Kumar et al. 2021). While operationalizing imperatives reduce, reuse and recycle are equally necessary to build CSCs, such uncertainties require proactive management strategies of organizations to support better sustainability performance and accelerate the shift towards circularity (Guest et al. 2022). Innovations in green products, processes, logistics, and services contribute to enhanced supply chain efficiency and competitiveness while simultaneously improving sustainability performance within durable goods sectors (Davies et al. 2024). Additionally, impact of social capital in sustainable performance by CSC management underlines how collaboration and engagement impact the supply chain partners in the cycles of resources (Fosso Wamba 2022). Moreover, firms, particularly sustainable start-ups, design and organize CSCs with a special focus on reducing negative sustainability impacts and creating value in different industries (Chen et al. 2024). Table 20.1 presents challenges like technological, financial, organizational, regulatory and collaborative dimensions. Overcoming these challenges necessitates the comprehensive strategy that aligns sustainability objectives with supply chain tactics, utilizes innovation, encourages involvement of stakeholders and cultivates the culture within circularity (Pego and Lourenço 2024).

20.3 Research Methodology

In this study, the qualitative method is used during the process of identifying challenges related to humanising CSCs with a view on sustainability. Purposive sampling selected a total of 34 experts to adequately represent industry experts from the manufacturing units (Grabowska et al. 2022). Participants selected for this investigation were chosen based on their extensive expertise accumulated over a decade of professional experience and having expertise in both SCM and sustainability or CE practices, as shown in Table 20.2. The data was retrieved through the process of

Table 20.1 Identifying challenges to humanising CSCs

Challenges	Description	References
Performance integration	The proficient assimilation of sustainability metrics and objectives into the established supply chain performance metrics and management systems. This means moving from traditional supply chains to Circular Economy (CE) principles	Sudusinghe and Seuring (2022), Bimpizas-Pinis et al. (2022), de Lima et al. (2022), Edwin Cheng et al. (2022)
Regulatory barriers	Complexity in addition to dynamic landscapes of regulations related to sustainability, waste management and CE practices. How to overcome barriers that do not quite support or incentivize circular initiatives enough	Nandha Gopan and Balaji (2023), Wang et al. (2022), Kazancoglu et al. (2021), Farooque et al. (2019), Mangla et al. (2018)
Technological limitation	Challenges in adopting innovations like IoT, blockchain and advanced analytics to enhance circularity. Addressing limitations in available technologies, infrastructure and digital capabilities need to enable CSCs	Mahroof et al. (2022), Dwivedi and Paul (2022), Khan et al. (2022), Ozkan-Ozen et al. (2020)
Financial constraints	Overcoming the financial barriers and costs associated with transitioning to CSCs. Problems related to new technologies securing funds for investment, processes and capacities that helps in achieve circularity	Saccani et al. (2023), Pla-Julián and Guevara (2019), Sehnem et al. (2019), Koh et al. (2017)
Resistance to change	Fostering a culture that embraces sustainability and CE principles. Overcoming organizational inertia and resistance to altering established supply chain practices and business models	Luthra et al. (2022), Friedman and Ormiston (2022), Iacovidou et al. (2021), Vermunt et al. (2019), Aminoff and Kettunen (2016)
Lack of standardization	Developing industry-wide standards, frameworks and metrics for measuring and reporting on CSC performance. Lack of consistency in approaches hinders collaboration and benchmarking	Khan and Abonyi (2022), Gebhardt et al. (2022), Nandi et al. (2021), Yadav et al. (2020)
Cross-sector collaboration	Partnerships and collaborations across industries and sectors will enable CSCs. Difficulties in finding new partners to work with, aligning objectives and cultural differences	Henry et al. (2022), Pedersen et al. (2021), DiVito et al. (2021), Soldatos et al. (2021), Klitsie et al. (2018)
Changing relationships	It involves adjusting the traditional supplier and customer relationships in order to make CSCs possible. Transactional relationships start giving way to more collaborative and deep partnerships with an eye on sustainability and shared value creation	Hernandez Marquina et al. (2024), Le (2022), Chari et al. (2022), Rogan et al. (2022), Hofstetter et al. (2021)
Developing open loops	Circular systems that go beyond the boundaries of individual organizations or supply chains incorporate the challenges on how to develop open-loop systems that attain the best retention and re-use of value and materials across a number of industries	Hussain and Malik (2020), Holgado and Aminoff (2019), Kalverkamp and Young (2019), Mishra et al. (2018), Miemczyk et al. (2016)

Table 20.2 Respondents demographic details

Age group	30–40–6
	40–50–13
	Above 50–15
Gender	Male—28
	Female—6
Geographic location	North zone—11
	South zone—9
	East zone—6
	West zone—8
Job responsibility	CSC consultants—9
	Sustainability managers—8
	Supply chain professionals—7
	Academics/Professors—4
	Regulatory experts—6

structured interviews which allowed for in-depth discussion, thus facilitating in-depth exploration regarding participants' insights on the key challenges to humanising CSCs. The guide questions had been designed so as to secure in-depth responses pertinent to the dimensions of sustainability and the exact challenges extracted from the literature (Jabłoński et al. 2020).

20.3.1 VIKOR Approach

Serafim Opricovic introduced VIKOR in 1998, which is elaborated as VIseKriterijumska Optimizacija I Kompromisno Resenje, as a technique for decision making based on multi criteria decision making techniques, wherein issues characterized by conflicting and non-commensurable criteria are foregrounded. This approach is designed to facilitate the attainment of a compromise solution that approaches the ideal outcome, achieved through the systematic ranking of alternatives and their selection from a designated pool. The technique is helpful where there are complex systems including many criteria, some of which might conflict with each other. In the context of this study, VIKOR is used to scrutinize and rank the challenges associated with the practice of CSCs (Khan et al. 2021). Such challenges are always varied in nature and therefore cut across environmental, social and economic aspects. VIKOR provides a step-by-step framework for ranking alternatives in decreasing order of proximity to the ideal solution, thus helping along with the decision-making processes in this complicated situation where trade-offs are inevitable (Singh et al. 2020).

20.3.2 *Fuzzy Set Theory (FST)*

Application of the FST into the fuzzification of decision-making processes increases the capability of classical set theory in handling uncertain and imprecise data. Developed by Zadeh (1965), FST theory represents the elements in a set with different degrees of membership rather than always using strict binary classification of membership. This is performed by applying membership functions that assign to every set element of values between 0 and 1, representative of the degree of its belongingness to the fuzzy set. Among the different applications, FST is especially useful in cases where information is incomplete, ambiguous, or changing, like in decision-making, pattern recognition and control systems. By accommodating the complexities of human cognition and subjective judgment, FST is able to model complex systems more robustly and flexibly, hence yielding better decision-making results in different areas (Khan et al. 2021).

20.4 Results and Discussion

Fuzzy VIKOR method was pioneered by Opricovic and Tzeng (2007), specifically for the purpose of ranking alternatives within a fuzzy decision-making context. Adoption of fuzzy VIKOR method will help improve the precision and reliability of the decision-making process within an intricate scenario marked with uncertainty by setting or establishing it as a useful tool for a wide array of applications in the purpose of evaluation.

20.4.1 *Stage 1: Creation of Decision Matrix*

This study has 3 criteria and 9 alternatives built on the literature and then ranked based on Fuzzy VIKOR method. Table 20.3 assigns weight to each sustainable criterion.

Table 20.4 explains the fuzzy scale used in the model.

Given the involvement of multiple experts in this evaluative process, the alternatives are assessed according to various criteria. Table 20.5 presents the results of the decision matrix, which reflects the arithmetic mean as determined by the industry experts.

Table 20.3 Assigning sustainable criteria characteristics

Sustainability criterion	Type	Weights
Economic	+	(0.333, 0.333, 0.333)
Social	+	(0.333, 0.333, 0.333)
Environmental	+	(0.333, 0.333, 0.333)

Table 20.4 Assigning of fuzzy scale for the terms

Code	Linguistic terms	L	M	U
1	Very Low	0	0	0.25
2	Low	0	0.25	0.5
3	Medium	0.25	0.5	0.75
4	High	0.5	0.75	1
5	Very High	0.75	1	1

Table 20.5 Calculating the decision matrix for the challenges based on criteria

Challenges	Economic	Social	Environmental
Performance integration	(0.325, 0.575, 0.750)	(0.300, 0.475, 0.700)	(0.325, 0.575, 0.800)
Regulatory barriers	(0.175, 0.375, 0.625)	(0.125, 0.350, 0.600)	(0.175, 0.375, 0.625)
Technological limitation	(0.375, 0.575, 0.800)	(0.225, 0.425, 0.675)	(0.250, 0.400, 0.600)
Financial constraints	(0.175, 0.300, 0.525)	(0.200, 0.375, 0.600)	(0.250, 0.425, 0.675)
Resistance to change	(0.150, 0.400, 0.650)	(0.125, 0.300, 0.525)	(0.225, 0.425, 0.625)
Lack of standardization	(0.175, 0.325, 0.525)	(0.150, 0.325, 0.550)	(0.300, 0.525, 0.700)
Cross-sector collaboration	(0.350, 0.600, 0.850)	(0.275, 0.450, 0.700)	(0.300, 0.525, 0.750)
Changing relationships	(0.125, 0.250, 0.500)	(0.250, 0.450, 0.675)	(0.125, 0.300, 0.550)
Developing open loops	(0.550, 0.800, 0.900)	(0.275, 0.500, 0.700)	(0.225, 0.400, 0.625)

20.4.2 Stage 2: Determining Positive Ideal Solution and Negative Ideal Solution

"If the criteria is positive, the positive ideal solution ($\tilde{f}^*$) and negative ideal solution $\left(\tilde{f}^\circ\right)$ can be obtained based on the following formula":

$$\tilde{f}_j^* = \max_i \tilde{f}_{ij} \quad i = 1, 2, \ldots, n \tag{20.1}$$

$$\tilde{f}_j^\circ = \min_i \tilde{f}_{ij} \quad i = 1, 2, \ldots, n \tag{20.2}$$

"If the criteria is negative, the positive ideal solution ($\tilde{f}^*$) and negative ideal solution $\left(\tilde{f}^\circ\right)$ can be obtained based on the substantial formula":

$$\tilde{f}_j^* = \min_i \tilde{f}_{ij} \quad i = 1, 2, \ldots, n \tag{20.3}$$

$$\tilde{f}_j^\circ = \max_i \tilde{f}_{ij} \quad i = 1, 2, \ldots, n \tag{20.4}$$

Table 20.6 shows the calculation of positive and negative ideal values.

Table 20.6 Calculation of positive and negative ideal solutions

Challenges	Positive ideal value	Negative ideal value
Economic	(0.375, 0.575, 0.800)	(0.175, 0.300, 0.525)
Social	(0.300, 0.475, 0.700)	(0.125, 0.350, 0.600)
Environmental	(0.325, 0.575, 0.800)	(0.175, 0.375, 0.600)

Table 20.7 Calculation of normalized decision matrix

Challenges	Economic	Social	Environmental
Performance integration	(−0.600, 0.000, 0.760)	(−0.696, 0.000, 0.696)	(−0.760, 0.000, 0.760)
Regulatory barriers	(−0.400, 0.320, 1.000)	(−0.522, 0.217, 1.000)	(−0.480, 0.320, 1.000)
Technological limitation	(−0.680, 0.000, 0.680)	(−0.652, 0.087, 0.826)	(−0.440, 0.280, 0.880)
Financial constraints	(−0.240, 0.440, 1.000)	(−0.522, 0.174, 0.870)	(−0.560, 0.240, 0.880)
Resistance to change	(−0.129, 0.516, 0.968)	(−0.435, 0.348, 1.000)	(−0.520, 0.160, 0.840)
Lack of standardization	(0.032, 0.613, 0.935)	(−0.478, 0.304, 0.957)	(−0.640, 0.000, 0.720)
Cross-sector collaboration	(−0.387, 0.258, 0.710)	(−0.739, 0.087, 0.739)	(−0.720, 0.000, 0.720)
Changing relationships	(0.065, 0.710, 1.000)	(−0.696, 0.087, 0.783)	(−0.400, 0.360, 1.000)
Developing open loops	(−0.452, 0.000, 0.452)	(−0.739, 0.000, 0.739)	(−0.520, 0.200, 0.840)

20.4.3 *Stage 3: Computation of Normalized Decision Matrix*

A normalized decision matrix can be calculated by the following formula:

$$\tilde{d}_{ij} = \left(\tilde{f}_j^* \ominus \tilde{f}_{ij}\right) / \left(r_j^* - l_j^\circ\right) \quad \text{Positive ideal solution} \tag{20.5}$$

$$\tilde{d}_{ij} = \left(\tilde{f}_{ij} \ominus \tilde{f}_j^*\right) / \left(r_j^\circ - l_j^*\right) \quad \text{Negative ideal solution} \tag{20.6}$$

where,

$$\tilde{f}_j^* = \left(l_j^*, m_j^*, r_j^*\right) \tag{20.7}$$

$$\tilde{f}_j^\circ = \left(l_j^\circ, m_j^\circ, r_j^\circ\right) \tag{20.8}$$

Table 20.7 illustrates the calculation of normalized values of the evaluation matrix.

20.4.4 Stage 4: Computation of the Values $\tilde{S}_i$ and $\tilde{R}_i$

The normalized matrix undergoes a transformation into a weighted normalized decision matrix, subsequently allowing for the calculation of values as delineated through the following formula:

$$\text{If } \tilde{R}_i = \left(R_i^l, R_i^m, R_i^r\right) \text{and } \tilde{s}_i = \left(s_i^l, s_i^m, s_i^r\right) \tag{20.9}$$

$$\tilde{S}_i = \left(\tilde{w}_j \otimes \tilde{d}_{ij}\right) \tag{20.10}$$

$$\tilde{R}_i = \left(\tilde{w}_j \otimes \tilde{d}_{ij}\right) \tag{20.11}$$

20.4.5 Stage 5: Calculation of the VIKOR Index (Q)

The value of Q can be calculated as follows.

$$\text{If } \tilde{Q}_i = \left(Q_i^l, Q_i^m, Q_i^r\right) \tag{20.12}$$

$$\tilde{Q}_i = v\frac{\left(\tilde{s}_i \ominus \tilde{s}^*\right)}{s^{\circ r} - s^{*l}} \oplus (1-v)\frac{\left(\tilde{R}_i \ominus \tilde{R}^*\right)}{R^{\circ r} - R^{*l}} \tag{20.13}$$

where,

$$\tilde{s}^* = \tilde{s}_i \tag{20.14}$$

$$s^{\circ r} = s_i^r \tag{20.15}$$

$$\tilde{R}^* = \tilde{R}_i \tag{20.16}$$

$$R^{\circ r} = R_i^r \tag{20.17}$$

"The variable v indicates the maximum group utility is equal to 0.5 in this study".

"The fuzzy numbers S, R and Q can be converted into crisp numbers using the subsequent formula":

$$\text{If } \tilde{A} = (l,m,r)\left(\tilde{A} \text{ is expreseed as a fuzzy number}\right) \tag{20.18}$$

$$Crisp\left(\tilde{A}\right) = \frac{2m + l + r}{4} \tag{20.19}$$

Table 20.8 illustrates the computation fuzzy values S, R and Q.

Table 20.9 delineates the crisp values of S, R, and Q, as well as the ranking of the alternatives based on R, S, and Q.

Table 20.8 Calculation of fuzzy values S, R and Q

Challenges	Fuzzy S	Fuzzy R	Fuzzy Q
Performance integration	(−0.200, 0.000, 0.253)	(−0.685, 0.000, 0.738)	(−0.847, 0.000, 0.847)
Regulatory barriers	(−0.133, 0.107, 0.333)	(−0.467, 0.286, 0.999)	(−0.720, 0.185, 1.000)
Technological limitation	(−0.147, 0.093, 0.293)	(−0.590, 0.122, 0.795)	(−0.769, 0.124, 0.902)
Financial constraints	(−0.080, 0.147, 0.333)	(−0.440, 0.284, 0.916)	(−0.662, 0.222, 0.975)
Resistance to change	(−0.043, 0.172, 0.333)	(−0.361, 0.341, 0.935)	(−0.634, 0.197, 1.000)
Lack of standardization	(0.011, 0.204, 0.319)	(−0.362, 0.305, 0.870)	(−0.578, 0.219, 0.964)
Cross-sector collaboration	(−0.129, 0.086, 0.246)	(−0.615, 0.115, 0.722)	(−0.804, 0.036, 0.842)
Changing relationships	(0.021, 0.236, 0.333)	(−0.343, 0.385, 0.927)	(−0.561, 0.278, 0.997)
Developing open loops	(−0.150, 0.067, 0.280)	(−0.570, 0.067, 0.676)	(−0.812, 0.000, 0.861)

Table 20.9 The crisp values S, R, Q and alternatives ranking

Challenges	Crisp value of R	Rank in R	Crisp value of S	Rank in S	Crisp value of Q	Rank in Q
Performance integration	0.013	1	0.013	2	0	1
Regulatory barriers	0.103	5	0.276	6	0.162	5
Technological limitation	0.083	4	0.112	4	0.095	4
Financial constraints	0.137	6	0.261	5	0.189	6
Resistance to change	0.158	7	0.314	8	0.19	7
Lack of standardization	0.184	8	0.28	7	0.206	8
Cross-sector collaboration	0.072	3	0.084	3	0.027	3
Changing relationships	0.207	9	0.338	9	0.248	9
Developing open loops	0.066	2	0.06	1	0.012	2

20.4.6 Stage 6: Proposing the Compromise Solution

In the final phase, decisions are made predicated on the values R, S, and Q for the alternatives arranged in descending order. There exist two conditions that necessitate decision-making and a collection of compromise solutions.

$$\text{Condition 1: Acceptable advantage}: Q\left(A^{(2)}\right)-Q\left(A^{(1)}\right)\geq 1/(m-1) \quad (20.20)$$

"where, m is number of alternatives. $A^{(1)}$ is the alternative with the first position and $A^{(2)}$is the alternative with second position in the ranking list by Q".

Condition 2: Acceptable stability in decision making: "The alternative $A^{(1)}$ must also be the best ranked by S or/and R".

If one of the conditions fail to be met, a collection of compromise solutions is proposed, which is comprised of:

Solution 1: "Alternatives $A^{(1)}, A^{(2)}, \ldots, A^{(M)}$ if Condition 1 is not satisfied; Alternative $A^{(M)}$ is determined by $Q(A^{(M)}) - Q(A^{(1)}) < 1/(m - 1)$ for maximum M (the positions of these alternatives are "in closeness")".
Solution 2: Alternatives $A^{(1)}$ and $A^{(2)}$ if only condition 2 is not satisfied.
Solution 3: The alternative exhibiting the minimal Q value will be designated as the optimal alternative if both stipulated conditions are fulfilled.

The findings of the conditions assessment are reflected in Table 20.10.

So, Performance integration, Technological limitation, Regulatory barriers, Financial constraints, Developing open loops, Cross-sector collaboration, Resistance to change, Lack of standardization and Changing relationships are selected in stage 6 as the final alternatives.

Performance integration (Rank 1 in R): This appears as Rank 1 in the context of performance integration, which means effective integration of sustainability issues into existing supply chain operations is of high importance (de Lima et al. 2024). A crisp value of 0.013 means it is relatively manageable in comparison with other challenges, hence establishing the baseline that alignment to the goals of sustainability and performance metrics goes hand-in-hand (Hernandez Marquina et al. 2022; Roy et al. 2022).

Regulatory Barriers (Rank 5 in R, 6 in S, 5 in Q): The regulatory barriers will find their ranking more or less always in the middle tier for all the criteria in this paradigm shift towards CSCs, opined (Sharma et al. 2019). The three crisp values obtained are 0.103 in R, 0.276 in S and 0.162 in Q, which indicate that while there are huge challenges in regulations, they could be greater in social and economic contexts. This consequently indicates that organizations must traverse intricate regulatory landscapes to guarantee adherence to legal mandates while concurrently advancing the sustainability objectives (Kazancoglu et al. 2020).

Technological Constraints (Rank 4 in R, 4 in S, 4 in Q): Because of close ranking on each and every parameter, technology constraint emerges as a very important challenge to the implementation of the CSCs (Agrawal et al. 2023a, b). The crisp values 0.083 for R, 0.112 for S and 0.095 for Q also represent the new investment

Table 20.10 Results of the conditions survey

Condition 1	Non acceptance
Condition 2	–
Selected solution	Solution 1

required in technology and innovation to make the Circular Practices real. According to Bressanelli et al. (2019), such constraints need to be dealt with to raise efficiency and effectiveness in supply chain operations.

Financial Constraints: It was ranked 6 in R and 5 in S, 6 in Q. This challenge will show almost the same ranking in the criteria thus indicating that the cost involved in the change in CSCs is of general concern (de Lima and Seuring 2023). The crisp values of 0.137 in R, 0.261 in S and 0.189 in Q indicate that though financial difficulties are important, they seem to be less crucial in the social context probably because of the benefits from circular practice in long-term economic potentials (Fulconis et al. 2019).

Resistance to change is ranked 7 in R, 8 in S and 7 in Q: Even though it is one of the lowest-ranked challenges for humanising CSCs, it is still one of the relevant concerns (Franco 2017). Such crisp values, at 0.158 in R, 0.314 in S and 0.19 in Q, point toward cultural and organizational inertias as barriers to adopting circular practices, particularly for the social and environmental dimensions. Looking at this resistance means raising a culture where sustainability is an integral part of one's life (Werning and Spinler 2020).

Lack of Standardization always ranked low for all criteria, which highlights the challenges organizations face in standardizing practices and metrics for CSCs. Crisp values range acutely low at 0.184 in R, 0.28 in S and 0.206 in Q, emphasizing more generally that standards should be developed at the industry level. Those will encourage collaboration and ensure effective measurements of circularity (Agrawal et al. 2023a, b).

Cross-Sector Collaboration: This is very highly ranked across all criteria, thus showing that the collaboration of the different sectors is very important in the success of CSC (Clarke and MacDonald 2019). The crisp values are 0.072 in R, 0.084 in S and 0.027 in Q, respectively, thus it indicates that creating a partnership and knowledge sharing among industries can be effective to the circular practices success (Rey-Garcia et al. 2021).

Changing Relationships (Rank 5 in R, 9 in S, 9 in Q): In the social and economic contexts, Crisp values are: 0.207 in R, 0.338 in S and 0.248 in Q. Changing relationships ranks lower as CSCs requires reframing of almost every nature of the traditional supplier and customer relationships (Calicchio Berardi and Peregrino de Brito 2021).

Development of Open Loops (Rank 2 in R, 1 in S, 2 in Q): Probably the high ranking of this challenge in the social context could mean that creating open loops, through which resources can be shared and reused again in other sectors. This can be seen as an essential opportunity to achieve improvements in sustainability (Batista et al. 2018). Crisp values of 0.066 in R, 0.06 in S and 0.012 in Q suggest that such open system development is able to make a significant contribution to the success of CSC.

Societal rankings differ and are discovered on varying principles that underscore the importance of including economical, societal and environmental consideration for catering into such complexities. Enterprises need to drive strategies entailing collaboration, technology investments and navigation of the regulatory environment while overcoming the cultural barriers to change concernedly (Ellis-Hill et al. 2022). In this way, they can make effective shifts to CSCs that reduce waste and contribute to broader sustainability objectives.

20.5 Conclusion

The significance of CSCs is a major drive that is related to sustainability because it takes into account the synergy of factors regarding environment, social and economy. The Fuzzy VIKOR approach aids the identification of challenges such as changing relationships, resistance to change and lack of standardisation. Fuzzy VIKOR methodology identifies resistance towards change as a potential problem to humanise CSCs, with impacts on circular practices due to organisational culture. The dearth of established measures and indicators do not allow advancements to measure progress and maintain circular practices. The shift is complex due to limitations in technology, financial resources, which requires further investment in innovative solutions. Engaging new partners in creating effective CSCs would be required in cultivating collaboration and adapting to the new relationships. Stakeholder collaboration across the sector is the way to deliver shared sustainability goals. A combination of strategies on planning, engagement and innovation will allow organisations to humanise CSCs. Overcoming challenges and operationalizing best practices will help to strengthen sustainability efforts towards a CE. Future research may consider circularity metrics, advanced innovative technologies and stakeholder involvement. Good case studies will provide lessons for organizations that strive to have humanised CSCs. The complexity of humanising the CSC is intrinsic to the creation of a sustainable platform that promotes social equity and economic viability.

References

Agrawal D, Dwivedi A, Patil A, Paul SK (2023a) Impediments of product recovery in circular supply chains: implications for sustainable development. Sustain Dev 31(3):1618–1637. https://doi.org/10.1002/sd.2472

Agrawal R, Surendra Yadav V, Majumdar A, Kumar A, Luthra S, Arturo Garza-Reyes J (2023b) Opportunities for disruptive digital technologies to ensure circularity in supply chain: a critical review of drivers, barriers and challenges. Comput Ind Eng 178:109140. https://doi.org/10.1016/j.cie.2023.109140

Aminoff A, Kettunen O (2016) Sustainable supply chain management in a circular economy: towards supply circles. In: Sustainable design and manufacturing 2016. Springer, pp 61–72. https://doi.org/10.1007/978-3-319-32098-4_6

Amiri M, Hashemi-Tabatabaei M, Ghahremanloo M, Keshavarz-Ghorabaee M, Zavadskas EK, Salimi-Zavieh SG (2022) Evaluating barriers and challenges of circular supply chains using a decision-making model based on rough sets. Int J Environ Sci Technol 19(8):7275–7296. https://doi.org/10.1007/s13762-021-03899-7

Bals L, Taylor KM, Rosca E, Ciulli F (2024) Toward a circular supply chain: the enabling role of information and financial flows in open and closed loop designs. Resour Conserv Recycl 209:107781. https://doi.org/10.1016/j.resconrec.2024.107781

Batista L, Bourlakis M, Smart P, Maull R (2018) In search of a circular supply chain archetype: a content-analysis-based literature review. Prod Plan Control 29(6):438–451. https://doi.org/10.1080/09537287.2017.1343502

Bimpizas-Pinis M, Calzolari T, Genovese A (2022) Exploring the transition towards circular supply chains through the arcs of integration. Int J Prod Econ 250:108666. https://doi.org/10.1016/j.ijpe.2022.108666

Bressanelli G, Perona M, Saccani N (2019) Challenges in supply chain redesign for the circular economy: a literature review and a multiple case study. Int J Prod Res 57(23):7395–7422. https://doi.org/10.1080/00207543.2018.1542176

Calicchio Berardi P, Peregrino de Brito R (2021) Supply chain collaboration for a circular economy: from transition to continuous improvement. J Clean Prod 328:129511. https://doi.org/10.1016/j.jclepro.2021.129511

Chari A, Niedenzu D, Despeisse M, Machado CG, Azevedo JD, Boavida-Dias R, Johansson B (2022) Dynamic capabilities for circular manufacturing supply chains: exploring the role of Industry 4.0 and resilience. Bus Strateg Environ 31(5):2500–2517. https://doi.org/10.1002/bse.3040

Chen L, Shen Q, Yu X, Chen X (2024) Knowledge spillovers along the sustainable supply chain of China's listed companies: the role of long-term orientation. J Innov Knowl 9(2):100478. https://doi.org/10.1016/j.jik.2024.100478

Clarke A, MacDonald A (2019) Outcomes to partners in multi-stakeholder cross-sector partnerships: a resource-based view. Bus Soc 58(2):298–332. https://doi.org/10.1177/0007650316660534

Czifra G, Molnár Z, Mĺkva M, Szabó P (2023) Dehumanization and humanization in the context of industry 4.0 and industry 5.0. Res Pap Fac Mater Sci Technol Slovak Univ Technol 31(52):10–20. https://doi.org/10.2478/rput-2023-0002

Dacre N, Yan J, Frei R, Al-Mhdawi MKS, Dong H (2024) Advancing sustainable manufacturing: a systematic exploration of Industry 5.0 supply chains for sustainability, human-centricity, and resilience. Prod Plan Control. https://doi.org/10.1080/09537287.2024.2380361

Davies J, Sharifi H, Lyons A, Forster R, Elsayed OKSM (2024) Non-fungible tokens: the missing ingredient for sustainable supply chains in the metaverse age? Transp Res Part E Logist Transp Rev 182:103412. https://doi.org/10.1016/j.tre.2024.103412

de Lima FA, Seuring S (2023) A Delphi study examining risk and uncertainty management in circular supply chains. Int J Prod Econ 258:108810. https://doi.org/10.1016/j.ijpe.2023.108810

de Lima FA, Seuring S, Sauer PC (2022) A systematic literature review exploring uncertainty management and sustainability outcomes in circular supply chains. Int J Prod Res 60(19):6013–6046. https://doi.org/10.1080/00207543.2021.1976859

de Lima FA, Seuring S, Genovese A (2024) How to enhance circular supply chains? aligning R-imperatives, uncertainty management and sustainability. Int J Oper Prod Manag 44(4):836–858. https://doi.org/10.1108/IJOPM-11-2022-0708

DiVito L, van Wijk J, Wakkee I (2021) Governing collaborative value creation in the context of grand challenges: a case study of a cross-sectoral collaboration in the textile industry. Bus Soc 60(5):1092–1131. https://doi.org/10.1177/0007650320930657

Dwivedi A, Paul SK (2022) A framework for digital supply chains in the era of circular economy: implications on environmental sustainability. Bus Strateg Environ 31(4):1249–1274. https://doi.org/10.1002/bse.2953

Earley R, Hornbuckle R (2017) Face-ing collaboration: a meditation on the faces of circular textile research. J Text Des Res Pract 5(2):85–109. https://doi.org/10.1080/20511787.2018.1478703

Edwin Cheng TC, Kamble SS, Belhadi A, Ndubisi NO, Lai K, Kharat MG (2022) Linkages between big data analytics, circular economy, sustainable supply chain flexibility, and sustain-

able performance in manufacturing firms. Int J Prod Res 60(22):6908–6922. https://doi.org/10.1080/00207543.2021.1906971
Ellen MacArthur Foundation (2015) Growth within: a circular economy vision for a competitive Europe. https://www.ellenmacarthurfoundation.org/publications. Accessed 23 Aug 2023
Ellis-Hill C, Pound C, Galvin K (2022) Making the invisible more visible: reflections on practice-based humanising lifeworld-led research—existential opportunities for supporting dignity, compassion and wellbeing. Scand J Caring Sci 36(4):1037–1045. https://doi.org/10.1111/scs.13013
Farooque M, Zhang A, Liu Y (2019) Barriers to circular food supply chains in China. Supply Chain Manage Int J 24(5):677–696. https://doi.org/10.1108/SCM-10-2018-0345
Fosso Wamba S (2022) Humanitarian supply chain: a bibliometric analysis and future research directions. Ann Oper Res 319(1):937–963. https://doi.org/10.1007/s10479-020-03594-9
Franco MA (2017) Circular economy at the micro level: a dynamic view of incumbents struggles and challenges in the textile industry. J Clean Prod 168:833–845. https://doi.org/10.1016/j.jclepro.2017.09.056
Friedman N, Ormiston J (2022) Blockchain as a sustainability-oriented innovation? Opportunities for and resistance to Blockchain technology as a driver of sustainability in global food supply chains. Technol Forecast Soc Chang 175:121403. https://doi.org/10.1016/j.techfore.2021.121403
Fulconis F, Pache G, Reynaud E (2019) Frugal supply chains: a managerial and societal perspective. Soc Bus Rev 14(3):228–241. https://doi.org/10.1108/SBR-06-2018-0059
Gebhardt M, Beck J, Kopyto M, Spieske A (2022) Determining requirements and challenges for a sustainable and circular electric vehicle battery supply chain: a mixed-methods approach. Sustain Prod Consum 33:203–217. https://doi.org/10.1016/j.spc.2022.06.024
Grabowska S, Saniuk S, Gajdzik B (2022) Industry 5.0: improving humanization and sustainability of Industry 4.0. Scientometrics 127(6):3117–3144. https://doi.org/10.1007/s11192-022-04370-1
Grosse EH, Sgarbossa F, Berlin C, Neumann WP (2023) Human-centric production and logistics system design and management: transitioning from Industry 4.0 to Industry 5.0. Int J Prod Res 61(22):7749–7759. https://doi.org/10.1080/00207543.2023.2246783
Guest D, Knox A, Warhurst C (2022) Humanizing work in the digital age: lessons from socio-technical systems and quality of working life initiatives. Hum Relat 75(8):1461–1482. https://doi.org/10.1177/00187267221092674
Henry LA, Rasche A, Möllering G (2022) Managing competing demands: coping with the inclusiveness–efficiency paradox in cross-sector partnerships. Bus Soc 61(2):267–304. https://doi.org/10.1177/0007650320978157
Hernandez Marquina MV, le Dain M-A, Zwolinski P, Joly I (2022) Sustainable performance of circular supply chains: a literature review. Procedia CIRP 105:607–612. https://doi.org/10.1016/j.procir.2022.02.101
Hernandez Marquina MV, le Dain M-A, Joly I, Zwolinski P (2024) Exploring determinants of collaboration in circular supply chains: a social exchange theory perspective. Sustain Prod Consum 50:1–19. https://doi.org/10.1016/j.spc.2024.07.017
Hofstetter JS, de Marchi V, Sarkis J, Govindan K, Klassen R, Ometto AR, Spraul KS, Bocken N, Ashton WS, Sharma S, Jaeger-Erben M, Jensen C, Dewick P, Schröder P, Sinkovics N, Ibrahim SE, Fiske L, Goerzen A, Vazquez-Brust D (2021) From sustainable global value chains to circular economy: different silos, different perspectives, but many opportunities to build bridges. Circ Econ Sustain 1(1):21–47. https://doi.org/10.1007/s43615-021-00015-2
Holgado M, Aminoff A (2019) Closed-loop supply chains in circular economy business models. In: Sustainable design and manufacturing 2019. Springer, pp 203–213. https://doi.org/10.1007/978-981-13-9271-9_19
Hussain M, Malik M (2020) Organizational enablers for circular economy in the context of sustainable supply chain management. J Clean Prod 256:120375. https://doi.org/10.1016/j.jclepro.2020.120375

Iacovidou E, Hahladakis JN, Purnell P (2021) A systems thinking approach to understanding the challenges of achieving the circular economy. Environ Sci Pollut Res 28(19):24785–24806. https://doi.org/10.1007/s11356-020-11725-9

Jabłoński M, Timmers P, Sarkis J (2020) Sustainability in business models in the network economy. Electron Mark 30(4):675–678. https://doi.org/10.1007/s12525-020-00444-1

Kalverkamp M (2018) Hidden potentials in open-loop supply chains for remanufacturing. Int J Logist Manage 29(4):1125–1146. https://doi.org/10.1108/IJLM-10-2017-0278

Kalverkamp M, Young SB (2019) In support of open-loop supply chains: expanding the scope of environmental sustainability in reverse supply chains. J Clean Prod 214:573–582. https://doi.org/10.1016/j.jclepro.2019.01.006

Kazancoglu I, Kazancoglu Y, Yarimoglu E, Kahraman A (2020) A conceptual framework for barriers of circular supply chains for sustainability in the textile industry. Sustain Dev 28(5):1477–1492. https://doi.org/10.1002/sd.2100

Kazancoglu I, Sagnak M, Kumar Mangla S, Kazancoglu Y (2021) Circular economy and the policy: a framework for improving the corporate environmental management in supply chains. Bus Strateg Environ 30(1):590–608. https://doi.org/10.1002/bse.2641

Khan AA, Abonyi J (2022) Information sharing in supply chains: interoperability in an era of circular economy. Cleaner Logist Supply Chain 5:100074. https://doi.org/10.1016/j.clscn.2022.100074

Khan SAR, Zkik K, Belhadi A, Kamble SS (2021) Evaluating barriers and solutions for social sustainability adoption in multi-tier supply chains. Int J Prod Res 59(11):3378–3397. https://doi.org/10.1080/00207543.2021.1876271

Khan SAR, Piprani AZ, Yu Z (2022) Digital technology and circular economy practices: future of supply chains. Oper Manag Res 15(3–4):676–688. https://doi.org/10.1007/s12063-021-00247-3

Klitsie EJ, Ansari S, Volberda HW (2018) Maintenance of cross-sector partnerships: the role of frames in sustained collaboration. J Bus Ethics 150(2):401–423. https://doi.org/10.1007/s10551-018-3859-5

Koh SCL, Gunasekaran A, Morris J, Obayi R, Ebrahimi SM (2017) Conceptualizing a circular framework of supply chain resource sustainability. Int J Oper Prod Manag 37(10):1520–1540. https://doi.org/10.1108/IJOPM-02-2016-0078

Kopnina H, Hughes AC, Zhang RS, Russell M, Fellinger E, Smith SM, Tickner L (2024) Business education and its paradoxes: linking business and biodiversity through critical pedagogy curriculum. Br Educ Res J 50(6):2712–2734. https://doi.org/10.1002/berj.4048

Kumar P, Singh RK, Kumar V (2021) Managing supply chains for sustainable operations in the era of industry 4.0 and circular economy: analysis of barriers. Resour Conserv Recycl 164:105215. https://doi.org/10.1016/j.resconrec.2020.105215

Le TT (2022) How humane entrepreneurship fosters sustainable supply chain management for a circular economy moving towards sustainable corporate performance. J Clean Prod 368:133178. https://doi.org/10.1016/j.jclepro.2022.133178

Longoni A, Luzzini D, Pullman M, Seuring S, van Donk DP (2024) Social enterprises in supply chains: driving systemic change through social impact. Int J Oper Prod Manag 44(10):1733–1749. https://doi.org/10.1108/IJOPM-10-2023-0835

Luthra S, Mangla SK, Shankar R, Prakash Garg C, Jakhar S (2018) Modelling critical success factors for sustainability initiatives in supply chains in Indian context using Grey-DEMATEL. Prod Plan Control 29(9):705–728. https://doi.org/10.1080/09537287.2018.1448126

Luthra S, Sharma M, Kumar A, Joshi S, Collins E, Mangla S (2022) Overcoming barriers to cross-sector collaboration in circular supply chain management: a multi-method approach. Transp Res Part E Logist Transp Rev 157:102582. https://doi.org/10.1016/j.tre.2021.102582

Mahroof K, Omar A, Kucukaltan B (2022) Sustainable food supply chains: overcoming key challenges through digital technologies. Int J Product Perform Manag 71(3):981–1003. https://doi.org/10.1108/IJPPM-12-2020-0687

Mangla SK, Luthra S, Mishra N, Singh A, Rana NP, Dora M, Dwivedi Y (2018) Barriers to effective circular supply chain management in a developing country context. Prod Plan Control 29(6):551–569. https://doi.org/10.1080/09537287.2018.1449265

Marcon M, Provensi T, Sehnem S, Campos LMS, Queiroz AAFSL (2023) The internalisation of the circular economy and ESG in Brazilian B Corps from the perspective of the stakeholder theory. Sustain Dev 31(5):3513–3527. https://doi.org/10.1002/sd.2601

Miemczyk J, Howard M, Johnsen TE (2016) Dynamic development and execution of closed-loop supply chains: a natural resource-based view. Supply Chain Manage Int J 21(4):453–469. https://doi.org/10.1108/SCM-12-2014-0405

Mishra JL, Hopkinson PG, Tidridge G (2018) Value creation from circular economy-led closed loop supply chains: a case study of fast-moving consumer goods. Prod Plan Control 29(6):509–521. https://doi.org/10.1080/09537287.2018.1449245

Nandha Gopan S, Balaji M (2023) Indian automotive supply chains: barriers to circular economy for sustainable development. Manag Decis 61(11):3589–3609. https://doi.org/10.1108/MD-03-2023-0435

Nandi S, Sarkis J, Hervani AA, Helms MM (2021) Redesigning supply chains using blockchain-enabled circular economy and COVID-19 experiences. Sustain Prod Consum 27:10–22. https://doi.org/10.1016/j.spc.2020.10.019

Opricovic S, Tzeng G-H (2007) Extended VIKOR method in comparison with outranking methods. Eur J Oper Res 178(2):514–529. https://doi.org/10.1016/j.ejor.2006.01.020

Ozkan-Ozen YD, Kazancoglu Y, Kumar Mangla S (2020) Synchronized barriers for circular supply chains in industry 3.5/industry 4.0 transition for sustainable resource management. Resour Conserv Recycl 161:104986. https://doi.org/10.1016/j.resconrec.2020.104986

Pedersen ERG, Lüdeke-Freund F, Henriques I, Seitanidi MM (2021) Toward collaborative cross-sector business models for sustainability. Bus Soc 60(5):1039–1058. https://doi.org/10.1177/0007650320959027

Pego A, Lourenço J (2024) An AI analysis on the circular economy value chain: a Portuguese perspective of evaluation business models. In: Digital sustainability: inclusion and transformation. Springer, pp 55–68. https://doi.org/10.1007/978-3-031-57650-8_5

Pla-Julián I, Guevara S (2019) Is circular economy the key to transitioning towards sustainable development? Challenges from the perspective of care ethics. Futures 105:67–77. https://doi.org/10.1016/j.futures.2018.09.001

Rašković M, Nagar S (2024) Driving social sustainability through storytelling. In: The Palgrave handbook of social sustainability in business education. Springer Nature, pp 209–225. https://doi.org/10.1007/978-3-031-50168-5_12

Ren Y-S, Ma C-Q, Chen X-Q, Lei Y-T, Wang Y-R (2023) Sustainable finance and blockchain: a systematic review and research agenda. Res Int Bus Financ 64:101871. https://doi.org/10.1016/j.ribaf.2022.101871

Rey-Garcia M, Mato-Santiso V, Felgueiras A (2021) Transitioning collaborative cross-sector business models for sustainability innovation: multilevel tension management as a dynamic capability. Bus Soc 60(5):1132–1173. https://doi.org/10.1177/0007650320949822

Rogan J, Fürstenberg F, Wieland A (2022) Shaping the transition from linear to circular supply chains. In: Circular economy supply chains: from chains to systems. Emerald, pp 69–87. https://doi.org/10.1108/978-1-83982-544-620221004

Roy T, Garza-Reyes JA, Kumar V, Kumar A, Agrawal R (2022) Redesigning traditional linear supply chains into circular supply chains: a study into its challenges. Sustain Prod Consum 31:113–126. https://doi.org/10.1016/j.spc.2022.02.004

Saccani N, Bressanelli G, Visintin F (2023) Circular supply chain orchestration to overcome circular economy challenges: an empirical investigation in the textile and fashion industries. Sustain Prod Consum 35:469–482. https://doi.org/10.1016/j.spc.2022.11.020

Sehnem S, Chiappetta Jabbour CJ, Farias Pereira SC, de Sousa Jabbour ABL (2019) Improving sustainable supply chains performance through operational excellence: circular economy approach. Resour Conserv Recycl 149:236–248. https://doi.org/10.1016/j.resconrec.2019.05.021

Shang C, Saeidi P, Goh CF (2022) Evaluation of circular supply chains barriers in the era of Industry 4.0 transition using an extended decision-making approach. J Enterp Inf Manag 35(4/5):1100–1128. https://doi.org/10.1108/JEIM-09-2021-0396

Sharma YK, Mangla SK, Patil PP, Liu S (2019) When challenges impede the process. Manag Decis 57(4):995–1017. https://doi.org/10.1108/MD-09-2018-1056

Singh M, Singh K, Sethi A (2020) An empirical investigation and prioritizing critical barriers of green manufacturing implementation practices through VIKOR approach. World J Sci Technol Sustain Dev 17(2):235–254. https://doi.org/10.1108/WJSTSD-08-2019-0060

Soldatos J, Kefalakis N, Despotopoulou A-M, Bodin U, Musumeci A, Scandura A, Aliprandi C, Arabsolgar D, Colledani M (2021) A digital platform for cross-sector collaborative value networks in the circular economy. Procedia Manuf 54:64–69. https://doi.org/10.1016/j.promfg.2021.07.011

Sudusinghe JI, Seuring S (2022) Supply chain collaboration and sustainability performance in circular economy: a systematic literature review. Int J Prod Econ 245:108402. https://doi.org/10.1016/j.ijpe.2021.108402

Tura N, Hanski J, Ahola T, Ståhle M, Piiparinen S, Valkokari P (2019) Unlocking circular business: a framework of barriers and drivers. J Clean Prod 212:90–98. https://doi.org/10.1016/j.jclepro.2018.11.202

Vadakkepatt GG, Winterich KP, Mittal V, Zinn W, Beitelspacher L, Aloysius J, Ginger J, Reilman J (2021) Sustainable retailing. J Retail 97(1):62–80. https://doi.org/10.1016/j.jretai.2020.10.008

Vermunt DA, Negro SO, Verweij PA, Kuppens DV, Hekkert MP (2019) Exploring barriers to implementing different circular business models. J Clean Prod 222:891–902. https://doi.org/10.1016/j.jclepro.2019.03.052

Wang JX, Burke H, Zhang A (2022) Overcoming barriers to circular product design. Int J Prod Econ 243:108346. https://doi.org/10.1016/j.ijpe.2021.108346

Werning JP, Spinler S (2020) Transition to circular economy on firm level: barrier identification and prioritization along the value chain. J Clean Prod 245:118609. https://doi.org/10.1016/j.jclepro.2019.118609

Yadav G, Luthra S, Jakhar SK, Mangla SK, Rai DP (2020) A framework to overcome sustainable supply chain challenges through solution measures of industry 4.0 and circular economy: an automotive case. J Clean Prod 254:120112. https://doi.org/10.1016/j.jclepro.2020.120112

Zadeh LA (1965) Fuzzy sets. Inf Control 8(3):338–353. https://doi.org/10.1016/S0019-9958(65)90241-X

Chapter 21
Beyond Efficiency: The Human Touch in Circular Supply Chains and Management

Manjeet Singh, Deepshikha Bhargava, Kirti Singh, and Amitabh Bhargava

Abstract The concept of a circular economy has emerged as one of the most promising alternatives to traditional linear economic models considered as "take, make, and dispose." Human Circular Supply Chain Management introduces the philosophy of a circular economy within supply chain management, offering new insights into supply chain sustainability that have driven up the research interest in this area. However, the current literature lacks an integrated overview about HCSCM. This conceptual ambiguity makes it difficult to outline the hypothesis of how HCSCM differs from other supply chain models focused on sustainability and therefore hampers further development in the field. Addressing the challenge, this study classifies diverse terminologies related to supply chain sustainability and presents a unified definition of HCSCM. It also presents a structured analysis of the literature review based on 261 research works that analyze the current status of HCSCM research. From this review, identified key and emerging research directions include design for circularity, procurement in HCSCM, biodegradable packaging, circular supply chain collaboration and coordination, HCSCM drivers and barriers, circular consumption, product liabilities and producer responsibility, and the role of technology in HCSCM.

Keywords Supply chain management · Circular economy · Circular supply chain · Circular supply chain management · Sustainable supply chain · Sustainability · Sustainable sourcing

M. Singh · D. Bhargava (✉) · K. Singh · A. Bhargava
Amity University, Greater Noida, Uttar Pradesh, India

K. Mathiyazhagan et al. (eds.), *Humanizing Circular Supply Chain Management*, Circular Economy and Sustainability,
https://doi.org/10.1007/978-3-031-86562-6_21

21.1 Introduction

It has also produced much debate and contention in academic literature regarding sustainability. Hence, the contentions in SCM literature are likely not to be an exemption either (Seuring and Müller 2008). Nevertheless, the habits of production, consumption, and trade are catastrophically unsustainable internationally (Preston 2012). In fact, at the present global level of consumption, most of the natural resources would have been exhausted long back, and this has only happened because of sourcing, manufacturing, distributing, consumption, reclamation, and regeneration of products. A Circular Economy (CE) is one such theory that is strong enough actually to bring in this paradigm shift. It is identified to be superior in comparison to the existing model of linear economic—take-make-and-dispose. Research and practice of the CE philosophy are emerging as a dynamic force for sustainability. Stewart and Niero (2018) and Hobson (2016) have recognized the humongous potential of the concept to help enterprises break through sustainability performance. Ellen MacArthur Foundation (2014) promotes CE as an industrial system that by design, is restorative and regenerative. All within biological and technical cycles, CE aims to ensure the best serviceable quality of materials, components, and products and their value. In this way, it safely widens natural wealth by giving an assurance of returning biological components or nutrients to the biosphere. Nevertheless, the majority of research demonstrates excitement and growing interest in SCM for CE (Ying and Li-jun 2012; Batista et al. 2018a, b; Bressanelli et al. 2018; De Angelis et al. 2018; Darom and Hishamuddin 2016; Aminoff and Kettunen 2016). However, it is still in its early stage that SCM research has advanced on how it should move forward in developing CE theories and practices in creating supply chains that support the potential and aim of a CE.

Based on the integration of human sustainability ideas in supply chain management, several names have been proposed and used interchangeably for sustainable supply networks in literature, including green supply chains, environmental supply chains, and closed-loop supply chains (Farooque et al. 2019; Ahi and Searcy 2015). These three—representing different levels of supply chain integration of sustainable thinking—do not rationally integrate circular thinking—the essence of CE philosophy—into SCM. Several review articles published recently on integrating CE into SCM have a somewhat restricted scope, as is the case of Batista et al. Writings on sustainability in SCM and CE remain dispersed. The principles of CE align strategically with some reflections, whereas others are more concentrated on SCM operations such as manufacturing, design, and procurement. A handful of studies have used the term "circular supply chain" to make this linkage with CE and SCM: (Farooque et al. 2019; Nasir et al. 2017; De Angelis et al. 2018; Canning 2006; Mishra et al. 2018; Genovese et al. 2017). A practical definition of human circular supply chain management (HCSCM) has just lately been published in academic journals. A more recent research study provided the following definition of HCSCM: "The coordinated forward and reverse flows of products, services, by-products, and related waste streams through the supply chain in creating value towards ensuring organizational economic, social, and environmental sustainability via extended life cycles with a purposeful business ecosystem perspective for organizations." These developments must use a multi-level analytical framework for studying human and organizational behavior.

21.2 Unveiling the Language of Sustainable Supply Chains: Exploring HCSCM

To the best of our knowledge, it would represent one of the pioneering attempts at conceptualization and defining an integrated, comprehensive view of HCSCM, teasingly different from other sustainability concepts emanating from supply chain literature. This chapter aims, in Sect. 21.2.1, to categorize the existing supply chain sustainability concepts and explore their connection with the philosophy of the CE. Based on this analysis, Sect. 21.2.2 provides a working definition for HCSCM.

21.2.1 Deconstructing Sustainability: A Framework for Human Supply Chain Practices

Elkington's (2004) developed the triple bottom line (TBL) theory, considering that an organization primarily rests for survival on three significant factors—the natural environment, society, and overall economic performance. The theory interlinks with most of the sustainability concepts in SCM literature. Other terms are the likes of "green supply chains" (Srivastava 2007; Chakraborty 2010; Seman et al. 2012; Malviya and Ravi 2015), "closed loop supply chains" (Souza 2013), "environmental supply chains" (Darom and Hishamuddin 2016), and "sustainable supply chain management" (Seuring and Müller 2008; Craig and Easton 2011; Leszczynska and Maryniak 2017; Helen et al. 2016). The three dimensions of this philosophy have differences in their emphasis. For example, Ahi and Searcy (2013) examined 12 definitions of GSCM out of 124 articles and 22 definitions of SSCM out of 56 articles. They say that most SSCM definitions explicitly capture all three dimensions of TBL. However, in any definition, the social themes were not explicitly stated in the published formulations of GSCM.

21.2.2 Human Circular Supply Chain Management: Redefining Resource Management

The regenerative dimension of HCSCM significantly contributes to the effectiveness of both SSCM and GSCM. This is pegged on the fact that there is an expansion of the scope of the definition of sustainability due to the incorporation of circular

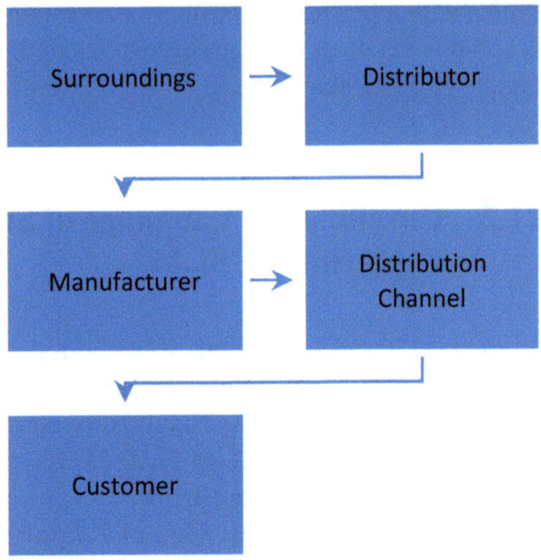

Fig. 21.1 Linear supply chain

thinking in EC into all supply chain operations. As earlier stated, HCSCM is concerned with both services and produced products, just like it is the case for CE. Companies work with partners not in their sector in HCSCM to maximize commodity value. With this clarity, supply chain managers may be able to set new standards of resource efficiency and, in the long run, profitability. This also helps reduce damage at the same time to the environment, society, and economy.

Figure 21.1 contrasts a traditional, linear supply chain to a human-centered, circular supply network approach. Most linear supply chain methods extract an item from the earth, transform and use it, and then it becomes waste—all too often deposited into landfills. A circular supply chain is one that is designed to maximize the efficiency of resource use by recovering, recycling, and reusing materials and by minimizing waste. For instance, closed-loop remanufacturing of photocopiers may recover 20–70% of materials, labor, and energy, and reduce 35–50% of generated wastes compared to virgin manufacturing (Toffel 2004). However, the value recovery potentiality in a closed-loop supply chain is constrained by some factors, such as product design, infrastructure, and consumer behaviors.is frequently constrained as operations are restricted to the producer's initial supply chain and do not involve additional auxiliary channel members or secondary supply networks (Moula et al. 2017). Reusing or recycling every undesired item inside the same supply chain is seldom feasible, therefore closed-loop supply chains nonetheless generate a lot of trash. A closed-loop supply chain goes one step further by reclaiming value from trash via partnerships with other companies operating in the same industry (open loop, same industry) or with companies operating in different industries (open loop, cross-sector) (Weetman 2017) (Figs. 21.1 and 21.2).

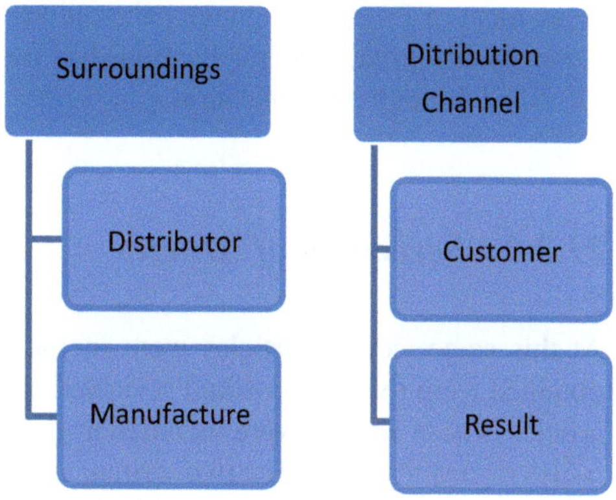

Fig. 21.2 Circular supply chains

21.3 Modus Operandi

A structured literature review was conducted in which an overview was given about the current position of HCSCM in academic research, whereby an analysis is provided. The search procedure and selection of the articles followed the same method as (Seuring and Müller 2008; Harland et al. 2006; Mayring 2003). The subsections below explain the approach used to source and screen the articles, analyze them, and describe sample characteristics.

21.3.1 Source of Article

It is also the case that Google Scholar, Scopus, and Web of Science are the three principal abstract and citation bibliographic databases. Indeed, in the case of Google Scholar, there is doubt expressed about the fitness for purpose for general research use on account of the substandard data, largely unaccounted for by Meho and Yang (2007) and Mongeon and Paul-Hus (2016). In addition to that, Web of Science and Scopus give the best depth and breadth of their historical coverage, correspondingly. The merits of Web of Science databases in accessing earlier materials hardly apply to this case since we are talking about studying a relatively recent event. Our focus points at Scopus. On average, there is an estimation that about 5% of papers covered by Web of Science are not duplicated in Scopus. On the other hand, Farooque et al. (2019) highlighted that Scopus and Web of Science's unique rate is around 50%. To ensure that not a single paper of any publisher is left out, complete text databases EBSCO, Elsevier, ProQuest, Sage, Springer, Taylor & Francis, and Wilson were used in conjunction with each other. In this review, all papers published up to 2018 were taken into consideration. The search was limited to only "Articles," "Articles in the press," and "Review articles" published in peer-reviewed journals to ensure the quality of the material and to reduce the number of articles selected to a manageable amount. In one sense, this can be considered a limitation because, with our limited language skills, the assessment of English-language publications was carried out by the team of authors alone. At this point, 2987 publications had been identified. A further 1748 items were identified through de-duplication.

21.3.2 Screening of Articles

At this stage, several articles were included and excluded based on the abstract obtained from the database and reviewed abstracts of each original sample of 1748 articles. These articles were included as they touch on some of the other CE-related issues regarding SCM. The two writers and researchers completed most of the analysis. The abstracts were reviewed by both researchers, and the outcomes were contrasted. All discrepancies in interpretation were resolved through dialogue until a consensus was achieved among the investigators. A backlog was created for any case for which an early judgment could not be made with certainty. The researchers worked through the backlog through thorough dialogue; where they had disagreements, they took the cautious approach of inclusion. This was bound to happen given the extensive work in HCSCM that takes one through many issues. It would have been somewhat challenging to develop precise inclusion or exclusion criteria, except that an article may foreground CE and look at SCM either at the corporate level or micro-level.

21.3.3 Analysis of Articles

This was essentially a data extraction process where the information from the 261 articles was preserved and abstracted. To lessen subjectivity in this process, the authors reviewed the data again and we convened often to settle any more disagreements that surfaced throughout the examination of the results. Our main study approach was content analysis (Krippendorff 2004). We had many meetings to talk about problems and ask questions about unclear areas to make sure we didn't overlook anything important. We collected the data using a straightforward matrix template, with questions derived from each study as follows:

- From a sustainability standpoint, what portion(s) of CE are integrated into SCM or the value chain in any other way?
- What component(s) of CE are integrated into SCM operations?
- The article discusses circular business models.
- How does technology fit into the integration of CE in SCM?
- Which industry sector was the focus of this?
- Which nation served as the research site?
- How was the analysis/research conducted?
- What were the findings, conclusions, and recommendations for the immediate and far future? (Fig. 21.3)

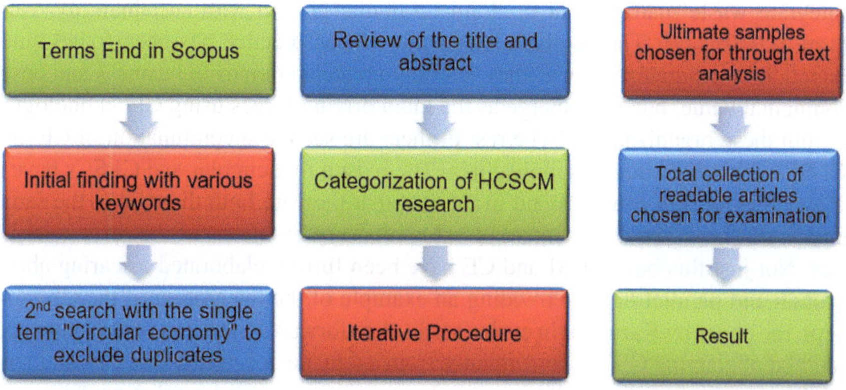

Fig. 21.3 Methodical means of reviewing material

21.4 Evaluation of Outcomes

Of the total number of papers published, the majority can be divided into two main categories as a result of HCSCM. In terms of sustainability, first classification emphasizes standard integration of the concept at a wide level of supply chain management and value chains. This represents that, as a whole, 60 papers were analysed. The body of research with CE philosophy integration at the supply chain management functional level comprises the second major category. Besides, one of the other HCSCM categories is Coupled 121, which, including subcategories from various functional domains and business models, accounts for more than 46% of the studied article total. These are subcategories with 57 and 12 papers respectively, which amount to 25% and 4% of the total number of documents researched respectively.

21.4.1 Supply Chain Management

To explore this area further, Masi et al. (2017) identified three categories of works under the category of circular supply chains that include: Ecoindustrial Parks (EIP), Environmental, Sustainable, and Green Systems, and Closed-loop Supply Chains. Even though EIP is a transient step toward a circular economy, the focus in this paper will be on the last two categories because they exactly fall under the sustainability feature of HCSCM.

Recent research, such as that by De Angelis et al. (2018), compared traditional, sustainable, and circular SCM approaches. Also, the idea of restorative or regenerative circularity in supply chains was discussed by Batista et al. (2018a), which also comes into focus by Winkler and Kaluza (2006). They believe that the Sustainable Supply Chain Networks should be designed and developed in such a way to integrate waste management to attain economic growth with sustainability. Park et al. (2010) say: "The combination of this SSCM practice has been found quite beneficial for organizations as they not only create a composite business and environmental value, but also emerge as the main driving forces using CE technologies within these organizations." The rest of them are seen as a combination of CE and CLSC within the circular supply chain. The reason that circular and CLSCs have a closer relationship can be found in the fact that they are both directed at the same end-in this case, the achievement of value recovery by means of the reverse logistics. Not just this but GSCM and CE have been further elaborated as caring about mutual and parallel fields. Providing an example of the integration of the CE concept, the paper by Kazancoglu et al. (2018). Puts forward a new and better appraised method of gauging GSCM performance from a total perspective where environmental, economic, logistic, and operational, organizational, and marketing flows are included.

21.4.2 Weaving Circularity Throughout the Supply Chain

A shift towards Circular Economy should entail fundamental transformations in the areas of business models and supply chains, as well as in the way the products and services are designed, manufactured, and consumed, the waste is handled, and finally the goods are reused, and recycled. The flows of logistics, however, are the factors that have an impact in every step of a supply chain. For this reason, the activities and chains of companies have undergone distinct micro-level CE practices in some companies. These practices were based on eco-design or green design, green purchasing, cleaner production, and EoL management as per the 3R principles. An equally interesting insight was uncovered by Masi et al. (2017) that organizations that were already familiar with CP were better prepared to subtly integrate it within the final performance measure of the green organization. Their research has revealed that apart from the predominant technical achievements, engineers are convinced that citizens should assume collective responsibility, by opting for sustainable, circular systems or through waste reduction. This is evident from the fact that Farooque et al. (2019), when as part of their study, grouped the already implemented cases of CE, concluded, that in high proportion, the parts as 'Recovery, Consumption and Use' were received the limelight. However, parts like 'Manufacturing, Distribution, and sales' were thinly covered under CE operation. Specifically, the novel concepts of recovery, consumption, and use, and so forth are emphasized, which therefore transform into the high potential of creeping-in areas in the social-technical system as well as the new sustainability values.

21.4.3 Circular Economy: Fueling Innovative Business Models

Again, linear economic models are unable to address the pervasive environmental challenges of today. These factors have led to the emergence of new business models founded in the CE philosophy. According to Nußholz (2017), a circular business model is "a company's value creation logic that is configured to resource efficiency by extending the useful life of products and parts through long-life design, repair, and remanufacturing and closing." Material loops (p. 12). Linder and Williander (2017) illustrate the creation logic in this context of a circular business model as "the utilization of economic value retained in goods after use in the production of new offerings" (p. 2). Various research contributions work towards the opening of the CBM; however, it was Roos (2014) who first conceptualized the innovative process development of CBM before posing specific questions to design an adequate business model in the circular value chain. Lüdeke-Freund et al. (2019) conducted a morphological analysis for 26 CBMs from existing literature in order to identify multiple options for business model design. In this paper, they suggest six significant patterns for closing resource loops. Further to this, Bocken et al. (2017) and Farooque et al. (2019) elaborated on how established businesses could experiment with their business model in order to meet goals of sustainability and circularity. There have been several independent business model frameworks developed within the extant literature. Lewandowski expanded the traditional business model canvas by incorporating take-back systems and adoption factors, creating a more comprehensive framework for designing circular economy (CE) business models. Additionally, a new framework called "Backcasting and Eco-design for the Circular Economy" (BECE) was developed to assist businesses in creating viable and sustainable business models that effectively implement CE principles in industrial settings.

Further, put in place for application from both product and service-oriented businesses is the BECE framework. The incentive worked the same when tested in favor of both types of businesses, for example, by Urbinati et al. (2017) and Heyes et al. (2018) also created a taxonomy of business models that would be applicable under the Circular Economy, which would allow for some firms that did adopt some of the innovations to take them toward being more cost-effective for adopting CE. Their business model canvas for the circular economy (CE) integrates circularity along two key dimensions: the Customer Proposition and Interface Value, which encompasses the customer value proposition and its delivery; and the Value Network, which focuses on supplier interactions and internal restructuring to support circular practices. There is an evidence-based EVPT and stepwise evaluation approach to include in a CE business model, as recently discussed by Manninen et al. (2018).

21.5 Shaping the Future: Research Directions for Sustainable and Inclusive Circular Systems

The review and analysis above indicate that the HCSCM area is new and still evolving; therefore, most relevant publications are at the level of conceptual or case study work, characteristic for a research domain at an embryonic level of development. Except for supply chain performance and EoL product management, a few research topics have received more attention than others. In fact, in all the supply chain functions, much research is yet to be done before realizing the full potential of HCSCM. This section concludes the current awkward work on various technical and process issues and is followed by related work on incentives to convert CE into realization. We, therefore, call for research in the following important, though much too little or no attention, directions critical for HCSCM:. The latter will be as follows: As CE is an emerging area of thought about sustainability, we believe that the subsequent developments in the areas of HCSCM shall dramatically improve SSCM and GSCM to an organization's better performance concerning sustainability.

21.5.1 Designing for Dematerialization: Minimizing Waste from the Start

Therefore, no one can deny that HCSCM requires a redesign of supply chains, products, and processes. Design for Circularity is the foundation of HCSCM. Therefore, some of the research opportunities that still exist include: supply chain design for EoL management, process innovation led by CE, and new methods and techniques for product design such as DFD (Ijomah et al. 2007), design for recycling (Gaustad et al. 2010) and design for remanufacturing (Farooque et al. 2019).

21.5.2 Sourcing for Sustainability: The Role of Procurement in Human Circular Supply Chains

Procurement is, in fact, one of the core strategic functions of many organizations. It is a function that is truly fundamental to a firm's sustainability performance. However, research covering procurement development of circular and thinking approaches is perhaps even more shocking than most other changes in thinking supply chain functions. Based on Connelly and Koenig, products under HCSCM must have new or more robust features-supports, durability, reliability, and reusability that tend to extend their life cycle, enable easy resource recoveries, and minimize wastages. An element that needs further research so that performance indicators are integrated into procurement and supplier management that are more oriented toward CE—in that the indicators are integrated into procurement and management of suppliers—can be done to diminish the impacts that products and services imprint on the environment through their life cycle.

21.5.3 Unlocking Circular Potential: The Power of Collaborative Networks in Sustainable Supply Chains

The remaining waste from one process or supply chain would be used as an input resource by another process or supply chain in a CE. Long-term collaboration is required among the different supply chains and with the supply chain partners. To maintain a flow of used goods, parts, and materials, critical areas for research on incentives and strategic value alignment were outlined by (Genovese et al. 2017). They also discussed how all these could materialize in terms of means for collaboration and coordination with suppliers, customers, and other stakeholders. Stewart and Niero (2018), Aminoff and Kettunen (2016), and Grimm et al. (2016), among others, advocated the retention of features such as knowledge management points and interactions between suppliers and customers.

21.5.4 Forces for Change: Catalysts and Challenges in Human Circular Supply Chain Management

There are challenges in the IT and electronics industries of China, the textile industry in Finland, and most recently, in the retailing industry in Finland. Yet, there are very few research studies conducted in contexts other than driver/bearer identification or relevance of applicability. Therefore, studies must be carried out on how industrial sector-specific and cross-cultural settings change the drivers and barriers of HCSCM. Also, to develop effective intervention policies to prevent or overcome these causes and barriers, context-specific research is needed to prioritize them in order of importance.

21.5.5 Technological Enablers for Sustainable Supply Chains

Enable technologies are one such technology that could help sustainable development. Their roles in HCSCM are all that is left. In that area, there has been rather insufficient study conducted. Recently, a special issue of the Journal of Cleaner Production entitled "Improving Natural Resource Management and Human Health to Ensure Sustainable Societal Development Based upon Insights Gained from Working within 'Big Data Environments'" has been published. Review of the 3Rs in China Circular Economy and Trash Avoidance. But no article was included where subjects were circular thought in big data analytics! Some research directions within the extensive data analytics domain for HCSCM include. With such promising technology, 3D printing has become a solution for the high efficiency and cheap cost of customized production. According to Helen et al. (2016) and Despeisse et al. (2017), the increasing variants of products are giving a proportionally short life span of customized products. Hence, researchers need to study the challenges of CE further. Improving the management of life cycle information was facilitated in HCSCM using the Internet of Things (IoT) and Radio Frequency Identification (RFID). These technologies also leave ample opportunity for better traceability improvements. Besides, the driving need for the CE principles to be more tightly embedded in an organization's information systems was also emphasized by Jensen and Remmen (2017).

21.6 Conclusion

In view of the increasing pace of CE principles, they are increasingly considered a better alternative to the traditional linear economy and hold tremendous potential to increase sustainability within enterprise boundaries. In turn, the integration of CE into supply chain management has generated increasing research interest. However, vagueness around HCSCM and strategic research objectives impeded the development of the field. Therefore, this study distinguishes HCSCM from broader concepts regarding sustainability and carries out a systematic literature review in order to gauge the status-quo in HCSCM. The study represents the potential of HCSCM in ways that, through applying principles of zero-waste generation, shifting supply chain functions toward restorative and regenerative processes can be focused on closed and open loops with effective business models. Based on these insights, the study offers future research directions and practical recommendations for academics, practitioners, and policymakers in integrating CE into supply chains, thus bettering the field of supply chain management and science. Most notably, this study has partial drawbacks pertaining mainly to the limitation in yielding English sources only, for which a number of important studies published in other languages were thus missed. Though the study reviewed a wide range of literature, including books, conference papers, and industry reports, only a selection could be included in the structured literature review in order to ensure quality in the analysis. The field of HCSCM is developing quickly. Therefore, this assessment of the literature need to be updated every few years to reflect developments in a relevant field of study. We hope that this assessment of the research will increase the urgency with which we must develop post-fossil carbon societies that are livable, sustainable, and egalitarian. In order to continue developing this exciting field of study, we thereby welcome any and all reader comments and input.

References

Ahi P, Searcy C (2013) A comparative literature analysis of definitions for green and sustainable supply chain management. J Clean Prod 52:329–341

Ahi P, Searcy C (2015) An analysis of metrics used to measure performance in green and sustainable supply chains. J Clean Prod 86(Suppl. C):360–377

Aminoff A, Kettunen O (2016) Sustainable supply chain management in a circular economy towards supply circles. In: Setchi R, Howlett RJ, Liu Y, Theobald P (eds) Sustainable design and manufacturing 2016. Springer International Publishing, Cham, pp 61–72

Batista L, Bourlakis M, Smart P, Maull R (2018a) In search of a circular supply chain archetypeea content-analysis-based literature review. Prod Plann Contr 29(6):438–451

Batista L, Gong Y, Pereira S, Jia F, Bittar A (2018b) Circular supply chains in emerging economiesea comparative study of packaging recovery ecosystems in China and Brazil. Int J Prod Res 57:7248–7268

Bocken NMP, Miller K, Weissbrod I, Holgado M, Evans S (2017) Business model experimentation for circularity: driving sustainability in a large international clothing retailer. Economic 2017(1):85–122

Bressanelli G, Perona M, Saccani N (2018) Challenges in supply chain redesign for the circular economy: a literature review and a multiple case study. Int J Prod Res:1–28

Canning L (2006) Rethinking market connections: mobile phone recovery, reuse and recycling in the UK. J Bus Ind Mark 21(5):320–329

Chakraborty S (2010) Concise chronological road map of evolving green supply chain management concepts: a review. IUP J Supply Chain Manag 7(4):7–25

Craig RC, Easton PL (2011) Sustainable supply chain management: evolution and future directions. Int J Phys Distrib Logist Manag 41(1):46–62

Darom NAM, Hishamuddin H (2016) Integration of economic and environmental aspects in sustainable supply chain management: a review. In: Proceedings of the 6th international conference on industrial engineering and operations management. Kuala Lumpur, Malaysia, pp 8–10

De Angelis R, Howard M, Miemczyk J (2018) Supply chain management and the circular economy: towards the circular supply chain. Prod Plann Contr 29(6):425–437

Despeisse M, Baumers M, Brown P, Charnley F, Ford SJ, Garmulewicz A, Knowles S, Minshall THW, Mortara L, Reed-Tsochas FP, Rowley J (2017) Unlocking value for a circular economy through 3D printing: a research agenda. Technol Forecasting Soc Change 115(1):75–84

Elkington's, J., 2004. Enter the triple bottom line. The triple bottom line: Does it all add up. 11 (12):1–16 Routledge

Ellen MacArthur Foundation (2012) Towards the circular economy. https:// www.ellenmacarthurfoundation.org/publications/towards-the-circulareconomy-vol-1-an-economic-and-business-rationale-for-an-accelerated-transition. Accessed 14 Mar 2018

Farooque M, Zhang A, Thürer M, Qu T, Huisingh D (2019) Circular supply chain management: a definition and structured literature review. J Clean Prod 228:882–900. https://doi.org/10.1016/j.jclepro.2019.04.303

Gaustad G, Olivetti E, Kirchain R (2010) Design for recycling. J Ind Ecol 14(2):286–308

Genovese A, Acquaye AA, Figueroa A, Koh SCL (2017) Sustainable supply chain management and the transition towards a circular economy: evidence and some applications. Omega 66:344–357

Grimm JH, Hofstetter JS, Sarkis J (2016) Exploring sub-suppliers' compliance with corporate sustainability standards. J Clean Prod 112(Part 3):1971–1984

Harland CM, Lamming RC, Walker H, Phillips WE, Caldwell ND, Johnsen TE, Knight LA, Zheng J (2006) Supply management: is it a discipline? Int J Oper Prod Manag 26(7):730–753

Helen R, Norbert B, Kulwant SP (2016) 3D printing services: classification, supply chain implications and research agenda. Int J Phys Distrib Logist Manag 46(10):886–907

Heyes G, Sharmina M, Mendoza JMF, Gallego-Schmid A, Azapagic A (2018) Developing and implementing circular economy business models in service oriented technology companies. J Clean Prod 177:621–632

Hobson K (2016) Closing the loop or squaring the circle? Locating generative spaces for the circular economy. Prog Hum Geogr 40(1):88–104
Ijomah WL, McMahon CA, Hammond GP, Newman ST (2007) Development of robust design-for-remanufacturing guidelines to further the aims of sustainable development. Int J Prod Res 45(18–19):4513–4536
Jensen JP, Remmen A (2017) Enabling circular economy through product stewardship. Proc Manuf 8:377–384
Kazancoglu Y, Kazancoglu I, Sagnak M (2018) A new holistic conceptual framework for green supply chain management performance assessment based on circular economy. J Clean Prod 195:1282–1299
Krippendorff K (2004) Content analysis: an introduction to its methodology. Sage
Leszczynska A, Maryniak A (2017) Sustainable supply chain—a review of research fields and a proposition of future exploration. Int J Sustain Econ 9(2):159–179
Linder M, Williander M (2017) Circular business model innovation: inherent uncertainties. Bus Strateg Environ 26(2):182–196
Lüdeke-Freund F, Gold S, Bocken NM (2019) A review and typology of circular economy business model patterns. J Ind Ecol 23(1):36–61
Malviya RK, Ravi K (2015) Green supply chain management (GSCM): a structured literature review and research implications. Benchmarking Int J 22(7):1360–1394
Manninen K, Koskela S, Antikainen R, Bocken N, Dahlbo H, Aminoff A (2018) Do circular economy business models capture intended environmental value propositions? J Clean Prod 171:413–422
Masi D, Day S, Godsell J (2017) Supply chain configurations in the circular economy: a systematic literature review. Sustainability 9(9):1602
Mayring P (2003) Qualitative inhaltsanalyse [Qualitative content analysis]. Qual Forsch 3:468–475
Meho LI, Yang K (2007) Impact of data sources on citation counts and rankings of LIS faculty: web of science versus Scopus and Google scholar. J Am Soc Inf Sci Technol 58(13):2105–2125
Mishra JL, Hopkinson PG, Tidridge G (2018) Value creation from circular economy-led closed loop supply chains: a case study of fast-moving consumer goods. Prod Plan Contr 29(6):509–521
Mongeon P, Paul-Hus A (2016) The journal coverage of Web of Science and Scopus: a comparative analysis. Scientometrics 106(1):213–228
Moula ME, Sorvari J, Oinas P (2017) Constructing a green circular society. https://helda.helsinki.fi/bitstream/handle/10138/231630/cbook2017(pdf).pdf?sequence=1. Accessed 15 Mar 2018
Nasir MHA, Genovese A, Acquaye AA, Koh SCL, Yamoah F (2017) Comparing linear and circular supply chains: a case study from the construction industry. Int J Prod Econ 183:443–457
Nußholz JLK (2017) Circular business models: defining a concept and framing an emerging research field. Sustainability 9(10):1810
Park J, Sarkis J, Wu Z (2010) Creating integrated business and environmental value within the context of China's circular economy and ecological modernization. J Clean Prod 18(15):1492–1499
Preston F (2012) A global redesign? Shaping the circular economy. Chatham
Roos G (2014) Business model innovation to create and capture resource value in future circular material chains. Resources 3(1):248–274
Seman NAA, Zakuan N, Jusoh A, Arif MSM, Saman MZM (2012) Green supply chain management: a review and research direction. Int J Manag Value Supply Chains 3(1):1–18
Seuring S, Müller M (2008) From a literature review to a conceptual framework for sustainable supply chain management. J Clean Prod 16(15):1699–1710. https://doi.org/10.1016/j.jclepro.2008.04.020
Souza GC (2013) Closed-loop supply chains: a critical review, and future research. Decis Sci J 44(1):7–38
Srivastava SK (2007) Green supply-chain management: a state-of-the-art literature review. Int J Manag Rev 9(1):53–80

Stewart R, Niero M (2018) Circular economy in corporate sustainability strategies: a review of corporate sustainability reports in the fast-moving consumer goods sector. Bus Strateg Environ 27(7):1005–1022

Toffel MW (2004) Strategic management of product recovery. Calif Manag Rev 46(2):120–141

Urbinati A, Chiaroni D, Chiesa V (2017) Towards a new taxonomy of circular economy business models. J Clean Prod 168:487–498

Weetman C (2017) A circular economy handbook for business and supply chains: repair, remake, redesign, rethink. Kogan Page, New York

Winkler H, Kaluza B (2006) Sustainable supply chain networks—a new approach for effective waste management. WIT Trans Ecol Environ 92:501–510

Ying J, Li-jun Z (2012) Study on green supply chain management based on circular economy. Phys Procedia 25(Suppl. C):1682–1688

Chapter 22
Enhancing Operational Excellence for Aeronautic Companies in MENA Region

Zoubida Benmamoun, Khaoula Khlie, Hayet Benhmida, and Mariam Elkhechafi

Abstract While implementing circular economy for supply chain management, operational efficiency becomes a major factor that affects the aeronautics firms that seek to reduce their wastage levels. This chapter focuses on the assessment of operating flows of aeronautic companies and uses both the Oliver Wight methodology and set of lean tools in the circular supply chain perspective. Through applying the principles of circular economy it is possible to evaluate current operational states against the intended objectives; this is achieved with the help of lean tools like Value Stream Mapping, 5S, Kaizen, and Kanban to point out the opportunities for improvement. The research builds knowledge regarding the match between sustainability operation strategies and supply chain excellence, as it is proven that having an integrated framework of practices is necessary. This chapter includes viable and specific prescriptions with recommendations for aeronautic organizations that seek to improve the applicability of circular supply chain techniques.

Keywords Circular supply chain · Operational efficiency · Lean tools · Aeronautics industry · Sustainability strategies

---

Z. Benmamoun (✉)
College of Engineering and Computing, Liwa University for Zoubida Benmamoun, Abu Dhabi, UAE
e-mail: zoubida.benmamoun@lc.ac.ae

K. Khlie · H. Benhmida
Faculty of Business, Liwa College, Abu Dhabi, UAE

M. Elkhechafi
LAREM Laboratory Groupe ISCAE, Casablanca, Morocco

K. Mathiyazhagan et al. (eds.), *Humanizing Circular Supply Chain Management*, Circular Economy and Sustainability,
https://doi.org/10.1007/978-3-031-86562-6_22

22.1 Introduction

Organizational efficiency is a basic goal for aeronautic firms, where accuracy, effectiveness, and constant enhancement are imperative factors for competitive benefit and sustenance. Operational excellence can thus be defined as a process of ongoing organizational development that is focused on the achievement of sustainable organizational objectives based on the efficient, effective and timely elimination of waste within the organization" (Dev et al. 2020; Chiarini and Kumar 2021).

The aeronautics field is inherently challenging due to supply chain relationships, compliance issues, as well as overall risk associated with transporting consumers and goods (Benmamoun et al. 2018). This definition of operational excellence goes further than the concept of efficiency because it helps to avoid waste and facilitate the effective use of resources, which is more effective than their consumption. The enhancement of supply chain management by incorporating circular economy practices is known as circular supply chain management (CSCM) and it is widely accepted as a process that can lead to sustainable development (Bag et al. 2020).

To comprehensively evaluate operational excellence, this chapter employs two complementary methodologies: lean manufacturing techniques including the Oliver Wight assessment tool and lean tools. It is implemented that the Oliver Wight methodology is a recognized approach that affords a business a structure for assessing the state of the organization and its operations. It focuses on ensuring that organizational activity is undertaken with the consideration of the strategic plan, and promoting the standardization of processes with good practice. It is claimed that it is useful for pointing out the differences between current working and best working, and in supporting the formalization of a scheme for improvement (Correll et al. 2021).

Similarly in this case, several of the lean tools are used to identify and improve the flow of value creating operations. The original principles of lean are rooted in the manufacturing context and are based on eliminating non-added value and its constant reduction through efficiency increments. Lean tools, Value Stream Mapping, 5S, Kaizen and Kanban are central to improving the discussions to lean methods in eliminating wastes, improving on process efficiency (Leksic et al. 2020). They are valuable because they offer applicable guidance on where waste might be lurking, how to create systems that cut out variation, and how to engage employees in the journey to using science to solve everyday problems.

The reveal of assessment results demonstrates to analyze considerable differences between the present state of functioning in comparison with set objectives, as a result, stresses the need for a systematic approach to achieving the company's goals in the sphere of operational performance. Therefore it provides important information but also key ideas and directions applicable in aeronautic companies that implement circular economy and operation excellence frameworks. The findings thus signal the need to adopt sustainability and efficiency, which are crucial elements that can be integrated in the smart context and further the objectives of circular supply chain management.

22.2 Literature Review

To enhance operational excellence for aeronautic companies in the MENA region, it is essential to consider various critical success factors and methodologies that have been identified in the literature. Abedelgadir and Mohammad (2022) call for the assessment of critical success factors for operational excellence in aviation. Knowledge of these factors can serve as a basis for enhancing the performance of the organization and achieve certain competitive advantage. Further, Revathi (2024) in discussing the strategic importance of quality management on operational improvement, emphasized that it is very relevant in achieving competitive advantage in various service and manufacturing industries.

Some of the matters aeronautic companies should pay most attention to include efficiency and operation performance. In addition, Resende (2023) proposes a facilitating structure on risk assessment in the exponent aviation environment, aiming at discussing how those integrated tool can improve the choices concerning risk management and reduction inside the industrial segment.

When considering the MENA region, this study gives an empirical review of operational excellence as outlined by Henríquez-Machado et al. (2021) to discuss sustainability through operational success from the perspective of an emergent country. It can be helpful to gain insight for aeronautic companies in the MENA region who are interested in using this paper's perspective to review their operational practices with regard to sustainability objectives. In addition, Sá et al. (2022) present the design of an excellence model incorporating the Shingo Model and sustainability, asserting that economic profitability is the starting point for operational excellence.

Efficiency in management activities as part of the concept of sustainable development is an important factor for aeronautic enterprises. Using the example of manufacturing, Wojtkowiak and Cyplik (2020) point out that operational excellence is in the heart of sustainable development concept: right-first-time processes and productivity, customer/market orientation. This perspective can help the aeronautic companies of the MENA to know how to implement sustainable practices into their overall business model.

However, to obtain operational effectiveness, the aeronautic companies can look toward other industries for useful lessons. Silva (2023) examines Lean Six Sigma in pharmaceutical manufacturing; he suggests that to find balance across the defined areas, and be quantifiably successful, a firm must demonstrate maturity in terms of its operation. For the purpose of the present paper, it is important to stress that this or a modified version of this approach can be adopted by aeronautic companies operating in the MENA region to improve their operational activities as well as select aspects of quality management.

Specifically, for aeronautic companies operating in the MENA where the environmental condition is different and competitive and technological advancement is an opportunity. Through infrastructure convergence, the aeronautic companies within the MENA region can improve their business operations and communication technology platforms.

Altogether, consolidating findings of the studies of critical success factors, quality management, risk analysis, sustainability, and technology can serve to improve the performance of businesses in the aeronautics industry of the MENA Countries. To achieve, and maintain operational excellence harmonization with a multidimensional model, which incorporates economic profitability, sustainability objectives, and technology progression will be on the spot in the ever-evolving aeronautic industry within the MENA region.

22.3 Methodology

This chapter employs two complementary methodologies to assess operational excellence in aeronautic companies: the Oliver Wight assessment as well as other lean tools. These methodologies are selected because each of these approaches offers a wide-ranging and realistic method of assessing and enhancing business activities. The Oliver Wight's assessment is one of the most widely recognized forms of assessments that help in develop in an end-to-end method in analyzing business processes and outputs. This occurs through focusing on the organization in its operational environment and the constant use of right method of working in the organization. The study is also done using the Oliver Wight ABCD Checklist for Operational Excellence where: Leadership, Planning and Control, People, Processes and Performance Measurement are analyzed and discussed to identify their strength and areas of improvement. Its aim is simply to determine that operational activities arc not optimally running, so that actionable plans can be created to close the identified gaps faster (Oliver Wight International 2010). Apart from Oliver Wight assessment the following lean tools are used to evaluate and improve processes within the operations. Lean has its roots in manufacturing and derives from the Toyota Production System and values the elimination of wastes and the focus on producing added value by eliminating sources of waste. Some lean tools that have been used in this assessment include Value Stream Mapping, 5S, Kaizen and Kanban.

Value Stream Mapping is employed to map and analyze the flow of material and information in the production process. This tool enables one to find out the system's constraints and inefficiencies because it shows the existing system and enables a user to plan for an enhanced future system (Ramani et al. 2021). The 5S methodology is used to understand and streamline the workplace, adding order and reducing the amount of non-added value activities. The 5S are named Sort, or organizing, Set in order –neatness, Shine –cleaning, Standardize and Sustain. This instrument assists in developing more cleanliness, order, and safety in workplace to have better output or quality enhancement (Balinado and Prasetyo 2020). Kaizen which implies continuously improving organizational culture is employed in the organization. This tool defines how to make gradual changes in approaches that, in turn, establishes sustainable and large changes over a long period of time (Rusdiana and

Soediantono 2022). Kanban is a method of work that employs the visualization of an organization's work process in order to limit the procurement of work and to improve the workflow of material and information. This tool employs cards or boards to provide signals showing the condition of the work and its status in passing through the various phases of production (Jebbor et al. 2023).

By combining the comprehensive evaluation framework of the Oliver Wight assessment with the practical, hands-on tools of lean methodology, this chapter provides a thorough and actionable assessment of operational excellence in aeronautic companies. The integration of these methodologies offers valuable insights into the current state of operations, identifies areas for improvement, and provides a clear path for achieving operational excellence within the context of circular supply chain management.

22.3.1 Assessment of the Maturity of Improvement Practices

Based on the findings of the first part of the article, we identified shortcomings regarding the measurement and the evaluation of the practice of these tools in the company. In response to this issue, we provide guidelines and operating conditions to companies that want to implement operational excellence, in order to give them a methodological framework and real examples of enterprises that succeeded in optimizing their processes.

The study is primarily based on the use of measurement references raised from literature, to determine adequate tools for the activities and the maturity level of the company.

Company benchmarking is the most valuable aspect of the study that is conducted as follows: identifying the maturity level of the integration of Lean Six-Sigma tools in the company, then assessing by the ABCD reference of operational excellence, and finally reviewing green practices for the value chain starting with green purchasing and procurement, green production, green distribution, ending with green return management.

22.3.1.1 Assessment of Lean Maturity Level in the Company

Maturity assessment is a prerequisite to successful implementation of an effective system. It makes it possible to verify the implementation of good practices and then their effectiveness. Maturity levels must take into account the analysis of the implementation of good practices, their coverage and their sustainability. The rating grid must illustrate the progress of the implementation.

Maturity assessment can also be: a manner to make a direct or indirect evaluation of the management, a monitoring mode of the process, a means for raising awareness to good practices for process pilots, but above all, it enables to identify sources of improvement.

Likert Scale to Assess the Integration of Lean Tools

The Likert scale (Inexistent, Existing, Practiced, Mastered, Exemplary) (Jebb et al. 2021) is chosen to examine the integration level of Lean tools of firms covered by the study, to position the integration of the company of each of the tools, and to specify its degree of integration in a five-level ranking.

A questionnaire was presented to the companies in order to identify and validate the tools deployed, and their level of integration in each participating company. The list of LSS tools is as follows:

- SIPOC: definition of the business process, supplier, input, process, output and customers.
- Voice of the Customer (VOC): definition of the customer requirements and expectations.
- CTQ Diagram for monitoring customer satisfaction.
- Quality function deployment QFD: product design that guarantee customer satisfaction.
- Production process map: visual representation of the workflow.
- Elimination of waste: eliminating non-value-added activities for the customers.
- Spaghetti diagram: visualization of physical flows of pieces or individuals.
- Value Stream Mapping VSM: mapping of physical flows of information.
- Control sheet: Graphical presentation of the data on problems.
- FMECA: Failure Mode, Effects and Criticality Analysis.
- Pareto chart: a graph representing the importance of the various drivers of a phenomenon.
- Brainstorming: a creative group problem-solving technique, under the guidance of a facilitator.
- Five Whys: interrogative and iterative technique, used to explore cause-and-effect relationships of a problem.
- Ishikawa Diagram: A structured representation of the causes of a problem.
- 5S: an approach which continuously optimizes working conditions, working time, organization, cleanliness and security of the work plan.
- SMED: a method of organization of work which seeks to reduce in a systematic way the time of serial tool changeover.
- JIT: an organization tool for a just in time management.
- POKA-YOKE: a concept aiming to avoid errors.
- KANBAN: a tool to reduce stock size.
- Benchmarking: a technique for comparative evaluation.
- Gantt chart: a scheduling tool that allows the different tasks of a project to be viewed in time.
- Standardized work: a concept of work process standardization.
- Visual Management: a management tool which provides clear and highly visible information.
- Kaizen: an improvement tool which brings small, gradual and successive changes.

22.3.1.2 Audit of Operational Excellence Using Oliver Wight Baseline

Using checklists for self-assessment helps companies focus on a number of management elements such as organization, motivation, productivity, creativity, delegation and excellence. That could be achieved by applying Olivier Wight baseline (Vartiak and Jankalova 2017). There are other useful initiatives to this effect.

However, this toolkit is selected because it is accessible, effective and allows not only to describe the current status of the business, but also to focus on aspects needed to become more competitive on international markets and to achieve exemplary levels of performance. It is an important means to assess the progress and the maturity of a company. Besides, it is a way to detect the deepest problems and to deal with issues immediately. In addition, it provides an overall vision at the same time to the following points: to assess the current status of the business, to clarify its objectives and purpose, to adapt the ABCD questionnaire to the company's requirements, to develop the action plan, to measure progress, to conduct monthly follow-ups (Wight 2000).

Points awarded are scored on a scale from 4 (excellent) to 0 (nothing), as presented in Table 22.1.

Structure of the Questionnaire

The ABCD questionnaire consists of five sections. At the level of each section, there is a number of social questions.

Preparation of the Audit

This audit was conducted with a clearly defined methodology. First, we contacted companies covered in this study to plan the ABCD evaluation meetings. Meetings were held with the heads of departments of both companies. Following these meetings, questionnaires were transferred in preparation for self-assessment, by identifying qualified human resources to answer the questions, putting in place computing

Table 22.1 Rating of the ABCD questionnaire "Oliver Wight"

Points	Level	Designation
4	Excellent	The activity is perfectly controlled, results are at the highest standard attainable
3	Very well	The activity is fully operational, original goals are achieved
2	Right	most processes and tools are in place, but not completely using technologies and/or not yet had all the expected results
1	Low	Men, processes, data and/or systems do not reach the minimum required level, so results are too weak or non-existent
0	Nothing	The activity is necessary but has not yet begun

and material resources required and planning the audit to deal with the entirety of the questions. The characteristics of classification of the toolkit are represented in Table 22.2.

22.3.1.3 Case Study of Two Aerospace Industrial Companies

Aerospace companies are engaged in the improvement of the quality of their products. They are obliged to progress, to strengthen their positions in their strategic spheres of activity, to be responsive to changes in market conditions and to their customers need, and to ensure and maintain their perennial development. That's why it is important that they restructure their strategies, by providing products rapidly with an optimal quality level and competitive prices.

To carry out this assignment, it is recommended to implement a quality management system reunited with the Lean Six Sigma approach.

Lean Maturity Assessment

The trend graph relating to the analysis of the scale results shows the gap between the level of integration of all the tools, and the target.

Figure 22.1 displays the Lean maturity assessment of the company n°1. The most suitable tool for this company is the 5S approach, given its importance and ease of use. Visual management, work standardization, brainstorming, voice of the Customer and check sheet are used too, but not in a systematic manner for all the processes. The other tools are either rarely or never used because they are difficult to understand by all company staff.

Figure 22.2 displays the Lean maturity assessment of the company n°2. The use of Lean tools is more common in this company. Value stream mapping and Spaghetti diagram are the most frequently used, as they provide a clear picture of the problem and staff are adequately trained on the two techniques. Most of the tools are used if needed, but not in a systematic manner for all the services. Generally speaking, company n°2 is more involved in the training and the use of Lean methods.

Table 22.2 Quality characteristics of the ABCD reference tool for assessment

Class A	Continuous improvement has become a way of life for employees, suppliers and customers. Improving quality and reducing costs and lead times creates a competitive advantage. There is a focus strategy for innovation
Class B	Most departments participate in the process, they are actively involved with suppliers and customers. Significant improvements were achieved in several fields
Class C	Processes are used in limited areas; some improvements were made on a departmental level
Class D	Processes are not established, or they are established but static

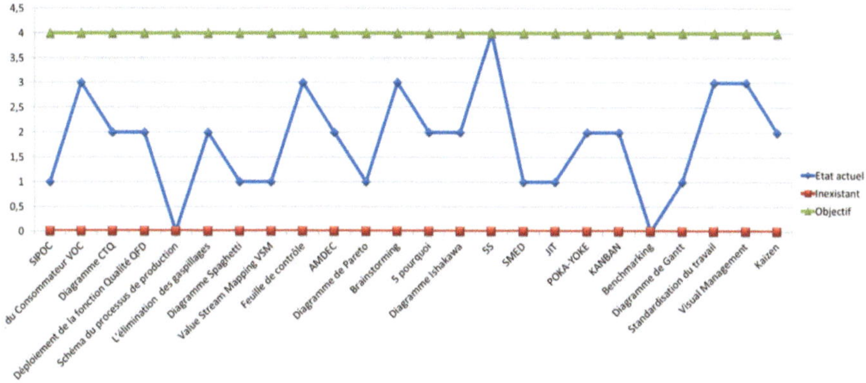

Fig. 22.1 Graph of lean maturity of the company n°1

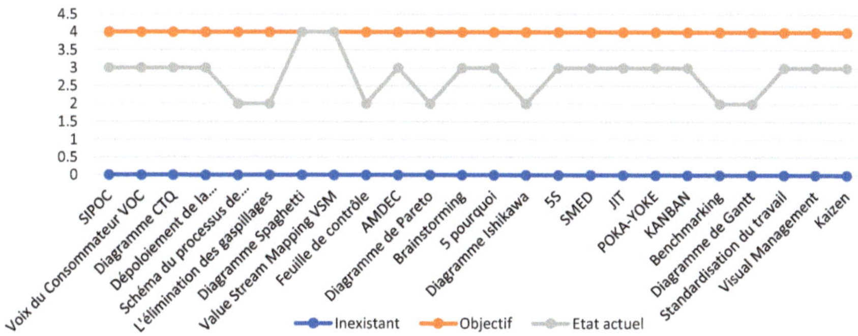

Fig. 22.2 Graph of lean maturity of the company n°2

Table 22.3 Scale of Oliver Wight baseline

Average	Performance level
>3.5	Class A
2.5–3.49	Class B
1.5–2.49	Class C
<1.5	Class D

Audit of Operational Excellence Using Oliver Wight Baseline

The results of the audit of operational excellence are presented in a table showing the sum of question scores by item and by section. Scores from 0 to 4 and the classes scale are represented in Table 22.3.

Following this audit, it seems that the companies are at different levels of maturity, as is highlighted in Table 22.4. The assessment results are shown in Figs. 22.3 and 22.4.

Table 22.4 Company rankings

Company 1	1.97	Class C
Company 2	2.52	Class B

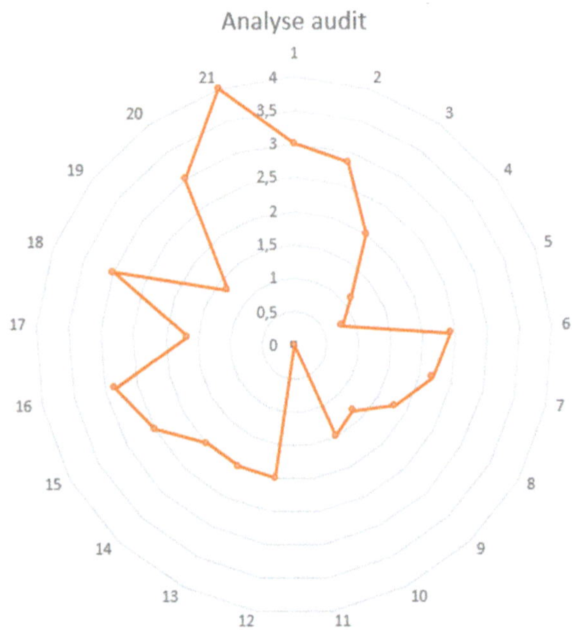

Fig. 22.3 OW audit analysis—company 1

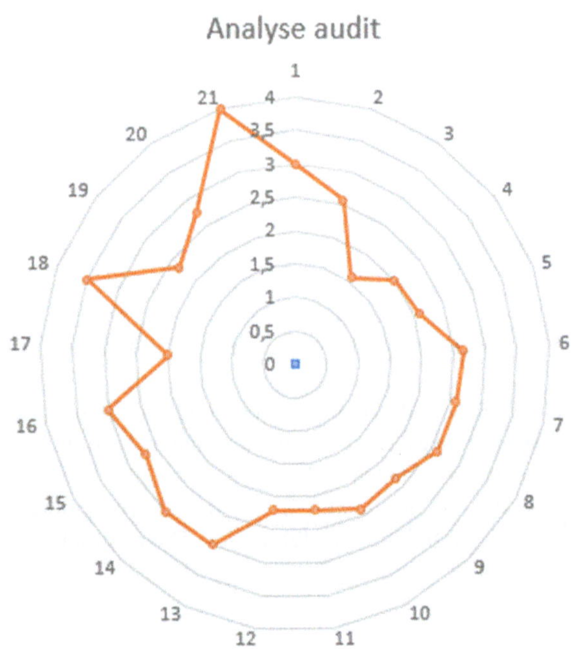

Fig. 22.4 OW audit analysis—company 2

Interpretation of Findings

The results achieved can be interpreted according to the specifications of each company.

The first company have several indirect customers and one single direct customer, so the requirement level is lower. The clients of the second company include the world's largest aircraft manufacturers which implies highly demanding requirements; thus, customers are making follow-up visits all year round.

It should be noted that even if these companies have initiated several continuous improvement projects, the current situation is not efficient against set targets, and there is still room for process improvement, and for organization in general.

22.3.1.4 Assessment of Green Practices

To identify relevant practices and key trends in industry that incorporate environmental aspects, we have chosen a case based on an approach using field studies.

Before performing field studies, we observed, in many companies with different activities, performance measures that take into account environment, links in the supply chain management, green procurement, green distribution and reverse logistics.

According to literature, to determine the integration level of these green initiatives in the companies, questionnaires have to be distributed to managers. For this study, we conducted the audit ourselves.

Assessment of Sustainable Practices

Sustainability indicators help to conduct a review of a company's progress towards the practice of sustainable development both internally and externally, this includes the firm's impact such as resource consumption and the effects of air pollution. The use of sustainability indicators can help decision-makers to set goals, to measure a company's progress and to compare alternatives, from a sustainability perspective, between the different feed materials, suppliers and technological improvements.

22.4 Conclusion

The assessment of operational excellence in the two aeronautic companies using the Oliver Wight methodology and various lean tools has provided significant insights into their current operational states and highlighted areas for improvement. The radar charts from the Oliver Wight audit reveal distinct performance patterns for each company, indicating variations in their adherence to best practices and the effectiveness of their operational processes. The lean maturity graphs further illustrate the current levels of lean implementation in both companies. While Company 1 applies all lean tools more evenly, it is seen to lag behind in some, namely Value Stream Mapping and 5 Why Analysis: there is a clear potential for improving the standardization and efficiency of these particular approaches. However, Company 2 identified a lean maturity level that is generally less than ideal; almost all of the areas analyzed in this regard are in the middle of the scores used during assessments. That is why, these findings reaffirmed the necessity to follow a systematic and systematic strategy in operation excellence. The Oliver Wight assessment is very effective in helping define a clear set of improvements that need to be made within the organization, in relation to the identified strategic objectives and improvement priorities. As for the second, the application of lean tools provides workable tools for ongoing, incremental improvement; it insists that there is never really a permanent end to the process of finding ways to eliminate waste and improve flow. Nevertheless, incorporating these methodological approaches into a circular supply chain management environment strengthens this focus on sustainability even more. Thus, by increasing the attention to minimizing wastage and making effective use of resources, it will be mutually beneficial for the aeronautic companies to work towards achieving operational excellence as well as boosting environmental targets. Therefore, complementing the Oliver Wight methodology with lean tools in the evaluation and enhancement of the aeronautic industry gives a strong approach. From this evaluation, it is imperative to understand that there is the need for specific development interventions and the opportunities that can stem from adopting the thinking of circular economy for supply chain processes. Thus, these methodologies indicate to the aeronautic companies how they can improve themselves in sustainable and efficient manner in their reacting efforts.

References

Abedelgadir M, Mohammad R (2022) Identification and ranking of the critical success factors for the achievement of operational excellence in the Sudanese aviation industry. J Oper Strategic Plan 5(2):144–172. https://doi.org/10.1177/2516600x221143846

Bag S et al (2020) Big data analytics as an operational excellence approach to enhance sustainable supply chain performance. Resour Conserv Recycl 153:104559

Balinado JRO, Prasetyo YT (2020) The impact of 5s lean tool to service operation: a case study in Toyota Dasmarinas-Cavite service operations. In: Proceedings of the 6th International Conference on Industrial and Business Engineering. ACM

Benmamoun Z, Hachimi H, Amine A (2018) Comparison of inventory models for optimal working capital; case of Aeronautics Company. Int J Eng 31(4):605–611

Chiarini A, Kumar M (2021) Lean Six Sigma and industry 4.0 integration for operational excellence: evidence from Italian manufacturing companies. Prod Plan Control 32(13):1084–1101

Correll JG, Snowden LC, Bentzley J (2021) Supercharged supply chains: discover unparalleled business planning and execution practices. John Wiley & Sons

Dev NK, Shankar R, Qaiser FH (2020) Industry 4.0 and circular economy: operational excellence for sustainable reverse supply chain performance. Resour Conserv Recycl 153:104583

Henríquez-Machado R, Muñoz-Villamizar A, Santos J (2021) Sustainability through operational excellence: an emerging country perspective. Sustain For 13(6):3165

Jebb AT, Ng V, Tay L (2021) A review of key Likert scale development advances: 1995–2019. Front Psychol 12:637547

Jebbor I, Benmamoun Z, Hachimi H (2023) Optimizing manufacturing cycles to improve production: application in the traditional shipyard industry. Processes 11(11):3136

Leksic I, Stefanic N, Veza I (2020) The impact of using different lean manufacturing tools on waste reduction. Adv Prod Eng Manage 15(1):81–92

Oliver Wight International, Inc. (2010) The Oliver Wight Class A Checklist for Business Excellence, vol 15. John Wiley & Sons

Ramani PV, Lingan KSD LK (2021) Application of lean in construction using value stream mapping. Eng Constr Archit Manag 28(1):216–228

Resende B (2023) Proposal of a facilitating methodology for fuzzy FMEA implementation with application in process risk analysis in the aeronautical sector. Int J Qual Reliabil Manage 41(4):1063–1088. https://doi.org/10.1108/ijqrm-07-2023-0237

Revathi M (2024) Enhancing operational excellence through quality management. Int Res J Adv Eng Manage 2(05):1816–1818. https://doi.org/10.47392/irjaem.2024.0268

Rusdiana IW, Soediantono D (2022) Kaizen and implementation suggestion in the defense industry: a literature review. J Ind Eng Manage Res 3(3):35–52

Sá JC et al (2022) The development of an excellence model integrating the Shingo model and sustainability. Sustain For 14(15):9472

Silva B (2023) Analysis of lean six sigma use in pharmaceutical production. Braz J Pharm Sci 59:e22949. https://doi.org/10.1590/s2175-97902023e22949

Vartiak L, Jankalova M (2017) The business excellence assessment. Procedia Eng 192:917–922

Wight OW (2000) The Oliver Wight ABCD checklist for operational excellence, vol 10. John Wiley & Sons

Wojtkowiak D, Cyplik P (2020) Operational excellence within sustainable development concept-systematic literature review. Sustainability 12(19):7933. https://doi.org/10.3390/su12197933

MIX
Papier aus verantwortungsvollen Quellen
Paper from responsible sources
FSC® C105338

If you have any concerns about our products,
you can contact us on
ProductSafety@springernature.com

In case Publisher is established outside the EU,
the EU authorized representative is:
Springer Nature Customer Service Center GmbH
Europaplatz 3, 69115 Heidelberg, Germany

Printed by Libri Plureos GmbH
in Hamburg, Germany